VOLUME TWO HUNDRED AND TWENTY FOUR

PROGRESS IN MOLECULAR BIOLOGY AND TRANSLATIONAL SCIENCE

Microbial Biofilms and Cancer - Part A

VOLUME TWO HUNDRED AND TWENTY FOUR

PROGRESS IN MOLECULAR BIOLOGY AND TRANSLATIONAL SCIENCE

Microbial Biofilms and Cancer - Part A

Edited by

DIBYAJIT LAHIRI
Department of Biotechnology, University of Engineering and Management, Kolkata, India

MOUPRIYA NAG
Department of Biotechnology, University of Engineering and Management, Kolkata, India

DEBASMITA BHATTACHARYA
Institute of Engineering and Management, India

GARGI BHATTACHARYA
Microbiologist, Ruby General Hospital, Kolkata, India

ELSEVIER

ACADEMIC PRESS
An imprint of Elsevier

Academic Press is an imprint of Elsevier
50 Hampshire Street, 5th Floor, Cambridge, MA 02139, United States
125 London Wall, London, EC2Y 5AS, United Kingdom

First edition 2026

ISBN: 978-0-443-43218-7
ISSN: 1877-1173

For information on all Academic Press publications
visit our website at https://www.elsevier.com/books-and-journals

Publisher: Zoe Kruze
Acquisition Editor: Leticia M. Lima
Editorial Project Manager: Sneha Apar
Production Project Manager: James Selvam
Cover Designer: Muthukumaran Thangaraj

Typeset by MPS Limited, India

Contents

Contributors

Reham Hamed Abdelaziz
Clinical and Chemical Pathology Department, Faculty of Medicine, Cairo University, Giza, Egypt; Microbiology, Immunology and Infectious Disease Department, College of Medicine and Health Sciences (CMHS), Arabian Gulf University, Manama, Bahrain

Kartik Adake
Department of Zoology, St. Xavier's College (Empowered Autonomous Institute), Mumbai; Department of Zoology, Savitribai Phule Pune University, Pune, Maharashtra, India

Kedar Ahire
Department of Zoology, Savitribai Phule Pune University, Pune, Maharashtra, India

Neethu Asokan
Department of Lifesciences, Sri Sathya Sai University for Human Excellence, Kalaburagi, Karnataka, India

Souradip Basu
Department of Biotechnology, University of Engineering and Management, Kolkata, India

Debasmita Bhattacharya
Department of Basic Science and Humanities, Institute of Engineering and Management, Kolkata, University of Engineering and Management, Kolkata, West Bengal, India

Rupsa Bhattacharya
Department of Biotechnology, Institute of Engineering and Management, University of Engineering and Management, Kolkata, West Bengal, India

Nagamani Bora
Division of Microbiology, Brewing and Biotechnology, School of Biosciences, University of Nottingham, Sutton Bonington Campus, Leicestershire, United Kingdom

Avijit Chakraborty
Department of Biotechnology, Institute of Engineering and Management, University of Engineering and Management, Kolkata, West Bengal, India

Sourav Chattaraj
Centre for Industrial Biotechnology Research, School of Pharmaceutical Science, Siksha 'O' Anusandhan Deemed to be University, Bhubaneswar, Odisha, India

Sayan Chatterjee
University School of Biotechnology, Guru Gobind Singh Indraprastha University, Dwarka, New Delhi, India

Saanya Chaturvedi
Department of Environmental Science and Engineering, Indian Institute of Technology (Indian School of Mines), Dhanbad, India

Karthiyaini Damodharan
Department of BioTechnology Vanniaperumal College for Women, Virudhunagar, Tamil Nadu, India

Suman Das
Department of Microbiology, Bankura Sammilani College, Bankura, West Bengal, India

Moumita Dev
Department of Food and Nutrition, Swami Vivekananda University, Barrackpore, Kolkata, West Bengal, India

Trisha Dutta
Department of Food Technology, Guru Nanak Institute of Technology, Kolkata, West Bengal, India

Arindam Ganguly
Department of Microbiology, Bankura Sammilani College, Bankura, West Bengal, India

Mousumi Ganguly
Department of Nutrition, Bankura Sammilani College, Bankura, West Bengal, India

Arpita Ghosh
Department of Microbiology, Vijaygarh Jyotish Ray College, Kolkata, West Bengal, India

Sahana Ghosh
Department of Microbiology, Vijaygarh Jyotish Ray College, Kolkata, West Bengal, India

Naiera M. Helmy Mohamed
Microbial Biotechnology Department, Biotechnology Research Institute, National Research Centre, Dokki, Cairo, Egypt

Shreyas Joshi
Tata Institute of Fundamental Research, Hyderabad, India

Riyan Kaibarta
Institute of Health Sciences, Presidency University, Kolkata, West Bengal. India

Sudhriti Maity
Department of Food Technology, Guru Nanak Institute of Technology, Kolkata, West Bengal, India

Saikat Mazumder
Department of Biotechnology, Institute of Engineering and Management, Kolkata, University of Engineering and Management; Department of Food Technology, Guru Nanak Institute of Technology, Kolkata, West Bengal, India

Debasis Mitra
Department of Microbiology, Graphic Era (Deemed to be University), Dehradun, Uttarakhand, India

Subhajit Nath
Department of Medical & Molecular Virology, University of Manchester, Manchester, United Kingdom

Soumyadip Pal
Department of Microbiology, Bankura Sammilani College, Bankura, West Bengal, India

Shreya Pandey
Department of Zoology, Daulat Ram College, University of Delhi, New Delhi, India

Sougata Rajak
Department of Biotechnology, Motilal Nehru National Institute of Technology Allahabad, Prayagraj, Uttar Pradesh, India

Pritam Dey Roy
Department of Microbiology, Vijaygarh Jyotish Ray College, Kolkata, West Bengal, India

Gustavo Santoyo
Genomic Diversity Laboratory, Institute of Biological and Chemical Research, Universidad Michoacana de San Nicolas de Hidalgo, Morelia, Mexico

Parul Thapar
Department of Nutrition and Dietetics, School of Allied Health Sciences, Manav Rachna International Institute of Research and Studies (Deemed to be University), Faridabad, Haryana, India

Panteha Torabian
Department of Biomedical and Chemical Engineering, Rochester Institute of Technology, Rochester, NY, United States

Preface

The relationship between microorganisms and human health has long fascinated scientists. Traditionally, microbes were viewed primarily as agents of infection, but advances in microbiome research have profoundly reshaped this perspective. Today, we recognize that microorganisms are integral to human physiology, influencing immunity, metabolism, and overall health. Among the many remarkable discoveries in this field is the realization that microbial biofilms highly organized communities of microorganisms embedded within a self-produced extracellular matrix play a far more significant role in disease than previously imagined. Increasing evidence now suggests that these resilient microbial communities contribute not only to chronic infections but also to the initiation, progression, and therapeutic resistance of several cancers.

The growing appreciation of this intricate relationship has inspired the development of this book, *Microbial Biofilms and Cancer: The Genetic and Molecular Interface*. Our goal is to bring together current knowledge from researchers across multiple disciplines to provide readers with a comprehensive understanding of how microbial biofilms influence cancer biology. While remarkable progress has been made in deciphering the genetic and molecular basis of cancer, the contribution of microbial communities has only recently begun to receive the attention it deserves. This volume aims to bridge that gap by presenting both established concepts and emerging discoveries that are transforming our understanding of cancer pathogenesis.

The chapters in this book explore the complex interplay between microbial biofilms and the host at multiple biological levels. Readers will find discussions on quorum sensing, microbial communication, chronic inflammation, immune modulation, genetic instability, epigenetic regulation, horizontal gene transfer, and host signaling pathways that collectively contribute to tumor development and progression. Particular attention has been given to clinically important microorganisms such as *Helicobacter pylori*, *Fusobacterium nucleatum*, *Porphyromonas gingivalis*, *Escherichia coli*, *Bacteroides fragilis*, and *Candida* species, all of which have been increasingly associated with various malignancies. By examining these interactions, the contributing authors provide valuable insights into the mechanisms through which microbial biofilms shape the tumor microenvironment and influence disease outcomes.

Beyond understanding the biology of cancer, this book also looks toward the future. Advances in genomics, metagenomics, artificial intelligence, bioinformatics, nanotechnology, and biosensor development are rapidly changing the landscape of biomedical research. These technologies are enabling scientists to identify microbial biomarkers with unprecedented precision, detect cancer-associated microbial signatures at earlier stages, and design innovative diagnostic and therapeutic strategies. The integration of computational approaches with molecular biology and microbiology represents an exciting frontier in precision oncology, and several chapters in this volume highlight these emerging opportunities.

One of the greatest strengths of this book lies in its interdisciplinary nature. The contributors represent diverse fields including microbiology, molecular biology, oncology, immunology, biotechnology, bioinformatics, and biomedical engineering. Their collective expertise reflects the collaborative spirit that is essential for addressing complex biomedical challenges. We believe that this volume will serve as a valuable resource for researchers, clinicians, postgraduate students, and healthcare professionals seeking a deeper understanding of the evolving role of microbial biofilms in cancer.

As editors, we are deeply grateful to all the authors who enthusiastically shared their knowledge and expertise to make this volume possible. Their dedication and scientific contributions have enriched the book immensely. We also sincerely thank the reviewers for their thoughtful suggestions, which have significantly improved the quality of the individual chapters. Our heartfelt appreciation extends to the editorial and production teams for their guidance, professionalism, and unwavering support throughout the publication process.

Science advances through curiosity, collaboration, and the willingness to challenge established ideas. We hope that this book not only provides readers with current scientific knowledge but also inspires new questions, encourages interdisciplinary collaborations, and stimulates future research in this rapidly evolving field. As our understanding of the relationship between microbial biofilms and cancer continues to expand, we are confident that the discoveries presented in this volume will contribute to the development of more effective strategies for cancer prevention, diagnosis, and treatment.

We sincerely hope that this book serves as a meaningful contribution to the scientific community and inspires the next generation of researchers to explore the fascinating interface between microbiology and oncology.

Dr. Dibyajit Lahiri
Dr. Moupriya Nag
Dr. Debasmita Bhattacharya
Dr. Gargi Bhattacharya

PART I

Pathogenesis, microenvironment alterations, and site-specific malignancies

Focus: Foundational mechanisms of biofilm formation, host-pathogen interactions, inflammation, cellular reprogramming, and site-specific oncogenesis

CHAPTER ONE

Introduction: Biofilms in cancer-an underestimated threat

Avijit Chakraborty* and Rupsa Bhattacharya
Department of Biotechnology, Institute of Engineering and Management, University of Engineering and Management, Kolkata, West Bengal, India
*Corresponding author. e-mail address: avijit.chakraborty@uem.edu.in

Contents

Progress in Molecular Biology and Translational Science, Volume 224
ISSN 1877-1173, https://doi.org/10.1016/bs.pmbts.2026.07.003

Abstract

Microbial biofilms are extremely organized groups of microorganisms entrenched in an extracellular polymeric substance (EPS) network which facilitates persistence, shielding against host immunity, and resistance to antimicrobial activity. Although traditionally related to chronic infections and complications in devices, recent evidence suggests that biofilms are also important in cancer formation and progression. Recent research has shown that microbial communities can be present in tumor micro-environment, and biofilms can be actively involved in carcinogenesis as opposed to being passive colonizers. In this chapter, a new association between biofilms and cancer is discussed through an understanding of their existence in various malignancies such as colorectal, oral, pancreatic, breast, gastric, lung, cervical, and liver cancers. The mechanisms of biofilms and tumor development are in detail discussed such as chronic inflammation, generation of genotoxic metabolites, immune modulation and destabilization of epithelial barriers. These mechanisms induce genomic instability, damage DNA, immunomodulation and the activation of oncogenic signaling pathways that promote the initiation and progression of tumors. Moreover, the chapter suggests the effect of biofilms on the tumor microenvironment, angiogenesis, metastasis, and cancer therapy resistance to chemotherapy, radiotherapy, and immunotherapy. Clinical implications are also discussed, and the problem includes difficulties in diagnosing, prognostic implications, and the effect of biofilm-related infections on patient outcomes. Although mounting evidence is obtained, the role of biofilms in oncology is under-recognized since it is challenging to detect, might be mistaken as contamination, and there is a lack of awareness in cancer research. The knowledge of the intricate relationships that exist between microbial biofilms and host tissues can provide fresh prospects to better diagnostics, specific antimicrobial therapy, and new treatment methods in cancers.

1. Background and context

Biofilms are now understood as highly organized, surface-associated microbial communities embedded within a self-produced extracellular polymeric substance (EPS), rather than as random aggregates of cells. This EPS matrix, primarily composed of polysaccharides along with proteins, lipids, and extracellular DNA (eDNA), provides structural integrity, mechanical stability, and protection from environmental stressors. Compared with planktonic (free-living) microorganisms, biofilm-associated cells display profound shifts in gene expression, metabolic coordination, and collective tolerance to antimicrobial agents and host immune defenses. Consequently, biofilms represent dynamic, structured ecosystems capable of complex biological behaviors. Central to this organization is quorum sensing (QS), a sophisticated cell-to-cell communication mechanism that regulates collective gene expression and coordinates group-level functions. Diverse QS systems, such as the *LuxI/LuxR* circuit in *Vibrio fischeri*, the *Agr* system in *Staphylococcus aureus,* and farnesol-mediated signaling in *Candida albicans*, illustrate how microorganisms synchronize activities including bioluminescence, virulence, morphogenesis, and biofilm maturation. Within biofilms, these signaling networks enable functional heterogeneity and division of labor, where subpopulations adopt specialized roles. For example, *Bacillus subtilis* differentiates into EPS producers, motile cells, and spore-forming cells under the control of intricate regulatory pathways involving the *TasA* amyloid protein and the master regulator *SinR*. Such differentiation parallels multicellular organization, conferring adaptive advantages in fluctuating and stressful environments.[1,2]

Historically, biofilms were first observed by Antonie van Leeuwenhoek in dental plaque, yet their ecological and clinical importance was only fully recognized centuries later. Investigations into biodeterioration in the 1970s and studies of indwelling medical devices (IMDs) in the early 1990s revealed biofilms as central contributors to persistent infections. Electron microscopy demonstrated bacteria embedded within extracellular matrices on explanted devices, fundamentally reshaping the understanding of chronic infections. Contemporary research has since expanded this perspective, revealing biofilms as both beneficial and detrimental depending on context. In the gut, resident bacteria and fungi form biofilms that can support host health, while pathogenic shifts may drive inflammation and therapeutic resistance. Similarly, chronic wounds exhibit polymicrobial biofilms, where interkingdom interactions between organisms such as *Candida albicans* and *Citrobacter freundii* complicate treatment outcomes.[3] In clinical environments, biofilms are strongly associated with device-related infections, periodontitis, cystic fibrosis, and catheter-associated complications. Fungal biofilms, particularly those formed by Candida species, are of special concern due to their robust adhesion, dimorphic transitions, and metabolic plasticity that confer resistance to antifungal therapies. Within implanted medical devices, biofilms often result in delayed symptom onset, evasion of host defenses, and diminished antibiotic efficacy. Device removal frequently becomes the primary therapeutic intervention, introducing surgical risks alongside psychological and economic burdens. These challenges underscore the urgent need for improved diagnostic strategies and targeted anti-biofilm therapies.[4]

Recent discoveries further highlight the remarkable complexity of microbial co-operation. Intercellular structures, including bacterial nanotubes and membrane-derived vesicles, facilitate the exchange of nutrients, proteins, and genetic material among neighboring cells. These mechanisms resemble primitive analogues of eukaryotic communication systems, suggesting that biofilms may have served as evolutionary platforms fostering early multicellularity. By promoting resource sharing, collective defense, and spatial organization, biofilms likely functioned as "evolutionary incubators," supporting the emergence of multicellular complexity in both prokaryotic and early eukaryotic lineages. Over evolutionary time, their role appears to have expanded from incubators to active promoters of multicellular life, emphasizing their foundational significance in biological systems. More recently, biofilm research has extended beyond infectious disease into oncology. With advancements in molecular biology, it became evident that diverse microbial communities exist in traditionally sterile tissues, including tumors. This paradigm shift is supported by the shared mechanistic link between chronic inflammation and carcinogenesis. Evidence of microbial aggregates and structured communities within tumor microenvironments suggests that tumors may serve as ecological niches for microbial colonization. Within this context, biofilms may influence cancer biology by modulating immune responses, altering extracellular matrix (ECM) architecture, and contributing to therapeutic resistance.[5]

Emerging evidence indicates associations between biofilms and malignancies such as colorectal and gastric cancers. The tumor microenvironment (TME), characterized by immune cells, fibroblasts, and ECM remodeling, promotes tumor heterogeneity and processes like epithelial-to-mesenchymal transition (EMT). This immunologically dynamic milieu often supports immune evasion through regulatory T cells and other suppressive mechanisms. Notably, parallels have been drawn between cancer cell metabolic adaptations, such as the Warburg effect, and the survival strategies of biofilm communities under hostile conditions. Biofilms may contribute to tumorigenesis through persistent inflammation, immune modulation, toxin production, and metabolic interference, suggesting that host–microbe interactions extend deeply into cancer progression.[6]

Collectively, current evidence positions biofilms as biologically sophisticated, adaptive systems that bridge microbiology, immunology, and evolutionary biology. They function not only as agents of chronic infection and antimicrobial tolerance but also as dynamic ecological communities capable of influencing host physiology and disease processes. Their evolutionary significance suggests a foundational role in the emergence of multicellular complexity, while their clinical relevance highlights their impact on persistent infections and treatment failure. The expanding recognition of biofilms within tumor microenvironments further redefines their importance, implicating them as potential modulators of cancer initiation, progression, and therapeutic response. Therefore, advancing biofilm research is essential for developing innovative diagnostic tools, anti-biofilm therapeutics, and integrated strategies addressing both infectious and non-infectious diseases, including cancer.

2. Understanding biofilms

2.1 Structure and composition

Biofilms are bacterial biopolymer matrices composed of proteins, polysaccharides, and DNA, which allow numerous species embedded within a self-produced EPS matrix to coexist. The matrix is critical for the construction and preservation of the biofilm, and it may be observed using sophisticated imaging techniques such as confocal microscopy. Exopolysaccharides provide a substantial contribution to this structure by functioning as a sticky framework and holding important components like mannose and galactose. Extracellular DNA, which is produced by cell autolysis or secretion, promotes biofilm development and antibiotic resistance by chelating ions. Furthermore, extracellular proteins present on bacterial surfaces are required for biofilm formation, assisting in cell adhesion and aggregation.[7] The EPS matrix constitutes approximately 80–90 % of the total biofilm biomass and serves as the structural and functional backbone of the biofilm. It is primarily composed of polysaccharides, proteins, lipids, and extracellular DNA (eDNA). Polysaccharides provide mechanical stability and adhesion to surfaces; proteins function as enzymes, structural scaffolds, and signaling mediators; lipids contribute to hydrophobic interactions; and eDNA enhances structural integrity and facilitates horizontal gene transfer. The EPS matrix acts as a diffusion barrier, restricting the penetration of antimicrobial agents and immune factors, thereby conferring enhanced protection to embedded cells.[2] Within this matrix resides a diverse microbial community that may include bacteria, fungi, algae, and even protozoa. These organisms interact through cooperative and competitive mechanisms, including quorum sensing, nutrient sharing, and metabolic cross-feeding. Microcolonies are often arranged in mushroom-like or pillar-shaped architectures separated by water channels that allow nutrient distribution and waste removal. This spatial organization promotes metabolic specialization and creates microniches characterized by gradients of oxygen, pH, and nutrient availability. Consequently, cells located in different regions of the biofilm exhibit distinct physiological states.[8] Biofilms are not random cell aggregates; they are dynamic, highly regulated ecosystems in which microbial interactions and matrix components together determine structural stability and functional resilience.

2.2 Stages of biofilm formation

Biofilm formation is a multistep process in which free-living (planktonic) microorganisms are adhered to a surface evolving into highly complex, surface-associated groups embedded in a self-generated EPS matrix. It starts with initial reversible attachment whereby the bacterial cells come into contact with a biotic or abiotic surface and stick on it loosely using physicochemical forces like van der Waals forces, hydrophobic forces, and electrostatic forces. Motility structures such as flagella and pili help in sensing the surface and temporary adhesion. It is then succeeded by irreversible attachment whereby cells are firmly attached to one another using adhesins, fimbriae and surface proteins. The bacteria start producing EPS components, such as polysaccharides, proteins, lipids, and, eDNA, at this stage, enhancing surface binding and triggering the transition of planktonic to sessile phenotype. In the third step, there is the formation of microcolony where attached cells multiply and form through small groups. The quorum sensing mechanisms are activated and cell-cell communication and coordinated gene expression is possible which regulates biofilm formation and matrix synthesis. As the biofilm further develops, it reaches the maturation phase which is defined by the development of the complex three-dimensional organization which is usually referred to as mushroom-like structures containing water channels with the ability to distribute nutrients, eliminate wastes and diffuse oxygen. In the mature biofilms, cells of the microorganisms are metabolically heterogeneous and become more resistant to antibiotic and host defense. Lastly, the cells are released in the biofilm in the dispersion stage as a result of environmental factors that include nutrient scarcity, oxygen pressure, or enzymatic damage of the matrix. These scattered cells transform back into planktonic and are also able to colonize new surfaces, which are involved in the proliferation and perpetuation of chronic infections.[9] Fig. 1 illustrates the sequential stages of the biofilm life cycle, highlighting the transition from initial bacterial attachment through mature biofilm development to dispersal, thereby providing a graphical overview of this dynamic process.

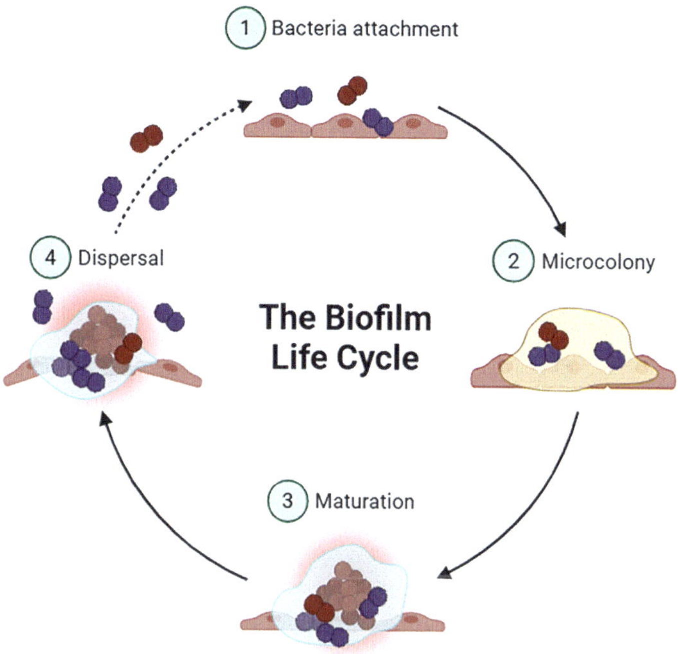

Fig. 1 The biofilm life cycle. *Created with Biorender.*

2.3 Key properties: Resistance, persistence, heterogeneity

Understanding the intrinsic properties of biofilms is essential for developing effective strategies to combat antimicrobial resistance (AMR) and persistent infections. There are several characteristics of biofilm which result in AMR and recurrent infections. One of the most important characteristics is resistance. Compared to planktonic cells, biofilm-associated cells significantly more resistant to host immune responses, antibiotics, as well as disinfectants.[10] The EPS matrix limits the antimicrobial penetration by acting as a physical and chemical barrier. Additionally, enzymes within the matrix may inactivate antibiotics, while altered gene expression and reduced metabolic rates further decrease drug susceptibility. Efflux pumps and stress-response pathways are often upregulated, enhancing tolerance. Another clinically significant characteristic of biofilms is phenotypic persistence. Within biofilms, a subpopulation of metabolically dormant persister cells survives antimicrobial exposure without acquiring genetic resistance. These cells contribute substantially to recurrent infections by repopulating the biofilm after treatment cessation.[11] Within biofilms, a subpopulation of cells known as persister cells enters a dormant or slow-growing state. These cells are not genetically resistant but are phenotypically tolerant to antibiotics, especially those targeting active cellular processes. After the completion of antibiotic treatment, these cells can reconstruct the biofilm, leading to recurrent infections. This property is particularly relevant in chronic infections such as urinary tract infections, wound infections, and device-associated infections. Biofilms also display remarkable heterogeneity. They are structurally and physiologically diverse, with gradients of oxygen, nutrients, and pH creating microenvironments within the same biofilm. Cells located on the surface may be metabolically active, while deeper layers may experience nutrient limitation and hypoxia. This spatial variation leads to differences in gene expression, metabolic activity, and antimicrobial susceptibility among cells in the same community. Such heterogeneity enhances survival under fluctuating environmental conditions. Together, these characteristics make biofilms highly adaptive systems that are difficult to eradicate and are major contributors to chronic and recurrent infections.[12]

3. Biofilms in the human body

Biofilms are not some simple aggregates of microorganisms in the environment, but highly adapted communities that can grow in different niches of the human organism. When the conditions are good, microorganisms stop to grow as planktons and develop structured, surface attaching population, embedded in. The mechanical stability, nutrient exchange and mechanism of microbial survival under immune pressure and antimicrobial treatment are provided by this structural organization. Biofilms are closely linked with chronic infections, complications of medical devices, and most recently, tumor-related microbial communities in the human host. Their existence has a significant bearing on disease development, treatment outcome, and clinical responses.[13]

3.1 Biofilms in chronic infections

Among the most clinically significant effects of biofilm formation, there are chronic infections. As opposed to acute infections where microbial growth and inflammatory reactions are often sudden, chronic infections are long-lasting and recurrently recur in the wake of treatment. One of the characteristics of most chronic infections is the development of biofilms on tissues or mucosal surfaces. Polymicrobial biofilms also develop on the necrotic tissue surfaces in chronic wounds, including diabetic foot ulcers, venous leg ulcers and pressure sores. These bio filaments interfere with wound healing, causing continuous low-grade inflammation, epithelial wound regeneration and respond to antibiotic treatment. Some of the microorganisms that are regularly found in the biofilms of wounds include *Staphylococcus aureus, Pseudomonas aeruginosa*, and other anaerobes, which show metabolic interaction and increased resistance to antibiotics.[14] Equally, the formation of biofilms in the lungs is related to the chronic respiratory infections, especially in the cases of patients with cystic fibrosis. Microbial colonization and biofilm maturation take place in a perfect surrounding of thick mucus secretions. Gradually, bacteria within biofilm also become very resistant to host immune reactions and extended antibiotic treatment leading to progressive destruction of the lungs. Another classical example of a chronic biofilm in a human body is the dental plaque. It is microbial communities comprised of multispecies that attach to tooth surfaces and the gingiva. Otherwise, plaque biofilms may result in dental caries and periodontal disease, and these diseases are characterized by chronic inflammation and tissue loss. Some of the factors that have contributed to the persistence of chronic biofilm infections include limited penetration of antibiotics through the EPS matrix, the altered metabolic condition of the sessile cells, quorum sensing-mediated control of virulence and the persistence of a subpopulation of persister cells. A combination of these processes enables microorganisms to endure adverse environments and resume infection despite apparently successful treatment.[15]

3.2 Medical device-associated biofilms

Medical devices that are implanted and indwelling have greatly enhanced the care of patients but have also provided new colonies of microbes. Biofilms that are related to medical devices constitute one of the primary issues in contemporary healthcare facilities. Biofilm formation is especially prone to form on urinary catheters, central venous catheters, prosthetic heart valves, orthopedic implants, contact lenses, and intrauterine devices. When introduced into the body, the devices are quickly covered with host proteins including fibrinogen and fibronectin creating a conditioning film, which facilitates microbial adhesion. These surfaces can be colonized by bacteria and fungi, and they start developing biofilms.[16] Catheter-associated urinary tract infections (CAUTIs) represent one of the prevalent medical-related infections across the globe. Microorganisms climb up the surfaces of catheters and form biofilms which can hardly be eradicated without removing the device. Equally, biofilms of the prosthetic joints might cause permanent inflammation and implant failure, which may require surgical revision. In cardiovascular medicine, prosthetic valve or pacemaker lead biofilms can cause infective endocarditis, which is a life-threatening condition. The biofilm matrix protects the microorganisms against immune response and systemic antibiotics enabling them to survive and periodically release planktonic cells into the blood.[17] Another common associated organism in the presence of indwelling devices is fungus biofilms and especially *Candida spp.* Such biofilms have a specific architecture and increased antifungal resistance making them difficult to treat. One of the critical characteristics of biofilms, which are associated with devices, is their resistance to antibiotic treatment. Systemic antimicrobial therapy frequently does not disappear up to effective concentrations in the biofilm, and microbial cells encrypted in the matrix display alternate phenotypes that diminish drug susceptibility. This leads to the common need to remove or replace the affected device which adds to the cost and morbidity of healthcare. Some of the preventative measures are surface modification, antimicrobial surfaces, aseptic methods and the creation of anti-biofilm material. However, biofilms that are device-associated are a clinical issue that persists.[18]

3.3 Microbial communities in cancers

Recent studies have found that biofilms and microbial communities can also have a role in tumor biology. Previously, tumors have been regarded as sterile environments since further development of sophisticated molecular tools has revealed the existence of different microbial communities within different types of tumors such as colorectal, pancreatic, breast and lung tumors.[19] Bacterial biofilms have also been noted to directly attach to the mucosal surface and penetrate tumor tissue in colorectal cancer. Chronic inflammation, generation of genotoxins, and host immunity regulation are some of the ways such microbial communities can help in carcinogenesis. Continuous inflammatory signaling may stimulate the damage of DNA, cell proliferation, and tumor growth. Microbes related to the tumors can also alter the outcome of treatment. Other bacteria may even metabolize chemotherapeutic agents and decrease their activity, whereas others are capable of regulating immune checkpoint responses. Moreover, the microenvironment of hypoxic and nutrient-varying ecotoxic state is similar to that of mature biofilms that maintain persistence of microbes in tumors.[6] Tumor-microbiome interactions are an idea that is growing rapidly. Biofilm microbes in tumors can modify local immune response, increase production of inflammatory cytokines and angiogenesis. They might also in certain instances cause metastasis through the effects on epithelial-mesenchymal transition pathways. Although the mechanisms through which biofilms are associated with cancer remain under study, it is becoming more apparent that microbial communities can no longer be seen as passive observers in the biology of tumors but as active participants. The insights into these interactions may open new opportunities of target therapies such as microbiome modulation, antimicrobial therapy, and immunotherapy.[20]

4. Emerging link between biofilms and cancer

Bacterial biofilms have been related to a variety of malignancies, including colorectal, breast, pancreatic, and lung cancers, and are frequently seen in tumour tissues and metastatic locations. Key bacteria including *Helicobacter pylori*, *Salmonella typhi*, *Fusobacterium nucleatum*, *Parvimonas micra*, *Campylobacter jejuni*, *Chlamydia psittaci*, and *Chlamydia pneumonia* are thought to have important roles in cancer formation. According to several studies, biofilms are strongly associated with cancers in the right colon, but frequency of biofilm formation is lower on the left side. Biofilms help bacteria move from causing specific illnesses to being engaged in oncogenesis, perhaps accounting for 11 % to 16 % of all cancer cases worldwide, notably gastrointestinal malignancies. Oncogenesis is a complicated process that requires several elements to work together. Microbiota can impact cancer susceptibility and development through a variety of methods. Biofilms can cause chronic inflammation, affect the immune response, create carcinogenic chemicals, disrupt host metabolism, and contribute to alterations in the tumour microenvironment (TME). The mucus barrier, especially Mucin 2 (MUC2), protects the colon by keeping intestinal microbes and epithelial cells apart. Breaches in this barrier allow microbiota to interact with epithelial cells, potentially causing inflammation and altering microbial composition. Furthermore, biofilms might cause treatment resistance, reducing the efficacy of cancer medicines.[21]

4.1 Evidence of microbial biofilms in tumors

There are several studies which have produced direct evidence in the existence of microbial biofilm in the tumor tissues. Modern imaging and molecular analysis have demonstrated the existence of dense aggregates of microbes within the extracellular matrices that can be compared to the classic biofilm structure. The adherence of bacterial biofilms to the mucosal surface along with the penetration of the epithelial layer has been observed in colorectal tumors. These polymicrobial communities are usually characterized by Fusobacterium, Bacteroides and *Escherichia coli.* Spatial arrangement of such microbial communities is a sign of ordered colonization and not by chance contamination. The microbial presence has also been reported in the pancreatic, breast and oral tumors. Indicatively, microbial DNA and bacterial cells have been observed in the tissues of pancreatic tumor indicating a possibility of these bacteria moving to other organs far away in the bloodstream through the gut or orally. On the same note, bacterial populations in the micro environment of the breast tumor have been found where they could play a role in tumor metabolism and immune response. Dense biofilm formations tend to occur on the surface of the mucosa that has been exposed to complicated oral microbiota in oral cancers. All these findings point in the same direction that it is possible to have organized microbial communities in tumors and not isolated microbial cells.[5,19]

4.2 Biofilms as drivers vs. passengers

Conversely, the passenger hypothesis posits that the presence of microbes in the tumor tissues is an after-effect of the presence of cancer. The tumor microclimates are also characterized by distorted metabolic states, diminished immunity, and deformed vascularization and this can also offer conducive living conditions to the microbes. Biofilms in this case might not directly cause cancer but can still affect disease progression due to the ability to mediate inflammation or by affecting therapeutic responses. The existing data suggests that the two mechanisms can take place concurrently. Other microbes can cause the formation of tumors and other can colonize the tumor microenvironment in the latter stages of the disease. The interplay that exists between the host tissues, microbial biofilms and immune responses is dynamic and knowledge of this interaction is vital in establishing the exact role of the microbes in cancer development.[21,22]

4.3 Cancer types in relation to biofilms

4.3.1 Colorectal cancer

One of the best-studied role of biofilm in cancer is colorectal cancer. Research has indicated that colorectal tumor patients have often bacterial biofilms on the mucosal lining. Such biofilms disrupt the epithelial protection, augment intestinal permeability, and enhance enduring inflammation. Some of these bacterial species in these biofilms secrete toxins that damage DNA or activate oncogenic signaling pathways, which provoke tumor formation and development. It is suggested that the contact of microbial communities with host epithelial cells is an important factor in determining the tumor microenvironment.[21]

4.3.2 Oral cancer

Dental plaque and other biofilms are common in the oral cavity, which is habitat to one of the most diverse microbial ecosystems in the human body. In oral malignancies, particularly oral squamous cell carcinoma (OSCC), microbial biofilms of pathogenic bacteria can lead to tissue injury and persistent inflammation. OSCC is the sixth most common malignant tumor, with 263,000 new cases and 127,000 deaths annually worldwide. OSCC represents over 90% of oral cancer cases, with rising incidences particularly in low-income countries and among females, especially those aged 45 or younger. Risk factors for OSCC include tobacco and alcohol use, betel nut consumption, poor nutrition, oral hygiene issues, dental injuries, and HPV infection. Genetic alterations may disrupt normal cellular processes and signaling pathways, involving oncogenes and tumor suppressor genes such as p53 and RB1. The oral microbiome, encompassing over 700 bacterial species, plays a crucial role in oral and systemic health, emphasizing the significance of microbial balance in the oral cavity. These biofilms have the potential to produce carcinogenic compounds, oxidative stress, and inflammatory cytokines that can lead to the transformation of malignant epithelial cells.[23]

4.3.3 Pancreatic cancer

Microbial populations have been identified in pancreatic tumor in a study which implies that there may be a connection between microbes and the development of pancreatic cancer. The bacteria of the mouth cavity or gastrointestinal tract can enter the pancreas via the blood or the biliary tract. These microbes can eventually develop biofilm-like colonies which can interact with tumor cells and immune factors. According to some studies, these microbial communities can have an impact on tumor progression and has an effect on chemotherapy and immunotherapy response.[24]

4.3.4 Breast cancer

Microbial populations have also been observed in the tissues of breast tumors. In spite of the fact that the breast has been regarded as a sterile organ, studies have revealed that bacteria may colonize breast tissue, and may even create organized communities. These microbes have the potential of affecting local immune responses, hormone metabolism and inflammatory pathways that may have an impact on tumor development and progression. Although the role of microbial biofilms in breast cancer is a developing field of research, there is an increasing amount of evidence that indicates the tumor microbiome can be a significant part of the breast cancer microenvironment.[25]

4.3.5 Gastric cancer

Gastric cancer is closely related to microbial infection especially through *Helicobacter pylori*. The bacterium is able to establish biofilm-like communities on the gastric mucosa, which allows the colonization of the stomach lining to continue. The permanent infection of *H. pylori* causes inflammatory reactions in the long term, destruction of gastric epithelial cells, the formation of carcinogenic substances, and finally the development of gastric adenocarcinoma.[26]

4.3.6 Lung cancer

Recent research has revealed the existence of microbial communities of the lung tumor and the surrounding tissues. Some bacterial species could also cause lung cancer through facilitating chronic inflammation and changes in the immune response system in the respiratory tract. Microbes forming biofilms can also be retained within the lung microenvironment with possible effects on tumor development and possible response to immunotherapy.[27]

4.3.7 Cervical cancer

Microbial flora of the female reproductive system can also have an effect on cervical cancer. The vaginal microbiota disturbances can facilitate continuous infections and inflammation and contribute to the oncogenic processes. Although human papillomavirus (HPV) is the major cause of cervical cancer, microbial biofilms in the cervical environment have a possible role in the development of the disease because they can modulate local immune response.[28]

4.3.8 Liver cancer

Recently, the gut-liver axis has been reported to be involved in the development of hepatocellular carcinoma. Microbial metabolites and bacterial products generated in the intestinal biofilms can be transported to the liver using the portal circulation. Those microbial products may induce inflammatory mechanisms and lead to the development of liver diseases, and eventually to the development of liver cancer (Table 1).[29]

Table 1 Microorganism involved in different types of cancer, their role in biofilm formation and mechanism contributing to cancer.

Cancer type	Major microorganisms involved	Role in biofilm formation	Mechanism contributing to cancer	References
Colorectal Cancer	*Fusobacterium nucleatum*, *Bacteroides fragilis*, *Escherichia coli*	Dense polymicrobial biofilms found on colon mucosa, particularly in proximal colon tumors	Biofilms disrupt epithelial barrier, induce chronic inflammation, produce genotoxins (e.g., colibactin), activate Wnt/β-catenin signaling	21,30,31
Oral Cancer	*Porphyromonas gingivalis*, *Fusobacterium nucleatum*, *Treponema denticola*	Dental plaque biofilms create a persistent inflammatory microenvironment	Activation of NF-κβ signaling, inhibition of apoptosis, immune evasion, production of carcinogenic metabolites	32,33
Pancreatic Cancer	*Fusobacterium nucleatum*	Biofilm-associated oral bacteria may translocate to the pancreas	Chronic inflammation, immune suppression, and modification of tumor microenvironment	24,34
Breast Cancer	*Staphylococcus epidermidis*, *Escherichia coli*, *Micrococcus luteus*	Biofilm-forming bacteria detected in breast tissue microbiome	Induction of DNA damage through toxins and oxidative stress	35,36

(continued)

Table 1 Microorganism involved in different types of cancer, their role in biofilm formation and mechanism contributing to cancer. *(cont'd)*

Cancer type	Major microorganisms involved	Role in biofilm formation	Mechanism contributing to cancer	References
Gastric Cancer	*Helicobacter pylori*	Biofilm formation enhances bacterial persistence on gastric mucosa	Chronic gastritis, secretion of virulence factors (CagA, VacA), epithelial cell damage	37
Gallbladder Cancer	*Salmonella enterica* serovar Typhi	Biofilm formation on gallstones enables chronic bacterial colonization	Persistent infection causes chronic inflammation and epithelial transformation	38
Lung Cancer	*Streptococcus*, *Veillonella* spp., *Haemophilus influenzae*, *Enterobacter spp.* and *Escherichia coli*	Respiratory microbial biofilms alter lung microbiome	Activation of inflammatory pathways such as ERK and PI3K signaling	39,40
Bladder Cancer	Uropathogenic *Escherichia coli*, *Klebsiella pneumoniae*, *Pseudomonas aeruginosa*, *Streptococcus*, *Mycobacterium*, *Bacteroidetes*	Biofilm formation during recurrent UTIs promotes persistent infection	Chronic inflammation and epithelial damage leading to tumorigenesis	41–43
Liver Cancer (Hepatocellular Carcinoma)	*Helicobacter hepaticus*, *Helicobacter pylori*, gut microbiota	Biofilm-forming bacteria from the gut may translocate to the liver via the portal vein	Chronic hepatic inflammation, activation of TLR signaling, production of bacterial toxins, oxidative stress leading to hepatocyte DNA damage	29

5. Mechanisms linking biofilms to carcinogenesis

Microbial biofilms have also been identified as significant factors towards the onset and progression of various cancers. In comparison to planktonic microorganisms, biofilm-associated microorganisms have changed expression, resistance to environmental pressure, and close interaction with host tissues. The latter properties enable biofilms to develop a persistent and dynamic microenvironment, capable of facilitating carcinogenesis. A number of biological processes identify how microbial biofilms could play a role in cancer progression, such as chronic inflammation, generation of genotoxic metabolites, immune modulation and epithelial barrier interference (Fig. 2).[44,45]

5.1 Chronic inflammation

Chronic inflammation is one of the major processes by which biofilms influence the pathogenesis of cancer. The repeated colonization of microbes in biofilms causes the sustained activation of the host immune system. Due to the resistance to immune clearance and antimicrobial activity of biofilms, they can stay longer on the tissues to trigger prolonged inflammatory signals. Chronic inflammation is credited with being one of

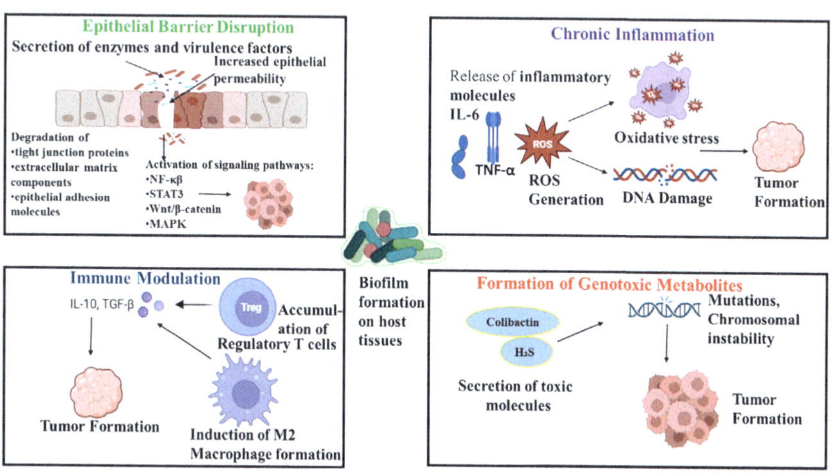

Fig. 2 This figure illustrates how microbial biofilms contribute to carcinogenesis through multiple mechanisms. Biofilm formation leads to epithelial barrier disruption, allowing microbial toxins to penetrate tissues. It also triggers chronic inflammation with ROS production and DNA damage, promotes immune modulation through Tregs and cytokines, and generates genotoxic metabolites like colibactin and hydrogen sulfide, causing mutations and tumor development. *Created with Biorender.*

the primary contributors of cancer. Microorganisms that are part of biofilms also activate host immune cells (macrophages, neutrophils, and dendritic cells) to release inflammatory factors (cytokines, chemokines, and reactive oxygen species or ROS). These inflammatory molecules are supposed to attack pathogens but may also have collateral damages to host tissues. Oxidative stress, DNA damage, and mutations in epithelial cells may be caused by prolonged exposure to ROS and reactive nitrogen species (RNS). Moreover, inflammatory cytokines like interleukin-6 (IL-6), tumor necrosis factor-alpha (TNF-α) and interleukin-1β (IL-1β) may trigger signaling pathways related to tumorigenesis. In relation to this, stimulation of the NF-*k*β and the STAT3 signaling cascades stimulate cell proliferation, suppresses apoptosis, and stimulates angiogenesis-processes that prefer tumor growth. In the long term, the inflammatory microenvironment that biofilms provide persistently can help trigger and develop malignancies. Besides, chronic inflammation may induce tissue remodeling and cell turnover. Recurrent episodes of destruction and the renewal process enhances the chances of formation of genetic mutations that may eventually result in malignant transformation. The mechanism has been noted in a number of cancers which are linked to microbial biofilms such as colorectal cancer, oral cancers and gastric cancer.[14,15]

5.2 Formation of genotoxic metabolites

The other significant process that connects biofilms with carcinogenesis is that microbial communities produce genotoxic metabolites. Microorganisms that are attached to a biofilm have varied metabolic activities and have the ability to synthesize a range of compounds that damage host DNA directly or disrupt cellular control processes. Certain types of bacteria occurring in biofilms generate toxins and metabolites which are genotoxins.[5,6] An example is where some kinds of bacteria are able to produce substances like colibactin, hydrogen sulfide and secondary bile acids. They have the potential to cause DNA two-strand breaks, chromosomal rearrangements and mutations in the host cells. Continued exposure to these genotoxic agents puts a person at high risk of developing cancer. The production of hydrogen sulfide, which is usually caused by sulfate-reducing bacteria in intestinal biofilms, has the potential of crippling the mitochondrial functionality and damaging DNA in epithelial cells.[30] Likewise secondary bile acids that form due to microbial metabolism of primary bile acids are also known to induce oxidative stress, inflammation and DNA damage that are known to cause carcinogenesis. The other role played by the biofilms is the metabolic cooperation that exists between the various microbial species. In the organized biofilm setting, organisms are able to trade metabolites, and signals such that they could produce complex metabolites that cannot be produced by single species. This synergy of metabolism could increase the presence of carcinogenic substances in the immediate microenvironment. Microbial enzymes may also transform dietary constituents or host molecules into substances that may cause cancer. As an illustration, the bacterial metabolism may convert nitrates and nitrites to N- nitroso compounds that are famous carcinogens. The long-term existence of such compounds in the tissues being colonized by biofilms enhances the probability of mutational occurrences and cancerous change.[45]

5.3 Immune modulation

The other contribution of biofilms to carcinogenesis is that it is able to regulate host immune responses. Immune evasion is one of the characteristics of biofilms. The extracellular polymeric substance matrix is a protective coating, which prevents immune attack and antimicrobial agents on microbial cells. Subsequently, microorganisms in biofilms may be allowed to survive in host tissues and constantly communicate with immune cells. Chronic exposure to antigens of microbes may cause dysregulation of the immune system.[46] The biofilms can change immune responses which would be effective antimicrobial response to a tolerogenic or immunosuppressive response. This change can be effected in a number of ways, such as the induction of regulatory T cells (Tregs), the inhibition of cytotoxic T lymphocytes, and the polarization of macrophages. Such as an example of biofilm-associated microbes can induce the M2 phenotype in macrophages, which is linked with tissue repair and immunosuppression instead of killing of microbes.[47] Anti-inflammatory cytokines like IL-10 and transforming growth factor-beta (TGF-β) are the anti-inflammatory cytokines that are produced by the M2 macrophage, inhibiting immune responses and facilitating tumor growth. Moreover, microbial elements, including lipopolysaccharides (LPS), peptidoglycan fragments, and extracellular DNA may be exposed to host pattern recognition receptors, i.e., Toll-like receptors (TLRs).[48] Chronic stimulation of these receptors can result in chronic immune stimulation and abnormal signaling pathways that induce tumorigenesis. The second significant event of immune modulation is the capacity of biofilm to provide a microenvironment to the survival and growth of tumors. Biofilms can be involved in the development of tumor cell growth, invasion and metastasis by inhibiting immune reactions and encouraging inflammatory signaling. There are also instances where the biofilms can be synergistic to tumor cells to form immunosuppressive tumor micro-environment, which helps to shield the cancer cells against destruction by the immune system.[5]

5.4 Disruption of epithelial barriers

Epithelial integrity plays a vital role in defending underneath tissues against microbial attacks and environmental injuries. Biofilms have the ability to interfere with these protective barriers, thus promoting carcinogenesis. Most biofilm forming microbes secrete enzymes and virulence factors that disintegrate components of epithelial junctions that include tight junction proteins, adherens junctions and extracellular matrix molecules. Enzymes like proteases, collagenases and hyaluronidases are also able to disrupt the integrity of the epithelial tissues where the microbes and their products can invade deeper tissues of the tissue.[46] The disruption of epithelial barriers results in the increased availability of microbial products like toxins, metabolites and inflammatory mediators to the host cells. This heightened exposure may cause oxidative stress and inflammatory response besides causing DNA damage all of which lead to cancer development. In addition, barrier disruption may alter the composition of the local microbiota, allowing opportunistic pathogens to colonize tissues and form more complex biofilm communities. Such dysbiosis can further amplify inflammation and metabolic disturbances associated with carcinogenesis. Biofilm-induced epithelial damage can also promote epithelial–mesenchymal transition (EMT), a process in which epithelial cells acquire migratory and invasive properties. EMT is an important step in tumor progression and metastasis. Microbial signals and inflammatory mediators released from biofilms can activate signaling pathways such as Wnt/β-catenin, NF-κB, and TGF-β, which are known to regulate EMT and tumor invasion.[48]

6. Biofilms and tumor progression

Besides their contribution to the onset of cancer, microbial biofilms also play an important part in tumor progression and severity of the disease. After a tumor has developed, the existence of biofilm of microbes within or around tumor tissue can modify a variety of biological functions that affect tumor growth, invasion, and metastasis. Biofilms form a multifaceted microenvironment, which comprises of microbial cells, extracellular polymeric substances (EPS), metabolic products, and host-derived molecules. This microenvironment is capable of interacting with tumor cells as well as the stromal components and eventually enhance cancer progressions. The most important ways that biofilms affect tumor progression are through the change of tumor microenvironment, angiogenesis, and metastasis.[6,21]

6.1 Effect on tumor microenvironment

The tumor microenvironment (TME) is a dynamic complex system that consists of tumor cells, immune cells, stromal cells, extra cellular matrix molecules, and different signaling molecules. There is growing evidence that microbial biofilms may also change the tumor microenvironment and cancer progression to a great degree. The tumor tissues colonized by biofilms produce a great variety of microbial metabolites, toxins, and signaling molecules interacting with host cells. These molecules are capable of triggering inflammatory reactions and modulating the behavior of the tumor cells and the stromal cells. Long-term presence of microbes usually leads to long-term activation of proinflammatory pathways including NF-*k*B, STAT3 and MAPK signaling, which are familiar with promoting tumor cell growth and survival. Immune cells can also be affected by biofilm within the tumor microenvironment.[49] The ongoing existence of microbial communities can cause an immune dysregulation process, which will reduce antitumor immunity. As an example, biofilm-derived molecules have the capability to inhibit the action of cytotoxic T cells and increase the mobilization of regulatory T cells and tumor-associated macrophages. These immune cells in most cases secrete anti-inflammatory cytokines and growth factors that encourage the growth of the tumor instead of destroying the cancerous cells. Metabolic interaction is another significant factor of the role of biofilm on the tumor microenvironment. Microorganisms in biofilms are able to modify nutrient availability by metabolism of host-derived nutrients to produce endogenous metabolic by-products including short-chain fatty acids, hydrogen sulfide, and secondary bile acids. These substances can impact tumor cell metabolism and pathways.[50] Certain metabolites encourage oxidative stress, damage to DNA, and gene expression alterations that are induced to aid tumor cell survival and proliferation. There is also another possibility of direct interaction between biofilms and tumor cells with microbial adhesion molecules and extracellular matrix components. The close physical interaction enables bacteria to alter cellular signaling cascades that control apoptosis, cell cycle and EMT. This leads to an increase in tumor permissiveness and the reduced immune surveillance of the tumor microenvironment. Besides, the polymeric matrix that is secreted by biofilms into the extracellular environment can modify the tumor microenvironment structure. The thick EPS matrix may be able to entrap inflammatory mediators, enzymes and microbial metabolites around tumor tissues, therefore, keeping the localized microenvironment in favor of tumor progression.[51]

6.2 Promotion of angiogenesis

Angiogenesis is the process of blood vessels developing out of the existing ones and is a crucial process in the development of tumours. Increased oxygen and nutrient supply is needed by the growing tumors to maintain high rates of cell proliferation. Understanding the mechanisms of biofilms in angiogenesis Biofilms may help promote angiogenesis by facilitating the release of pro-angiogenic factors into the tumor microenvironment. The microbial biofilms tend to cause chronic inflammation resulting in the synthesis of a number of molecules that promote angiogenesis. vascular endothelial growth factor (VEGF) can be induced by signaling pathways induced by inflammatory cytokines including interleukin-6 (IL-6), tumor necrosis factor-alpha (TNF-α) and interleukin-8 (IL-8).[5,52] VEGF is among the most significant angiogenic regulators and triggers the growth and the migration of endothelial cells which constitute new blood vessels. Microbial metabolites formed in the biofilms can also facilitate angiogenesis other than inflammatory cytokines. Host cell receptors may be activated by certain bacterial products and lead to pro-angiogenic signaling cascades. These signals induce the growth of endothelial cells and augment the permeability of the vascularities, which allow the tumor tissue to develop new blood vessels. Another effect of biofilms would be on hypoxia conditions in tumors. Poor blood supply and high rates of cell growth usually result in low oxygen levels in tumor tissues. Hypoxia provokes the expression of Hypoxia-inducible factor-1 alpha (HIF-1α), a transcription factor, which increases angiogenic genes, such as VEGF.[53] The hypoxic conditions can also be aggravated by the microbial metabolism in biofilms that use oxygen and change the metabolic activity of the local area. This increased hypoxia can induce more robust angiogenic reactions, and thus speed up the vascularization of tumors. Moreover, the microbial elements like lipopolysaccharides and other pathogen-associated molecular patterns may induce immune message transduction which indirectly stimulates angiogenesis. These are the molecules that trigger the immune cells to transport growth factors and matrix-remodeling enzymes that aids in the formation of new blood vessels. These mechanisms consider that biofilms are able to promote tumor vascularization substantially. When the blood vessels increase in size, they do not only help in supplying nutrients and oxygen to the tumor cells, but also enable the survival and growth of the tumor.

6.3 Facilitation of metastasis

One of the most harmful phenomena in cancer progression is known as metastasis, the process of spreading cancer cells that are located in the main tumor to other organs. The contribution of biofilms to metastasis is that it facilitates a number of cellular and molecular processes that allow tumor cells to loose, migrate and invade adjacent tissues. The induction of EMT is one of such mechanisms. When the cells become EMT, the cell-cell adhesive characteristics are lost and mesenchymal-like characteristics are developed enabling the cells to move and invade adjoining tissues. Inflammatory signaling pathways NF-$k\beta$, TGF-β, and Wnt/β-catenin pathways Biofilm-associated microorganisms have the capability to activate EMT.[5] The stimulation of these pathways results in a decrease of epithelial markers such as E-cadherin and an increase of mesenchymal markers such as vimentin and N-cadherin. Biofilms also have the potential of supporting metastasis by dismantling extra cellular constituents. A variety of biofilm-forming bacteria synthesize enzymes (proteases, collagenases and hyaluronidases) which disintegrate structural proteins of the extracellular matrix. This destruction undermines the tissue barriers and allows tumor cell migration into the tissues and blood vessels. Enhanced vascular permeability is also another cause of metastasis. The inflammatory reaction caused by biofilm may destroy endothelial junctions of blood vessels, giving tumor cells easy entry to the circulatory system. When the tumor cells penetrate the blood or lymphatic system it is possible to have secondary tumors in other body organs. In addition, microbial metabolites and toxins produced by biofilms may influence tumor cell motility and survival.[54] Certain bacteria can activate signaling pathways that enhance cytoskeletal reorganization, cell migration, and resistance to apoptosis. These effects increase the likelihood that tumor cells will survive during the metastatic process. Biofilms may also protect circulating tumor cells from immune attack. The immunomodulatory effects of biofilms can suppress immune responses and create conditions that allow tumor cells to evade immune surveillance during metastasis.[55]

7. Biofilms and therapy resistance

A major difficulty with microbial biofilm in cancer-related settings is that it greatly contributes to therapeutic resistance. Biofilms give microorganisms structural defense, metabolic plasticity, and greater communication systems that enable them to endure severe environments, such as antimicrobial and anticancer therapy. In tumor tissues or mucosal surfaces, biofilms can also be the major barrier to the efficacy of traditional treatments when they occur there or on the loci. This resistance is not only antibiotic-resistant but can also affect the result of chemotherapy, radiotherapy, and immunotherapy. The role of biofilms in therapy resistance is a crucial tool that could improve treatment interventions and come up with new therapeutic options.[55]

7.1 Antibiotic resistance

Biofilms are the well-known ones in terms of exceptional antibiotic resistance. It can be seen that microorganisms that are embedded on biofilms can survive antibiotic levels which are up to 100–1000 times the concentration needed to kill planktonic bacterial cells. Such resistance has a number of structural, physiological and genetic causes. A major way to cause antibiotic resistance in biofilms is through the protective barrier of the EPS matrix. This is a polysaccharide-based, protein-based, lipid-based, and extracellular DNA-based matrix, which prevents the deeper layers of the biofilm to be penetrated by antibiotics. Consequently, the bacteria, which are found in the inner parts of the biofilm, might be subjected to sub-lethal levels of antimicrobial agents. The other significant aspect is the metabolic heterogeneity in biofilms. Cells in the deeper layers tend to have a nutrient limitation and low levels of oxygen and have lower growth rates or are dormant. Most antibiotics have the highest efficacy with the actively dividing cell of the bacterium, thus the dormant or slow-growing cell in the biofilm may survive antibiotic exposure. It is also in biofilms that persister cells are resistant to antibiotics (relatively only a small number of bacteria) that do not have genetic resistance. These persister cells are capable of surviving the treatment and re-population of the biofilm upon withdrawal of antibiotic treatment. Moreover, biofilms aid in horizontal gene transfer within communities of microbes. The proximity between bacteria cells will encourage the exchange of plasmids and other genetic elements with antibiotic resistance genes. This increases the rate of the development of resistance in microbial populations. In cancer, chronic biofilm infections may be complicated by causing chronic infections particularly in immunocompromised patients during either chemotherapy or surgery. These infections can also be treated with prolonged antibiotic treatment which can also contribute to the emergence of antimicrobial resistance.[44,56]

7.2 Chemotherapy tolerance

Biofilms can also complicate the success of chemotherapy which is a major treatment of most cancers. There are chemotherapy agents that are used to attack rapidly dividing cancer cells but the biofilms in the tumor tissues can change drug distribution cellular signaling and immune responses thus lowering the effectiveness of the therapy. The change in the tumor microenvironment is one of the mechanisms by which biofilms affect chemotherapy tolerance.[6] Microorganisms attached to biofilms secrete metabolites, toxins, and inflammatory mediators and they can influence the behavior of tumor cells. Such signals can stimulate cancer cell survival pathways such as NF-*k*B, STAT3, and PI3K/Akt signaling pathways which promote cell survival against apoptosis undergoing chemotherapy. Also, biofilm microbial enzymes have the potential to metabolize or chemically alter some chemotherapeutic drugs, altering their efficacy. They contain some bacteria that can break down or destroy anticancer compounds before they can get to tumor cells.[57] Such microbial metabolism can have a great impact on the pharmacokinetics and bioavailability of chemotherapy agents. Biofilms have also the ability of being a physical barrier to limit the drug permeation into tumor tissues. Like they do with the antibiotics, the EPS matrix can entrap or neutralize the chemotherapeutic molecules, and they cannot reach the target cells in adequate concentrations. In addition, biofilm-induced chronic inflammation can enhance cancer cell survival and growth. The anti-apoptotic pathways and expression of drug resistance genes in tumor cells can be triggered by inflammatory cytokines issued in reaction to microbial biofilms. All these processes make the cancer cells less sensitive to chemotherapeutic agents.

7.3 Radiotherapy interference

The other significant treatment modalty involves radiotherapy which is applied in destroying cancerous cells by exposing them to ionizing radiations that destroy cellular DNA.[58] Nevertheless, the existence of microbial biofilms in tumor tissues may have the capability of disrupting the efficacy of radiation treatment. Oxygen availability in tumor tissues is one of the most important factors which determine radiotherapy effectiveness. Oxygen supplements radiation damage on DNA by the production of ROS. Hypoxic or low oxygen tumor regions are usually more resistant to radiotherapy. Biofilms may assist in the development of hypoxia in tumors in a number of ways.[59] The microbial metabolism uses oxygen and nutrients and this can contribute to decreasing the oxygen already present in the hypoxic tumor tissues. This depletion in oxygen concentration may reduce the efficacy of radiation induced DNA damage in cancer cells. Also, biofilms can induce chronic inflammation, which changes the tissue architecture and activity of blood vessels. The results of inflammatory reactions can be abnormal vascular architecture which inhibits oxygen transportation to tumor tissues.[60] Consequently, the tumor cells in these hypoxia areas might develop resistance to radiation treatment. Certain microbial metabolism processes occurring in biofilms can also have an effect on cellular stress responses and mechanisms of DNA repairs. These metabolites can induce protective signaling in tumor cells, which can more effectively repair radiation induced DNA damage and survive radiotherapy treatment. In addition, immune modulation caused by biofilm could inhibit the immunological effects of radiotherapy. Radiation therapy also destroys directly cancer cells in addition to activating immune responses towards cancer cells.[45] Biofilms may compromise this positive immune response in case they inhibit immune response or cause an immunosuppressive environment.

7.4 Immune therapy implication

Although biofilm-mediated immune modulation is described in Section 5.3, its clinical significance becomes particularly evident during cancer immunotherapy. Biofilm-induced immunosuppressive tumor microenvironments substantially reduce the efficacy of immune checkpoint inhibitors, adoptive T-cell therapies, and cancer vaccines. Persistent microbial biofilms promote T-cell exhaustion, impair antigen presentation, and limit cytotoxic lymphocyte infiltration, thereby diminishing antitumor immune responses. In addition, biofilm-associated dysbiosis may alter the gut and intratumoral microbiome, reducing patient responsiveness to immune checkpoint blockade targeting the PD-1/PD-L1 and CTLA-4 pathways. These mechanisms collectively contribute to primary and acquired resistance to immunotherapy despite adequate therapeutic administration. Immunotherapy has become a prospective measure of cancer treatment in modern medicine. Immune checkpoint inhibitor, adoptive T-cell therapy and cancer vaccines are all forms of therapy that seek to improve the capacity of the immune system to detect and destroy tumor cells.[61] Nonetheless, microbial biofilms can affect the efficacy of immunotherapy in a number of ways. Biofilms are also famous on their capacity to alter the host immune responses. Biofilms have the potential to protect bacterial cells by the extracellular environment and microbial elements in the biofilm against immune detection and suppressing immune cell activity.[9] The immune system can also be indirectly influenced by this immune evasion mechanism on its capacity to identify tumor cells. Persistent biofilm infections can lead to immune exhaustion or chronic immune activation. In such conditions, immune cells such as T lymphocytes may become less effective at attacking tumor cells. Biofilm-induced immune suppression may involve increased production of anti-inflammatory cytokines such as interleukin-10 (IL-10) and transforming growth factor-beta (TGF-β), which inhibit antitumor immune responses. Biofilms may also promote the recruitment of regulatory T cells and myeloid-derived suppressor cells within the tumor microenvironment.[62] These immune cells suppress cytotoxic T cell activity and create an immunosuppressive environment that favors tumor growth. Another important factor is the interaction between the microbiome and immune checkpoint therapy. Certain microbial communities have been shown to influence the response to immune checkpoint inhibitors. Biofilm-forming microbes may alter microbial composition and immune signaling pathways, potentially reducing the effectiveness of these therapies. Additionally, biofilms may contribute to chronic inflammation, which can both promote tumor progression and interfere with immune-based treatments. The persistent inflammatory environment created by biofilms may disrupt the delicate balance required to be effective.[63]

8. Clinical implications

There are clinical implications in the increasing appreciation of microbial biofilm in cancer biology. Biofilms are being identified in growing numbers in tumor tissues, and are thought to mediate cancer occurrence, progression, and response to treatment. Nevertheless, their clinical relevance is not just limited to fundamental pathogenesis. Biofilms can make cancer diagnosis complex, disease prognosis is a possibility and can also impact the overall patient outcome. Since biofilms are a structural complexity and usually hard to diagnose through conventional clinical techniques, they pose peculiar challenges to the diagnosis and management of cancer-related infections and tumor progression. These implications are important in understanding how to enhance clinical strategies that could be used in detecting, treating and providing better care to the patients at an earlier stage.[5,6,21]

8.1 Diagnostic challenges

Detection and proper diagnosis of biofilms is one of the significant clinical problems related to biofilms. The conventional microbiological diagnostic methods are mainly used to detect planktonic microorganisms as opposed to microorganisms found within biofilm structures. Consequently, most of the biofilm-related infections can go undetected or misdiagnosed in the clinical practice. Biofilm's complexity in structure renders it hard to isolate microorganisms by the conventional methods of culture. Due to the impact of many biofilm-associated bacteria, they can enter a low-growing or dormant metabolic state and this also makes them less detectable in regular cultures.[18] As such, laboratory findings can give false information about lack of infection in the presence of biofilms in the tissues. The identification of biofilms in cancer patients is made even more complex in cases where the patient is on chemotherapy or immunosuppressive therapy. The compromised immune response and the complicated tumor microenvironment may conceal the infection or inflammatory symptoms of the presence of biofilms. The heterogeneity of the microbial communities in biofilms is another limitation to diagnosis. Biofilms usually have several microbial species that are engaged in an organized structure. Traditional diagnostic methods which are oriented to a single-pathogen detection can miss polymicrobial biofilm communities which play a combined role in the progression of disease. More sophisticated methods of diagnosis are also under discussion in order to overcome these drawbacks.[6] Polymerase chain reaction (PCR), metagenomic sequencing, and fluorescence in situ hybridization (FISH) are molecular methods that enable researchers to identify molecular evidence of a microbial population and observe microbial community in tissue directly. Confocal laser scanning microscopy and electron microscopy are also imaging methods that can be used to determine biofilm structures in clinical samples. Nevertheless, biofilm detection is not yet integrated into clinical diagnostics despite such technological developments, which makes it difficult because of its cost as well as the complexity of its technical aspects and the absence of standard operating procedures. Thus, better strategies of diagnosis of biofilm-related conditions are one of the most important aspects of the research in the spheres of oncology and infectious disease medicine.[64]

8.2 Prognostic significance

Prognostic implications of the presence of microbial biofilms in tumor tissues also can be important. Some studies have proposed that microbial communities attached to biofilm can mediate tumor malignancy, disease advancement and treatment reaction. Tumor development and survival are facilitated by a pro-inflammatory microenvironment commonly established by biofilms.[5] Continuous inflammation due to microbial biofilms may activate signaling pathways that promote cell growth, angiogenesis and apoptotic resistance. Consequently, the biofilm colonization, and consequently, tumors can show an earlier aggressive biological behavior in relation to tumours without microbial invasion. The biofilm presence in cancer has been linked to high tumor invasiveness and further tissue penetration in some cancers, including colorectal cancer and oral cancers. There are chances that biofilm-related bacteria are capable of producing enzymes and metabolites that destroy the extracellular matrix and ease the invasion of tumors into the adjacent tissues.[21,46] This procedure could result in higher stages of the tumor at the diagnosis point. The therapeutic interventions can also be affected by biofilms. As explained in the above sections, microbial biofilms may disrupt chemotherapy, radiotherapy, and immunotherapy. In patients with tumors containing biofilm-related microbial communities, the therapeutic effect might be decreased, which will have implications on the prognosis of the disease. Moreover, the composition of the microbial community in biofilms can be used as the potential biomarker of the disease progression. Detection of specific bacterial species or microbial signature in the tumor tissues may give good information on the behavior of the tumor, the chances of developing metastasis and also the chances of resistance to treatment. Microbial profiling and biofilm detection could prove useful methods in the future to predict disease outcome and personalized therapy. The detection of biofilm-related microbial biomarkers can assist clinicians with risk stratification and patients with custom-made therapy.[6,35]

8.3 Impact on patient outcomes

Biofilms are capable of having significant impacts on the overall patient outcomes in cancer management. They can lead to ongoing infections, complications in the course of the treatment, and limited efficacy of the treatment processes because they can be present in the tumor tissues and other medical devices. Cancer patients are usually typified by weakened immunity which is either as a result of the disease or chemotherapy and radiotherapy treatment. This condition of immunocompromised exposes them to infections including those due to biofilm forming microorganisms. A biofilm is able to grow on a number of body surfaces such as mucousal surfaces, medical equipment implants, and surgical sites.[5,6] Due to the extreme resistance of biofilms to antimicrobial therapies, biofilm-related infections may prove hard to eliminate. Recurring infections can also be treated with a long course of antibiotics, frequent hospitalization, or other medical interventions. This may lead to higher healthcare expenses and reduced quality of life of the patients of such complications. Biofilms can also affect the outcome of the surgery in cancer patients. Examples include biofilm growth on implants or prosthetics after surgery and consequent infection and retarded healing. In other instances, infected equipment can be required to be pulled out or changed and this can also complicate treatment. The other factor is the effect of biofilms on the efficacy of cancer therapy. Because biofilms modify the tumor microenvironment and enhance therapy resistance, they have the potential to decrease the efficacy of anticancer treatment. This can result in partial tumor removal, increased recurrence, and the risk of metastasis.[65] Furthermore, chronic inflammation caused by biofilms can increase the progression of the disease and cause systemic complications. The constant inflammatory reactions may impair the physiological health of the patient, thereby lowering the tolerance levels of the patient to harsh cancer therapies. Clinically, the understanding of the role of biofilms in cancer-related infections can be used to enhance better management practices of patients. Early detection of biofilms, antimicrobial therapy aimed at a particular target, and creation of biofilm-disrupting agents could improve the treatment outcomes considerably.[66]

9. Why biofilms remain underestimated

Although evidence of microbial biofilms causes and progression of cancer continues to grow, their potential in oncology has been widely undervalued. Over a long period of time, cancer research mainly involved genetic mutations, environmental carcinogens and host factors whereas little was done on the role of microbes. Despite the recent research focusing on the significance of microbial communities in the human body and disease, biofilms in the tumor tissues continue to be ignored in clinical practice and cancer research. Various factors are known to drive this underestimation such as the challenges in identifying biofilms, confusion of the presence of microbes as contamination and poor understanding by oncology professionals.[67]

9.1 Detection difficulties

Among the biggest reasons why biofilms have not been taken seriously is that the identification of the same is quite challenging in the clinical and research laboratory. Biofilms are highly organized microbial communities which are entrenched in an extra cellular polymeric structure (EPS) encased by a tight adhesion to the biological surfaces. This organization renders them very dissimilar to planktonic microorganisms that are simpler to isolate and identify by conventional microbiological methods. The classical diagnostic procedures, including culture-based, do not always identify microorganisms contained in the biofilm. A number of the microbes in the biofilm are in slow-growing or inactive condition because of the low nutrient supply and reduced metabolic rates in the lower layers of the biofilm. Since the traditional methods of culture make use of the active growth of living organisms, such microbes can still escape detection during laboratory studies.[68] The other difficulty is that biofilms often have several species of microbes that constitute polymicrobial communities. Typical standard diagnostic methods are aimed at detecting single pathogens as opposed to more complicated microbial consortia. Consequently, one may fail to capture all the diversity and organization of the biofilm communities.[9] Moreover, biofilms can be localized in certain areas of tissues and are not necessarily uniformly distributed throughout a tumor or an infected area. False-negative outcomes can thus be caused by sampling errors when collecting biopsy or tissues. It is not always easy to tell the difference between planktonic bacteria and structured biofilms, even with the presence of microbial DNA, without special imaging procedures. Fluorescence in situ hybridization (FISH), metagenomic sequencing, confocal laser scanning microscopy, and electron microscopy are advanced technologies that have enhanced the detection and visualization of biofilms at the tissues. These methods are however not frequently offered in the majority of clinical laboratories because of their high cost, technical demands, and non-standardized procedures.[45] This means that a considerable number of biofilms are left undiagnosed and not properly reported in clinical research.

9.2 Misinterpretation as contamination

The other significant aspect that has led to the underestimation of the biofilms is the common misreading of the presence of microbes to be lab or environmental contamination.[69] Whenever bacteria are identified in tumor samples, then they are usually thought to have been added during the collection, processes or analysis of the samples. In the past, internal tissues were regarded as sterile places. As such, the presence of bacteria in tumor tissues or biopsy samples were regularly explained by contamination by either the skin, surgical equipment, or laboratory equipment. This premise has made scientists and medics not to pay enough attention to detect findings of microbes that can actually be reflecting biologically important biofilm communities. Biofilms have the ability to survive in stable populations of microbes on tissues without the emergence of symptoms of acute infection. Their presence may not be immediately clinically relevant because they may tend to cause either subtle or chronic inflammatory reactions as opposed to visible indications of infection. Consequently, microbial signals in the samples of tumor may be rejected instead of any further examination. Moreover, a biofilm is occasionally of low concentration in comparison with host cells in tumor tissues. When microbial DNA or small colonies of bacteria have been observed by means of molecular techniques, researchers can dismiss them as irrelevant or attribute it to background contamination. Nevertheless, recent findings with new sequencing and imaging instruments have shown that some bacteria can develop organized biofilms in tumor tissues and can contribute actively to cancer development. These results indicate that the presence of microbes in tumors cannot automatically be deemed as contamination but should rather be considered and assessed closely in regard to whether it holds any biological significance. Better sampling methods, strict contaminations control and thoroughness of interpretation of microbial data is required to differentiate actual biofilm-associated microbes and laboratory contaminants.[5,9,70]

9.3 Limited oncology unawareness

Another significant factor why biofilms have not been extensively studied in cancer research is limited awareness of oncology researchers and clinicians. Over the decades, the main attention in oncology has been put on genetic mutations, oncogenes, tumor suppressor genes, and pathways of the host cellular mechanisms linked to carcinogenesis. Although these are the key considerations of cancer biology, the possibility of microbial communities playing an important role is only recently beginning to be recognized as a topic of serious interest. Biofilm biology and how microbial communities can communicate with tumor tissues may not be well known to many oncologists and cancer researchers.[71] This has led to the neglect of microbial factors in the study of cancer pathogenesis or treatment resistance. The other factor is the historical divide that exists between the disciplines of microbiology and oncology. Traditionally, microbial infections and cancer biology were researched as relatively independent fields. Nonetheless, there is a growing indication of the fact that these areas are strongly interconnected. Microbial biofilms have the potential to have an impact on inflammation, immune responses, metabolic processes, and cellular signaling pathways directly involved in cancer progression. Moreover, oncology training can offer clinical training exposure to microbiome studies and biofilm pathology. In the absence of adequate knowledge on these concepts, clinicians might not pay attention to microbial biofilms when assessing cancer progression, resistance to treatment, or inexplicable inflammatory reactions. This gap in knowledge has also been added by funding priorities and research focus.[68] It is only more recently that research efforts on the interaction between the human microbiome and cancer have started to take place in large-scale research initiatives. With the growing amount of such studies, there is a growing awareness of the importance of biofilms in oncology. To develop this field, the need to involve microbiologists, oncologists, immunologists and bioinformatics researchers in more interdisciplinary collaboration will be necessary. The combination of microbiome analysis and biofilm detection into cancer research can assist in delivering novel insights into the mechanisms of tumor development and resistance to treatment that were unknown previously.[72]

10. Conclusion

Biofilms have become known as significant factors in cancer pathogenesis, growth, and treatment failures. Their organized microbial communities facilitate chronic presence in the host tissues, facilitating chronic inflammation, the generation of genotoxic metabolites, immune regulation, and epithelial derailment. Through these processes, microenvironment is established which is conducive to tumor initiation and growth. The other effect of biofilms on tumor progression is that biofilms change the tumor microenvironment, enhancing angiogenesis and metastasis. Moreover, they have protective extra cellular matrix and metabolic plasticity that adds resistance to antibiotics and other anticancer therapies like chemotherapy, radiotherapy and immunotherapy. In clinical terms, biofilms are also known to create diagnostic difficulties, have the potential to impact the disease prognosis, and could have an adverse effect on patient outcomes. Biofilms have been underestimated owing to their increasing importance because of their detection challenge, continued confusion with contamination, and lack of awareness on their part among oncologists. Altogether, the importance of microbial biofilms in the biology of cancer is vital to advance diagnostic methods and come up with new treatment plans that can destabilize the biofilm-related processes in cancer.

Acknowledgements

AC would like to thank All India Council for Technical Education (AICTE) for providing AICTE Doctoral Fellowship. RB would like to thank Department of Health Research (DHR) for providing DHR Women Scientist fellowship.

References

1. Di Martino P. Extracellular polymeric substances, a key element in understanding biofilm phenotype. *AIMS Microbiol.* 2018;4(2):274–288. https://doi.org/10.3934/microbiol.2018.2.274
2. Baxter KJ, Sas E, Clark KB, Walsh M, Pradeep N, Batool A, et al. Biofilms: from the cradle of life to life support. *Npj Biofilms Microbiomes.* 2026;12(1):11. https://doi.org/10.1038/s41522-025-00875-8
3. Harbool SS, Ghannoum M. Microbial biofilms: where are we and where are we going? *J Pediatric Pharmacol. Ther.* 2025;30(4):529–538. https://doi.org/10.5863/JPPT-25-01211
4. Penesyan A, Paulsen IT, Kjelleberg S, Gillings MR. Three faces of biofilms: a microbial lifestyle, a nascent multicellular organism, and an incubator for diversity. *Npj Biofilms Microbiomes.* 2021;7(1):80. https://doi.org/10.1038/s41522-021-00251-2
5. Gupta A, Bhattacharya D, Elayaperumal S, Sivamani Y, Lahiri D, Nag M. Biofilms: a cause for the development of cancer: a review. *Microbe.* 2025;6:100236. https://doi.org/10.1016/j.microb.2025.100236
6. Choi E, Murray B, Choi S. Biofilm and cancer: interactions and future directions for cancer therapy. *Int J Mol Sci.* 2023;24(16):12836. https://doi.org/10.3390/ijms241612836
7. Shree P, Singh CK, Sodhi KK, Surya JN, Singh DK. Biofilms: understanding the structure and contribution towards bacterial resistance in antibiotics. *Med Microecol.* 2023;16:100084. https://doi.org/10.1016/j.medmic.2023.100084
8. Moënne-Loccoz Y, Mavingui P, Combes C, Normand P, Steinberg C. *Microorganisms and biotic interactions. Environmental Microbiology: Fundamentals and Applications.* Springer Netherlands; 2015:395–444. https://doi.org/10.1007/978-94-017-9118-2_11.
9. Sharma S, Mohler J, Mahajan SD, Schwartz SA, Bruggemann L, Aalinkeel R. Microbial biofilm: a review on formation, infection, antibiotic resistance, control measures, and innovative treatment. *Microorganisms.* 2023;11(6):1614. https://doi.org/10.3390/microorganisms11061614
10. Liu HY, Prentice EL, Webber MA. Mechanisms of antimicrobial resistance in biofilms. *Npj Antimicrob Resist.* 2024;2(1):27. https://doi.org/10.1038/s44259-024-00046-3
11. Li X-Z, Plésiat P, Nikaido H. The challenge of efflux-mediated antibiotic resistance in gram-negative bacteria. *Clin Microbiol. Rev.* 2015;28(2):337–418. https://doi.org/10.1128/CMR.00117-14
12. Niu H, Gu J, Zhang Y. Bacterial persisters: molecular mechanisms and therapeutic development. *Signal Transduct Target Ther.* 2024;9(1):174. https://doi.org/10.1038/s41392-024-01866-5

13. Muhammad MH, Idris AL, Fan X, Guo Y, Yu Y, Jin X, et al. Beyond risk: bacterial biofilms and their regulating approaches. *Front Microbiol.* 2020;11. https://doi.org/10.3389/fmicb.2020.00928
14. Sahoo K, Meshram S. Biofilm formation in chronic infections: a comprehensive review of pathogenesis, clinical implications, and novel therapeutic approaches. *Cureus.* 2024. https://doi.org/10.7759/cureus.70629
15. Guillaume O, Butnarasu C, Visentin S, Reimhult E. Interplay between biofilm micro-environment and pathogenicity of Pseudomonas aeruginosa in cystic fibrosis lung chronic infection. *Biofilm.* 2022;4:100089. https://doi.org/10.1016/j.bioflm.2022.100089
16. Khatoon Z, McTiernan CD, Suuronen EJ, Mah T-F, Alarcon EI. Bacterial biofilm formation on implantable devices and approaches to its treatment and prevention. *Heliyon.* 2018;4(12):e01067. https://doi.org/10.1016/j.heliyon.2018.e01067
17. Tegegne DT, Abbott IJ, Poźniak B. Catheter-associated urinary tract infections: understanding the interplay between bacterial biofilm and antimicrobial resistance. *Int J Mol Sci.* 2025;26(18):9193. https://doi.org/10.3390/ijms26189193
18. Grari O, Ezrari S, El Yandouzi I, Benaissa E, Ben Lahlou Y, Lahmer M, et al. A comprehensive review on biofilm-associated infections: Mechanisms, diagnostic challenges, and innovative therapeutic strategies. *Microbe.* 2025;8:100436. https://doi.org/10.1016/j.microb.2025.100436
19. Chen Y, Wu F-H, Wu P-Q, Xing H-Y, Ma T. The role of the tumor microbiome in tumor development and its treatment. *Front Immunol.* 2022;13. https://doi.org/10.3389/fimmu.2022.935846
20. Chen Y, Fang Y, Lyu Z, Tian Y, Niu S, Li Y-R, et al. Microbiome modulation of tumorigenesis and immune responses. *J Biomed Sci.* 2026;33(1):4. https://doi.org/10.1186/s12929-025-01208-9
21. Kushwaha M, Nukala V, Singh AK, Makharia GK, Mohan A, Kumar A, et al. Emerging implications of bacterial biofilm in cancer biology: Recent updates and major perspectives. *Gut Microbes Rep.* 2024;1(1). https://doi.org/10.1080/29933935.2024.2339270
22. Xing J, Fang Y, Zhang W, Zhang H, Tang D, Wang D. Bacterial driver–passenger model in biofilms: a new mechanism in the development of colorectal cancer. *Clin Transl Oncol.* 2022;24(5):784–795. https://doi.org/10.1007/s12094-021-02738-y
23. Sukmana BI, Saleh RO, Najim MA, AL-Ghamdi HS, Achmad H, Al-Hamdani MM, et al. Oral microbiota and oral squamous cell carcinoma: a review of their relation and carcinogenic mechanisms. *Front Oncol.* 2024;14. https://doi.org/10.3389/fonc.2024.1319777
24. Li JJ, Zhu M, Kashyap PC, Chia N, Tran NH, McWilliams RR, et al. The role of microbiome in pancreatic cancer. *Cancer Metastasis Rev.* 2021;40(3):777–789. https://doi.org/10.1007/s10555-021-09982-2
25. Xu Y, Wang M-C. The intratumoral microbiota in breast cancer: from basic research to clinical translation. *Gut Microbes.* 2025;17(1). https://doi.org/10.1080/19490976.2025.2560695
26. Rizzato C, Torres J, Kasamatsu E, Camorlinga-Ponce M, Bravo MM, Canzian F, et al. Potential role of biofilm formation in the development of digestive tract cancer with special reference to helicobacter pylori infection. *Front Microbiol.* 2019;10. https://doi.org/10.3389/fmicb.2019.00846
27. Yan C, Chen Y, Tian Y, Hu S, Wang H, Zhang X, et al. The emerging role of microbiota in lung cancer: a new perspective on lung cancer development and treatment. *Cell Oncol.* 2025;48(5):1267–1298. https://doi.org/10.1007/s13402-025-01103-3
28. Feng X, Song W, Ren X, Xu Z, Li C, Feng M, et al. Microbial influences on HPV infection and cervical carcinogenesis: emerging evidence from the vaginal microbiome. *Front Microbiol.* 2025;16. https://doi.org/10.3389/fmicb.2025.1733315

29. Xiao K, Li K, Xiao K, Yang J, Zhou L. Gut microbiota and hepatocellular carcinoma: metabolic products and immunotherapy modulation. *Cancer Med.* 2025;14(9). https://doi.org/10.1002/cam4.70914
30. Bautista J, Lamas-Maceiras M, Hidalgo-Tinoco C, Guerra-Guerrero A, Betancourt-Velarde A, López-Cortés A. Gut microbiome–driven colorectal cancer via immune, metabolic, neural, and endocrine axes reprogramming. *Npj Biofilms Microbiomes.* 2026;12(1):21. https://doi.org/10.1038/s41522-025-00883-8
31. Lichtenstern CR, Lamichhane-Khadka R. A tale of two bacteria – Bacteroides fragilis, Escherichia coli, and colorectal cancer. *Front Bacteriol.* 2023;2. https://doi.org/10.3389/fbrio.2023.1229077
32. Starska-Kowarska K. The role of porphyromonas gingivalis in oral carcinogenesis and progression by remodelling the tumour microenvironment: a narrative review. *Cancers.* 2025;17(21):3478. https://doi.org/10.3390/cancers17213478
33. Zhang W, Wang S, Wang H, Tang Y-J, Tang Y, Liang X. Who is who in oral cancer? *Exp Cell Res.* 2019;384(2):111634. https://doi.org/10.1016/j.yexcr.2019.111634
34. D'Antonio DL, Zenoniani A, Umme S, Piattelli A, Curia MC. Intratumoral fusobacterium nucleatum in pancreatic cancer: current and future perspectives. *Pathogens.* 2024;14(1):2. https://doi.org/10.3390/pathogens14010002
35. Allen-Taylor D, Boro G, Cabato PM, Mai C, Nguyen K, Rijal G. Staphylococcus epidermidis biofilm in inflammatory breast cancer and its treatment strategies. *Biofilm.* 2024;8:100220. https://doi.org/10.1016/j.bioflm.2024.100220
36. Nazeer HA, Kannan S, Balakrishanan J, Nair VK, Kavitha Y, Bhosale NK. The gut microbiota and breast cancer: a comprehensive review of emerging links and therapeutic implications. *Med Microecol.* 2025;26:100155. https://doi.org/10.1016/j.medmic.2025.100155
37. Duan Y, Xu Y, Dou Y, Xu D. Helicobacter pylori and gastric cancer: mechanisms and new perspectives. *J Hematol & Oncol.* 2025;18(1):10. https://doi.org/10.1186/s13045-024-01654-2
38. Di Domenico EG, Cavallo I, Pontone M, Toma L, Ensoli F. Biofilm producing Salmonella Typhi: chronic colonization and development of gallbladder cancer. *Int J Mol Sci.* 2017;18(9):1887. https://doi.org/10.3390/ijms18091887
39. Kanteres T, Angelakopoulou IM, Angelakopoulou EA, Tsolaki F, Tagarakis G, Manika K, et al. Lung microbiome dynamics in health and lung cancer. *Microb Genomics.* 2025;11(10). https://doi.org/10.1099/mgen.0.001509
40. Yiminniyaze R, Zhang Y, Zhu N, Zhang X, Wang J, Li C, et al. Characterizations of lung cancer microbiome and exploration of potential microbial risk factors for lung cancer. *Sci Rep.* 2025;15(1):15683. https://doi.org/10.1038/s41598-025-98424-y
41. Maslennikova IL, Nekrasova IV, Starčič Erjavec M, Karimova NV, Kuznetsova MV. Uropathogenic Escherichia coli biofilms: antibiotic pressure and interaction with human neutrophils. *Int J Mol Sci.* 2025;26(19):9484. https://doi.org/10.3390/ijms26199484
42. Yang F, Zhang W, Wang R, Mao S, Liu Y-Q, Jiang Y-S, et al. *Pseudomonas aeruginosa* exacerbates bladder cancer progression by activating cancer-driven immunosuppression. *Cancer Immunol. Res.* 2026;14(3):387–398. https://doi.org/10.1158/2326-6066.CIR-25-0873
43. Zhang W, Yang F, Mao S, Wang R, Chen H, Ran Y, et al. Bladder cancer-associated microbiota: recent advances and future perspectives. *Heliyon.* 2023;9(1):e13012. https://doi.org/10.1016/j.heliyon.2023.e13012
44. Rather MA, Gupta K, Mandal M. Microbial biofilm: formation, architecture, antibiotic resistance, and control strategies. *Braz J Microbiol.* 2021;52(4):1701–1718. https://doi.org/10.1007/s42770-021-00624-x
45. Zhao A, Sun J, Liu Y. Understanding bacterial biofilms: from definition to treatment strategies. *Front Cell Infect Microbiol.* 2023;13. https://doi.org/10.3389/fcimb.2023.1137947

46. Montanari E, Bernardo G, Le Noci V, Anselmi M, Pupa SM, Tagliabue E, et al. Biofilm formation by the host microbiota: a protective shield against immunity and its implication in cancer. *Mol Cancer*. 2025;24(1):148. https://doi.org/10.1186/s12943-025-02348-0
47. Uruén C, Chopo-Escuin G, Tommassen J, Mainar-Jaime RC, Arenas J. Biofilms as promoters of bacterial antibiotic resistance and tolerance. *Antibiotics*. 2020;10(1):3. https://doi.org/10.3390/antibiotics10010003
48. Sanjabi S, Zenewicz LA, Kamanaka M, Flavell RA. Anti-inflammatory and pro-inflammatory roles of TGF-β, IL-10, and IL-22 in immunity and autoimmunity. *Curr Opin Pharmacol*. 2009;9(4):447–453. https://doi.org/10.1016/j.coph.2009.04.008
49. Goenka A, Khan F, Verma B, Sinha P, Dmello CC, Jogalekar MP, et al. Tumor microenvironment signaling and therapeutics in cancer progression. *Cancer Commun*. 2023;43(5):525–561. https://doi.org/10.1002/cac2.12416
50. Yu X, Zhang Y, Yao Y, Cho WC, Shang A. Biofilm-mediated immune dysregulation in chronic pulmonary diseases: mechanisms and clinical implications. *Front Microbiol*. 2026;16. https://doi.org/10.3389/fmicb.2025.1736384
51. Pérez Escriva P, Correia Tavares Bernardino C, Letellier E. De-coding the complex role of microbial metabolites in cancer. *Cell Rep*. 2025;44(3):115358. https://doi.org/10.1016/j.celrep.2025.115358
52. Saman H, Raza SS, Uddin S, Rasul K. Inducing angiogenesis, a key step in cancer vascularization, and treatment approaches. *Cancers*. 2020;12(5):1172. https://doi.org/10.3390/cancers12051172
53. Shibuya M. Vascular endothelial growth factor (VEGF) and its receptor (VEGFR) signaling in angiogenesis: a crucial target for anti- and pro-angiogenic therapies. *Genes & Cancer*. 2011;2(12):1097–1105. https://doi.org/10.1177/1947601911423031
54. Loh C-Y, Chai J, Tang T, Wong W, Sethi G, Shanmugam M, et al. The E-Cadherin and N-Cadherin switch in epithelial-to-mesenchymal transition: signaling, therapeutic implications, and challenges. *Cells*. 2019;8(10):1118. https://doi.org/10.3390/cells8101118
55. Luo Y-C, Huang X-T, Wang R, Lin Y-J, Sun J-X, Li K-F, et al. Advancements in understanding tumor-resident bacteria and their application in cancer therapy. *Military Med Res*. 2025;12(1):38. https://doi.org/10.1186/s40779-025-00623-1
56. Gupta N, Chauhan K, Singh G, Chaudhary S, Rathore JS. Decoding antibiotic resistance in Pseudomonas aeruginosa: embracing innovative therapies beyond conventional antibiotics. *Microbe*. 2025;6:100233. https://doi.org/10.1016/j.microb.2025.100233
57. Yao Y, Zhu Y, Chen K, Chen J, Li Y, Li D, et al. Microbiota in cancer: current understandings and future perspectives. *Signal Transduct Target Ther*. 2026;11(1):39. https://doi.org/10.1038/s41392-025-02335-3
58. Baskar R, Lee KA, Yeo R, Yeoh K-W. Cancer and radiation therapy: current advances and future directions. *Int J Med Sci*. 2012;9(3):193–199. https://doi.org/10.7150/ijms.3635
59. Twhigg L, Ng HM, Glyn T, Wall C, Purcell R. Fibre, microbes and radiotherapy: unravelling the gut's impact on radiotherapy in cancer. *ESMO Gastrointest Oncol*. 2025;9:100174. https://doi.org/10.1016/j.esmogo.2025.100174
60. Tavakolpournegari A, Moosavi SS, Matinahmadi A, Zayani Z, Bidooki SH. Exploring how reactive oxygen species contribute to cancer via oxidative stress. *Stresses*. 2025;5(4):69. https://doi.org/10.3390/stresses5040069
61. Garg P, Pareek S, Kulkarni P, Horne D, Salgia R, Singhal SS. Next-generation immunotherapy: advancing clinical applications in cancer treatment. *J Clin Med*. 2024;13(21):6537. https://doi.org/10.3390/jcm13216537
62. Sarkar S, Chowdhury S, Kar K, Chakraborty P, Das A. Immunological interventions in bacterial infections: from mechanisms to clinical applications. *Clin Immunol. Commun*. 2025;8:124–137. https://doi.org/10.1016/j.clicom.2025.11.003

63. Pei B, Ma X, Wu N, Wang C, Yang W. Effect of microbiome group on immune checkpoint inhibitors in treatment of gastrointestinal tumors. *Chin J Cancer Res.* 2023;35(3):252–265. https://doi.org/10.21147/j.issn.1000-9604.2023.03.05
64. Shakoori AR. *Fluorescence in situ hybridization (FISH) and its applications. Chromosome Structure and Aberrations.* Springer India; 2017:343–367. https://doi.org/10.1007/978-81-322-3673-3_16.
65. Mishra S, Gupta A, Upadhye V, Singh SC, Sinha RP, Häder D-P. Therapeutic strategies against biofilm infections. *Life.* 2023;13(1):172. https://doi.org/10.3390/life13010172
66. Cavallo I, Sivori F, Mastrofrancesco A, Abril E, Pontone M, Di Domenico EG, et al. Bacterial biofilm in chronic wounds and possible therapeutic approaches. *Biology.* 2024;13(2):109. https://doi.org/10.3390/biology13020109
67. Che S, Yan Z, Feng Y, Zhao H. Unveiling the intratumoral microbiota within cancer landscapes. *IScience.* 2024;27(6):109893. https://doi.org/10.1016/j.isci.2024.109893
68. Wang X, Liu M, Yu C, Li J, Zhou X. Biofilm formation: mechanistic insights and therapeutic targets. *Mol Biomed.* 2023;4(1):49. https://doi.org/10.1186/s43556-023-00164-w
69. Xu Y, Dhaouadi Y, Stoodley P, Ren D. Sensing the unreachable: challenges and opportunities in biofilm detection. *Curr Opin Biotechnol.* 2020;64:79–84. https://doi.org/10.1016/j.copbio.2019.10.009
70. Fayaz T, Renuka N, Ratha SK. Antibiotic occurrence, environmental risks, and their removal from aquatic environments using microalgae: advances and future perspectives. *Chemosphere.* 2024;349:140822. https://doi.org/10.1016/j.chemosphere.2023.140822
71. Parida S, Sharma D. The microbiome and cancer: creating friendly neighborhoods and removing the foes within. *Cancer Res.* 2021;81(4):790–800. https://doi.org/10.1158/0008-5472.CAN-20-2629
72. Kandalai S, Li H, Zhang N, Peng H, Zheng Q. The human microbiome and cancer: a diagnostic and therapeutic perspective. *Cancer Biol & Ther.* 2023;24(1). https://doi.org/10.1080/15384047.2023.2240084

CHAPTER TWO

Quorum sensing and host cell reprogramming

Karthiyaini Damodharan*
Department of BioTechnology Vanniaperumal College for Women, Virudhunagar, Tamil Nadu, India
*Corresponding author. e-mail address: karthiyainid1@gmail.com

Contents

Progress in Molecular Biology and Translational Science, Volume 224
ISSN 1877-1173, https://doi.org/10.1016/bs.pmbts.2026.07.014

Abstract

Quorum sensing (QS) is a mechanism that allows bacteria to coordinate their behavior based on population density through the production and detection of small signaling molecules. This process regulates several important functions, including virulence, biofilm formation, and metabolic adaptation.

In recent years, it has become clear that QS signals are not limited to communication among bacteria but can also influence host cells during infection. At the same time, host cells respond to microbial invasion by undergoing coordinated changes in gene expression, chromatin structure, and metabolism, a process commonly referred to as host cell reprogramming.

Studies now show that QS molecules such as acyl-homoserine lactones, autoinducing peptides, and quinolone signals can interact with host signaling pathways, including NF-κB, MAPK, and PI3K/Akt. These interactions affect immune responses, epithelial barrier function, and cellular metabolism, often favouring pathogen survival.

This chapter summarizes current understanding of quorum sensing mechanisms, highlights their roles in major pathogens, and discusses how these signals influence host cellular processes. These insights may help guide the development of new anti-virulence strategies.

1. Introduction

For much of the twentieth century, bacteria were regarded as autonomous, unicellular organisms whose behaviours were largely independent and reactive. This perception has been fundamentally revised with the discovery that microbial populations engage in sophisticated cell–cell communication systems that coordinate collective behaviour.[1,2] Among these systems, quorum sensing (QS) has emerged as a central regulatory mechanism through which bacteria synchronize gene expression in a population-density dependent manner.[3]

QS controls a wide range of collective behaviours, including the production of virulence factors, formation of biofilms, adaptation to environmental stress and interspecies competition. While initially studied as a mechanism of bacterial communication, it is now evident that QS also plays an important role in interactions between microbes and host organisms.[4]

The concept of quorum sensing originated from studies of bioluminescent marine bacteria in the 1970s, where researchers observed that light production occurred only when bacterial populations reached a critical density.[5] This phenomenon was later attributed to the accumulation of diffusible chemical signal molecules called **autoinducers** that, upon reaching threshold concentrations, activate transcriptional regulators controlling specific gene networks.[6] Since those early discoveries, quorum sensing has been identified across diverse Gram-negative and Gram-positive bacteria, as well as in fungi, revealing it as a widespread and evolutionarily conserved communication strategy.[2]

Importantly, quorum sensing is not confined to intra-species communication. Accumulating evidence demonstrates that QS molecules participate in cross-kingdom signaling, influencing host tissues during infection.[7] Pathogenic bacteria use quorum sensing not merely to coordinate among themselves, but to manipulate host cellular pathways in ways that favour colonization, persistence, and immune evasion.

Simultaneously, advances in molecular biology, transcriptomics, and systems immunology have revealed that host cells are not passive targets during infection. Instead, they undergo profound functional changes, a process often referred to as **host cell reprogramming**.[8] This reprogramming encompasses transcriptional remodeling, epigenetic modification, metabolic rewiring, and altered signaling cascades. While host cell reprogramming may initially serve defensive purposes, pathogens have evolved mechanisms to hijack or redirect these responses to promote their survival.

The intersection between quorum sensing and host cell reprogramming represents a frontier area in microbiology and immunology. Increasing evidence suggests that bacterial autoinducers can directly modulate host signaling pathways, including NF-κB, MAPK, and PI3K/Akt pathways, thereby influencing inflammatory responses, apoptosis, oxidative stress regulation, and epithelial barrier integrity.[9] In chronic infections, quorum sensing–regulated biofilms further reshape host tissue environments, contributing to persistent inflammation and immune dysfunction (Fig. 1).[10]

One of the most studied quorum sensing systems is the LuxI/LuxR paradigm in Gram-negative bacteria, which relies on acyl-homoserine lactone (AHL) signals.[6] Certain AHL molecules have been shown to penetrate mammalian cell membranes and alter cytokine production profiles.[7] Similarly, Gram-positive bacteria employ autoinducing peptides

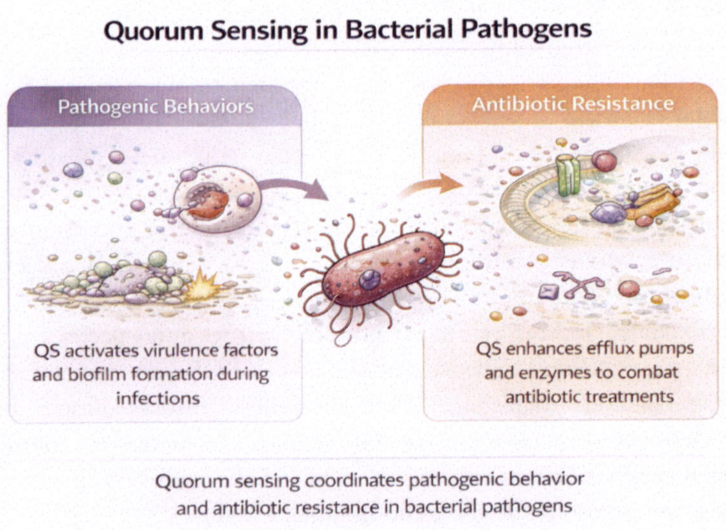

Fig. 1 Quorum sensing regulates pathogenic behavior and antibiotic resistance in bacterial pathogens.

(AIPs) that regulate virulence through two-component systems,[11] some of which indirectly affect host immune recognition. More recently, the universal signaling molecule autoinducer-2 (AI-2) has been implicated in interspecies and potentially host-microbiome communication,[12] suggesting broader ecological implications.

Host cell reprogramming during infection is multifaceted. At the transcriptional level, pathogens can induce or suppress gene networks controlling inflammation, cell survival, and metabolism. Epigenetically, histone modifications and DNA methylation patterns may be altered in infected cells, leading to sustained changes in gene expression even after pathogen clearance.[8] Metabolically, immune cells often shift toward glycolysis (aerobic glycolysis or the "Warburg-like effect") during activation, yet some pathogens exploit or redirect these shifts to establish intracellular niches.

In chronic infections such as cystic fibrosis–associated lung disease, biofilm-forming bacteria such as *Pseudomonas aeruginosa* use quorum sensing to regulate extracellular matrix production and toxin secretion.[13] These QS-regulated factors reshape epithelial cell responses, impair ciliary function, and modulate macrophage polarization. Similarly, in *Staphylococcal* infections, QS-controlled toxins and immune-modulatory proteins influence neutrophil recruitment and cytokine release patterns.[11]

Understanding the molecular mechanisms underlying QS-mediated host reprogramming has significant translational implications. Traditional antimicrobial therapies target bacterial growth or viability, however, increasing antibiotic resistance needs alternative strategies. Targeting quorum sensing often termed as **"quorum quenching"** that offers a promising anti-virulence approach that may attenuate pathogenicity without imposing strong selective pressure for resistance.[3]

The integration of systems biology, transcriptomic profiling, metabolomics, and advanced imaging techniques has accelerated discoveries in this field. Multi-omics analyses now allow simultaneous monitoring of bacterial gene expression and host cellular responses, providing unprecedented insight into the bidirectional communication between microbes and host tissues. This chapter explores the conceptual and mechanistic landscape linking quorum sensing to host cell reprogramming. It first outlines the fundamental principles of quorum sensing, including signal production, detection, and feedback regulation (Fig. 2A), followed by the canonical LuxI/LuxR signaling system in Gram-negative bacteria (Fig. 2B).

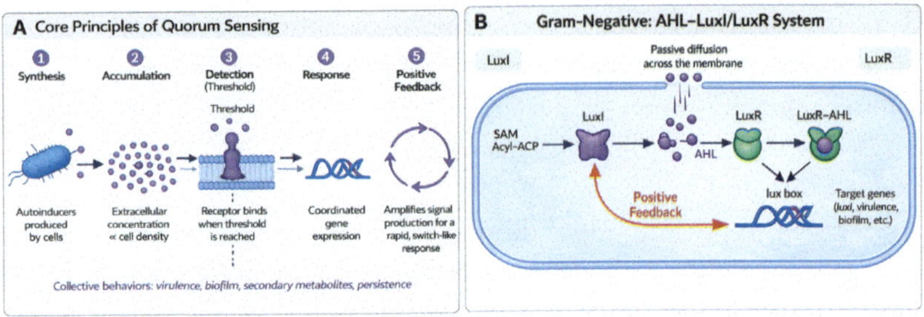

Fig. 2 Molecular basis and regulatory architecture of quorum sensing systems. (A). **Core principles of quorum sensing.** QS involves five key steps: (1) synthesis of autoinducers, (2) accumulation proportional to cell density, (3) detection once a threshold concentration is reached, (4) coordinated gene expression, and (5) positive feedback that amplifies signal production. These processes regulate collective behaviors such as virulence, biofilm formation, secondary metabolite production, and persistence. (B). **Gram-negative QS: AHL–LuxI/LuxR system.** Autoinducer synthase LuxI produces acyl-homoserine lactone (AHL) from S-adenosylmethionine (SAM) and acyl-ACP. AHL diffuses across the membrane and binds the transcriptional regulator LuxR. The LuxR–AHL complex activates target genes (e.g., *lux* operon, virulence, biofilm genes) and often induces positive feedback on *luxI* expression. (C) **Hierarchical QS networks (regulatory cascades).** QS systems can be organized in hierarchical layers, where a primary regulator (e.g., LasR) controls secondary systems (e.g., RhlR, PqsR), which in turn regulate downstream genes. This architecture enables coordinated outputs such as virulence, biofilm maturation, secondary metabolite production, and stress responses, while providing robustness and adaptability. (D) **Gram-positive QS: AIP two-component system.** Autoinducing peptides (AIPs) are synthesized as precursors, processed, and exported. Detection occurs via membrane-bound histidine kinase receptors, triggering phosphorylation cascades (His → Asp) that activate or repress target gene expression. (E) **Autoinducer-2 (AI-2): interspecies communication.** AI-2 is produced by LuxS via the activated methyl cycle and forms furanone derivatives. It serves as a universal signal detected by diverse bacteria through transporter-mediated uptake or receptor systems. AI-2 enables interspecies communication and contributes to stress adaptation, motility, and host–microbe interactions. (F) **Pseudomonas quinolone signal (PQS) system.** PQS is synthesized from anthranilate via pqsABCDE enzymes. It regulates virulence gene expression and biofilm formation and participates in iron chelation through PQS–Fe complexes. The PQS system integrates quorum sensing with metabolic functions. (G) **Integration with global regulatory networks.** QS interacts with environmental inputs (nutrients, oxygen, stress, host signals) and regulatory layers such as sigma factors, small RNAs (sRNAs), and second messengers (e.g., c-di-GMP). This integration coordinates outputs including virulence, biofilm formation, metabolism, motility, and persistence. (H) **Population heterogeneity in QS responses.** Within structured communities like biofilms, microenvironmental gradients (e.g., oxygen, nutrients) create heterogeneous QS activation states. Surface cells exhibit high QS activity, intermediate layers show mixed responses, and deeper regions display low or inactive QS, promoting resilience and division of labor. (I) **Evolutionary dynamics: cooperation vs. cheating**. QS-regulated behaviors can be exploited by "cheater" cells that do not produce signals but benefit from communal outputs. Spatial structure and feedback mechanisms stabilize cooperation by favoring signal-producing cells and limiting cheater spread.

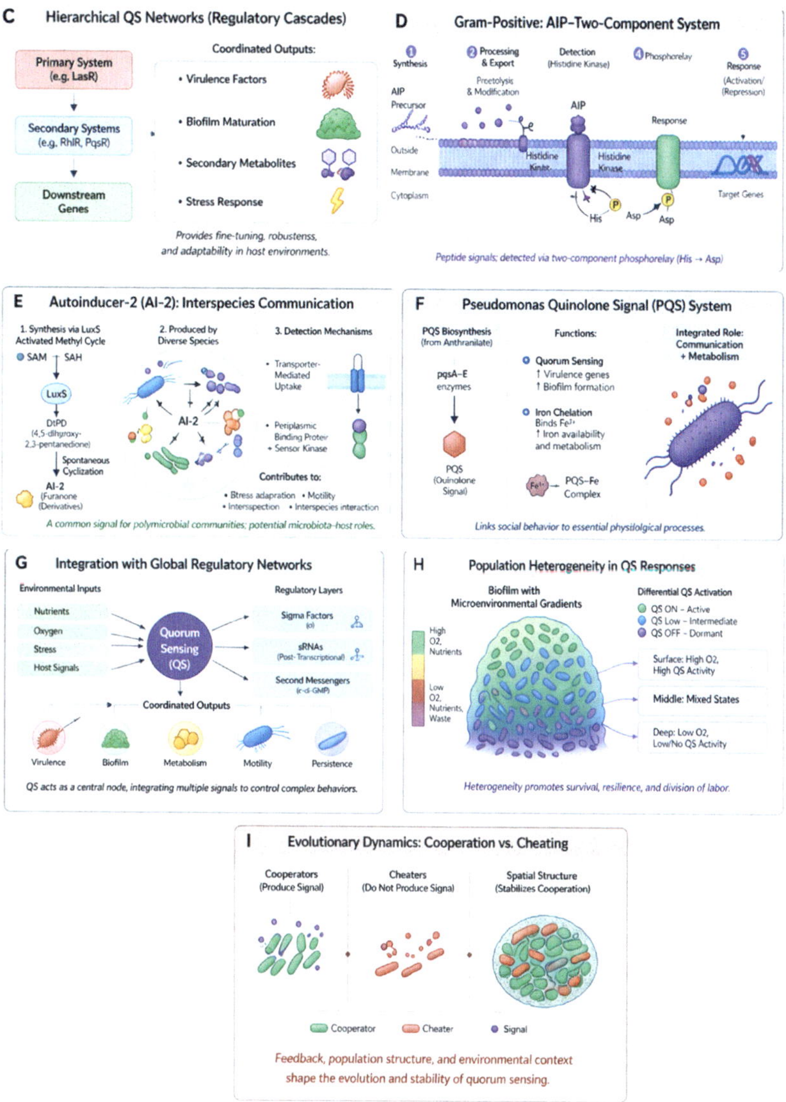

Fig. 2 *(continued)*

Subsequently, hierarchical quorum sensing networks that integrate multiple regulatory layers are discussed (Fig. 2C), along with peptide-based signaling systems in Gram-positive organisms (Fig. 2D). The chapter then examines interspecies communication mediated by autoinducer-2 (AI-2) (Fig. 2E) and quinolone-based signaling systems such as the *Pseudomonas* quinolone signal (PQS) (Fig. 2F).

Finally, broader systems-level integration, population heterogeneity, and evolutionary dynamics of quorum sensing are addressed (Fig. 2G–I). This structured progression provides a comprehensive foundation for understanding how quorum sensing interfacs with host cellular pathways to shape infection outcomes.

2. Molecular basis of quorum sensing

Quorum sensing (QS) is a cell density dependent regulatory mechanism that enables bacteria to coordinate gene expression at the population level through the production, release, accumulation, and detection of small signaling molecules termed autoinducers.[1]

Quorum sensing functions as a transcriptional regulatory network in which extracellular chemical signals provide feedback control over collective behaviours. Although the basic mechanism of quorum sensing are signal synthesis, accumulation, detection, and response is conserved, the molecular architectures and chemical characteristics of signaling molecules differ significantly between Gram-negative and Gram-positive organisms, as well as among polymicrobial communities.[2]

2.1 Core principles of quorum sensing circuits

Quorum sensing (QS) is a regulatory mechanism that allows bacteria to coordinate gene expression based on population density. This process relies on the production, release, accumulation, and detection of small signaling molecules known as autoinducers.[1]

Although QS systems vary among bacterial species, they generally follow a similar pattern (Fig. 2A). Signaling molecules are synthesized and released into the surrounding environment, where they accumulate as the bacterial population grows. Once a threshold concentration is reached, these molecules are detected by specific receptors, triggering changes in gene expression. In many cases, positive feedback mechanisms further amplify signal production, allowing a coordinated response across the population.[2,4]

QS systems are not isolated pathways; they are integrated with broader regulatory networks that respond to environmental conditions such as nutrient availability, oxygen levels, stress, and host-derived signals. This integration allows bacteria to regulate their behavior in complex and changing environments.

2.2 Quorum sensing in gram-negative bacteria

2.2.1 Acyl-homoserine lactone (AHL) systems

The canonical QS system in Gram-negative bacteria is the LuxI/LuxR paradigm, first described in marine bioluminescent species. This system relies on **acyl-homoserine lactones (AHLs)** as signaling molecules (Fig. 2B).[6]

2.2.1.1 Signal biosynthesis

AHLs are produced by LuxI-type enzymes using S-adenosylmethionine (SAM) as the lactone donor and acyl-acyl carrier protein (acyl-ACP) intermediates derived from fatty acid biosynthesis. Differences in acyl chain length and chemical modifications, such as oxo or hydroxy substitutions, contribute to species-specific signaling characteristics.[2,6]

2.2.1.2 Signal detection

AHL molecules can move freely across bacterial membranes and accumulate in the surrounding environment. As their concentration increases and reaches a critical level, they bind to LuxR-type transcriptional regulators inside the cell. This interaction stabilizes the protein and promotes its dimerization, allowing the complex to bind specific DNA sequences known as lux boxes and regulate gene expression.[1]

2.2.1.3 Positive feedback

In many systems, the LuxR-AHL complex activates transcription of the luxI gene itself, creating a positive feedback loop that amplifies signal production and synchronizes gene expression across the population.

2.2.2 Hierarchical QS networks

In several pathogens, QS systems are organized hierarchically rather than as single circuits. A well-characterized example includes multi-tiered systems where one QS regulator activates subordinate systems, producing a regulatory cascade (Fig. 2C).

Such hierarchical organization allows fine-tuning of:

- Virulence factor expression
- Biofilm maturation
- Secondary metabolite production
- Stress response pathways

This layered architecture increases robustness and adaptability in fluctuating host environments.[13]

2.3 Quorum sensing in Gram-positive bacteria

In contrast to Gram-negative systems that utilize diffusible small molecules, Gram-positive bacteria predominantly employ autoinducing peptides (AIPs) as signaling molecules (Fig. 2D). These peptides exhibit considerable structural diversity and undergo complex biosynthetic processing before becoming biologically active.

2.3.1 Peptide signal biosynthesis and processing

AIPs are synthesized as precursor peptides encoded by genes such as agrD in staphylococci. These precursor peptides typically contain an N-terminal leader sequence and a C-terminal pro-peptide region that forms the mature signal.
The maturation process involves:

- **Proteolytic cleavage** of the precursor peptide
- **Cyclization and post-translational modification**, often generating thiolactone or lactone rings (e.g., in Staphylococcus aureus AIPs)
- **Export via membrane-associated transporters**, such as AgrB, which functions both as a processing enzyme and exporter

In Staphylococcus aureus, AgrB cleaves AgrD and facilitates the formation of a cyclic thiolactone ring structure, producing the active AIP. This cyclic structure is critical for receptor specificity and signaling fidelity.

In contrast, Streptococcus species produce **competence-stimulating peptides (CSPs)** that are typically linear or minimally modified peptides. These are processed by dedicated secretion systems (e.g., ComAB transporter) and do not form thiolactone rings, reflecting functional and structural divergence among Gram-positive QS signals.

2.3.2 Two-component signal transduction systems

Unlike AHLs, peptide signals cannot freely diffuse across membranes due to their size and polarity. Instead, they are detected by membrane-bound histidine kinase receptors that form part of two-component systems (TCS). A typical TCS includes:

1. A sensor histidine kinase (e.g., AgrC, ComD)
2. A cytoplasmic response regulator (e.g., AgrA, ComE)

Upon binding of the extracellular AIP:

- The histidine kinase undergoes **autophosphorylation** at a conserved histidine residue

- The phosphate group is transferred to an **aspartate residue** on the response regulator
- The activated response regulator binds DNA and modulates transcription

This phosphorylation cascade enables rapid signal amplification and integration with other regulatory networks.

2.3.3 Functional and structural diversity of AIPs

AIPs exhibit substantial diversity across species:

- **Staphylococci:** cyclic thiolactone-containing peptides with high specificity (cross-inhibition between strains)
- **Streptococci:** linear CSPs involved in competence and stress adaptation
- **Enterococci:** peptide pheromones regulating conjugation and plasmid transfer

This diversity allows fine-tuned regulation of species-specific behaviors, including virulence expression, genetic competence, and horizontal gene transfer.

2.4 The autoinducer-2 (AI-2) system: Interspecies communication

Autoinducer-2 (AI-2) represents a distinct class of signaling molecules implicated in interspecies communication. Synthesized via the LuxS enzyme as part of the activated methyl cycle, AI-2 is chemically derived from 4,5-dihydroxy-2,3-pentanedione (DPD), which spontaneously cyclizes into furanone derivatives (Fig. 2E).

Unlike species-specific AHL or AIP systems, AI-2 is produced by diverse Gram-negative and Gram-positive bacteria, suggesting a role in polymicrobial community coordination.

AI-2 detection mechanisms vary:

- In some bacteria, AI-2 is internalized via specific transporters.
- In others, periplasmic binding proteins and sensor kinases mediate detection.[12]

AI-2 signaling contributes to:

- Biofilm formation
- Motility regulation
- Stress adaptation
- Interspecies competition

Importantly, AI-2 has been proposed as a mediator of microbiota-host communication, although the mechanistic basis of such interactions remains under investigation.

2.5 The Pseudomonas quinolone signal (PQS) system

Certain bacteria employ quinolone-based signaling molecules, such as the Pseudomonas quinolone signal (PQS). These molecules function both as QS signals and as iron chelators, linking communication to metabolic and environmental sensing (Fig. 2F).
PQS participates in:

- Regulation of virulence genes
- Biofilm maturation
- Oxidative stress responses
- Secondary metabolite production

The integration of metabolic sensing with communication underscores the evolutionary advantage of multifunctional signal molecules.[14]

2.6 Systems-level regulation and network integration

Modern transcriptomic and proteomic analyses reveal that QS regulons may encompass hundreds of genes, representing up to 10 % of a bacterial genome in certain species. QS-regulated genes include those encoding:

- Toxins
- Proteases
- Adhesins
- Secretion systems
- Efflux pumps
- Metabolic enzymes

Importantly, QS often interfaces with global regulators such as alternative sigma factors, small RNAs (sRNAs), cyclic-di-GMP signaling pathways, and stress response networks. Post-transcriptional control via RNA-binding proteins further modulates QS output.

This multilayered regulation ensures that QS-dependent behaviors occur only under optimal environmental and physiological conditions (Fig. 2G).[4]

2.7 Quorum sensing and population heterogeneity

Although QS is classically described as a synchronized population-wide switch, single-cell analyses reveal phenotypic heterogeneity within bacterial communities. Subpopulations may exhibit differential QS activation, contributing to bet-hedging strategies that enhance survival under host immune pressure or antibiotic exposure (Fig. 2H).

Such heterogeneity is particularly relevant in biofilms, where nutrient gradients, oxygen limitation, and host factors create microenvironments influencing QS dynamics.[10]

2.8 Evolutionary and ecological considerations

QS systems are subject to evolutionary pressures that balance cooperative benefits against exploitation by "cheater" cells that do not produce signals but benefit from communal behaviors. Mechanisms that stabilize cooperation include spatial structuring, feedback amplification, and coupling of QS to essential metabolic processes (Fig. 2I).

From an ecological perspective, QS contributes to niche adaptation, interspecies competition, and microbiome homeostasis. These ecological dimensions are increasingly recognized as central to understanding pathogen behavior within host environments.[15]

3. Quorum sensing systems in major human pathogens

Quorum sensing systems reach their greatest biological and clinical significance in pathogenic bacteria, where they orchestrate virulence programs, biofilm development, immune evasion strategies, and metabolic adaptation during host colonization.[3,4] In many organisms, QS is not a peripheral regulatory mechanism but rather a central coordinator of infection-stage transitions. This section examines representative Gram-negative and Gram-positive pathogens in which QS has been extensively characterized and directly linked to disease progression.[16] The major quorum sensing systems across clinically relevant pathogens and their roles in virulence regulation and biofilm formation are summarized in Fig. 3.

3.1 Gram-negative pathogens

3.1.1 Pseudomonas aeruginosa: A hierarchical QS network

Pseudomonas aeruginosa possesses one of the most sophisticated and extensively studied QS architectures. Its system consists of at least three interconnected signaling circuits.[13]

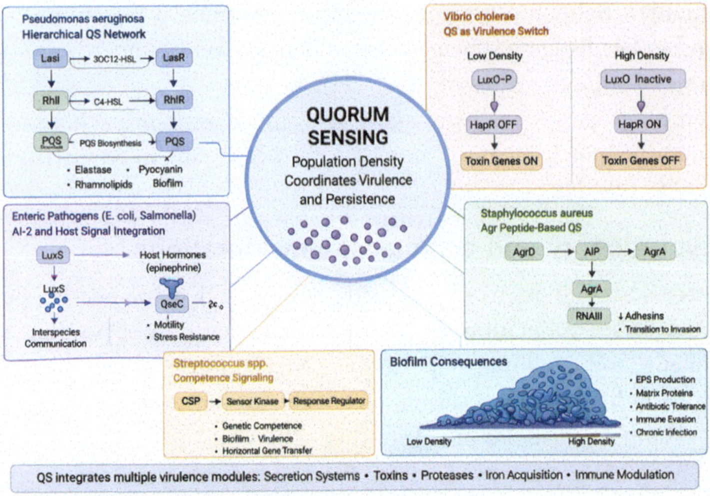

Fig. 3 Quorum sensing systems in major human pathogens and their roles in virulence, biofilm formation, and disease progression. This figure illustrates the major quorum sensing (QS) mechanisms across clinically relevant bacterial pathogens and their integration into virulence regulation and biofilm development. In *Pseudomonas aeruginosa*, hierarchical QS systems (Las, Rhl, and PQS) coordinate the expression of virulence factors and biofilm-associated genes. In *Vibrio cholerae*, QS functions as a density-dependent regulatory switch, where LuxO and HapR control the transition between virulence expression and repression. Enteric pathogens such as *Escherichia coli* and *Salmonella* utilize AI-2–mediated signaling that integrates interspecies communication with host-derived hormonal signals via the QseC sensor kinase. In *Staphylococcus aureus*, the Agr peptide-based QS system regulates the shift from adhesion to toxin production through RNAIII-mediated control. In *Streptococcus* species, competence-stimulating peptides regulate genetic competence, virulence, and horizontal gene transfer. Collectively, QS-mediated biofilm formation promotes extracellular matrix production, antibiotic tolerance, immune evasion, and chronic infection persistence.

- **Las system (LasI/LasR)**
- **Rhl system (RhlI/RhlR)**
- **PQS system (Pseudomonas quinolone signal)**

These systems operate in a hierarchical manner, with the Las system generally positioned at the apex of the regulatory cascade.

3.1.1.1 Las system

The LasI enzyme synthesizes N-3-oxo-dodecanoyl homoserine lactone (3OC12-HSL), which binds to the transcriptional regulator LasR. The LasR–AHL complex activates:

- Elastase (LasB)
- Exotoxin A
- Alkaline protease
- Components of the type II secretion system
- rhl and pqs operons (hierarchical activation)

3.1.1.2 Rhl system

The RhlI enzyme produces N-butyryl homoserine lactone (C4-HSL), which interacts with RhlR. This system regulates:

- Rhamnolipid biosynthesis
- Pyocyanin production
- Secondary virulence factors
- Biofilm maturation genes

3.1.1.3 PQS system

The PQS signal (2-heptyl-3-hydroxy-4-quinolone) links quorum sensing with iron metabolism and oxidative stress responses.[14] PQS contributes to:

- Outer membrane vesicle formation
- Biofilm architecture stabilization
- Redox homeostasis

Clinically, QS in *P. aeruginosa* is critical in cystic fibrosis lung infections, where dense biofilms form in mucus-rich environments.[17] QS-regulated factors impair epithelial barrier integrity, disrupt ciliary function, and modulate macrophage inflammatory responses. Importantly, QS mutants often exhibit attenuated virulence in animal infection models.[18]

3.1.2 Vibrio cholerae: QS as a virulence switch

Unlike *P. aeruginosa*, *Vibrio cholerae* uses QS to repress virulence at high cell density.[19] Its regulatory network includes:

- CAI-1 (species-specific signal)
- AI-2 (interspecies signal)
- LuxO response regulator
- HapR transcription factor

At low cell density:

- LuxO is phosphorylated.
- Small regulatory RNAs repress HapR.
- Virulence genes (cholera toxin, toxin-coregulated pilus) are expressed.

At high cell density:

- Accumulated autoinducers inhibit LuxO phosphorylation.
- HapR is expressed.
- Virulence genes are repressed.

This density-dependent virulence repression promotes host colonization initially but facilitates detachment and environmental dissemination later. Thus, QS functions as a lifecycle regulator coordinating transitions between host-associated and aquatic environments.

3.1.3 Enteric pathogens and AI-2 signaling

Several enteric pathogens, including pathogenic *Escherichia coli* and *Salmonella* species, produce AI-2 via LuxS. AI-2 contributes to:

- Motility regulation
- Biofilm formation
- Acid stress resistance
- Colonization efficiency

In enterohemorrhagic *E. coli* (EHEC), quorum sensing interfaces with host-derived signals such as epinephrine and norepinephrine through the QseC sensor kinase, demonstrating integration between bacterial QS and host neuroendocrine signaling. Such cross-talk highlights the complexity of pathogen sensing within the intestinal microenvironment.[12]

3.2 Gram-positive pathogens

3.2.1 Staphylococcus aureus: The Agr system

In Staphylococcus aureus, quorum sensing is mediated by the accessory gene regulator (Agr) system, which utilizes structurally specialized autoinducing peptides (AIPs). The Agr locus encodes four key components: AgrD (precursor peptide), AgrB (processing/export protein), AgrC (sensor kinase), and AgrA (response regulator).
The biosynthesis of AIP involves a highly coordinated process:

- AgrD is synthesized as a precursor peptide
- AgrB cleaves the precursor and catalyzes **thiolactone ring formation**, generating a cyclic AIP
- The mature AIP is exported to the extracellular environment

Upon reaching threshold concentration, AIP binds to the membrane-bound histidine kinase AgrC, triggering autophosphorylation. The phosphate is then transferred to AgrA, which activates transcription from P2 and P3 promoters.

This leads to:

- **Positive feedback amplification** of AIP production
- Expression of **RNAIII**, a central regulatory RNA molecule
- Upregulation of secreted toxins (e.g., hemolysins, proteases)
- Downregulation of surface adhesins

This switch-like transition facilitates progression from colonization to invasive infection.

Notably, structural variation in AIPs across S. aureus strains results in **cross-inhibition**, where AIPs from one strain can inhibit signaling in another, representing an additional layer of competitive regulation.

3.2.2 Streptococcus species and competence regulation

In *Streptococcus* species, quorum sensing regulates genetic competence through competence-stimulating peptides (CSPs). These peptides are synthesized as precursor molecules and exported via ABC transporters such as ComAB.

Unlike Staphylococcal AIPs, CSPs are typically **linear peptides** that do not undergo thiolactone cyclization. Upon extracellular accumulation, CSPs bind to the sensor kinase ComD, initiating phosphorylation of the response regulator ComE.
Activated ComE induces:

- Expression of competence genes
- DNA uptake machinery
- Stress response pathways

This system promotes horizontal gene transfer, facilitating acquisition of antibiotic resistance and virulence determinants.

The biochemical simplicity of CSPs compared to AIPs reflects functional specialization toward competence rather than toxin regulation, yet both systems demonstrate the central role of peptide-mediated signaling in Gram-positive bacterial adaptation.

3.3 QS and biofilm-associated infection

Across a wide range of pathogens, quorum sensing is closely associated with biofilm formation. QS regulates key processes such as the production of extracellular polymeric substances (EPS), secretion of matrix proteins, expression of surface attachment factors, and the transition from motile to sessile growth.[20]

Biofilms provide several advantages to bacterial communities, including increased tolerance to antibiotics, protection from host immune responses, metabolic diversity, and the ability to establish persistent infections. Within these structured communities, spatial gradients of nutrients and oxygen influence QS activity, leading to the formation of subpopulations with distinct phenotypes.[10]

This heterogeneity makes biofilms particularly difficult to eradicate and is a major factor contributing to chronic and recurrent infections.

Across pathogens, quorum sensing is intimately linked to biofilm development. QS regulates:[20]

- Extracellular polymeric substance (EPS) production
- Matrix protein secretion
- Surface attachment factors
- Motility-to-sessility transition

Biofilms confer:

- Enhanced antibiotic tolerance
- Immune evasion
- Metabolic heterogeneity
- Chronic infection persistence

Within biofilms, spatial gradients influence QS activation patterns, generating subpopulations with distinct phenotypes.[10] This heterogeneity complicates therapeutic eradication and contributes to recurrent infection.

3.4 QS as a virulence integrator

Rather than acting as a simple on–off switch for individual genes, QS integrates multiple virulence modules, including:

- Type III and type VI secretion systems
- Toxin secretion
- Protease production
- Immune-modulatory molecules
- Iron acquisition systems

This integration ensures coordinated expression only when bacterial density is sufficient to overcome host defenses. Disruption of QS frequently attenuates virulence without abolishing bacterial viability, reinforcing its importance as an anti-virulence therapeutic target.[16] A comparative summary of major quorum sensing systems in clinically relevant pathogens is presented in Table 1.

Table 1 Major quorum sensing systems in clinically relevant pathogens.

Pathogen	QS system	Signal molecule (s)	Key regulatory components	Major QS-regulated functions	Clinical relevance	References
Pseudomonas aeruginosa	Las, Rhl, PQS	3OC12-HSL, C4-HSL, PQS	LasI/LasR, RhlI/RhlR, PqsABCDH	Elastase, pyocyanin, rhamnolipids, biofilm maturation, redox regulation	Chronic lung infections, cystic fibrosis	3,17
Staphylococcus aureus	Agr system	Autoinducing peptides (AIPs; cyclic thiolactone)	AgrD (precursor), AgrB (processing/export), AgrC (sensor kinase), AgrA (response regulator), RNAIII	Toxin secretion, adhesion repression, virulence switching, immune modulation	Skin infections, sepsis, device-associated infections	11
Vibrio cholerae	CAI-1, AI-2 system	CAI-1, AI-2	CqsA/CqsS, LuxS, LuxO, HapR	Density-dependent repression of virulence genes, biofilm dispersal, environmental adaptation	Cholera pathogenesis	3
Enterohemorrhagic *E. coli* (EHEC)	AI-2, QseC system	AI-2	LuxS (AI-2 synthesis), QseC (sensor kinase), QseB (response regulator)	Motility, toxin production, host hormone sensing (epinephrine/norepinephrine), colonization	Gastrointestinal infections	7

(continued)

Table 1 Major quorum sensing systems in clinically relevant pathogens. (*cont'd*)

Pathogen	QS system	Signal molecule (s)	Key regulatory components	Major QS-regulated functions	Clinical relevance	References
Salmonella enterica	AI-2 system	AI-2	LuxS (AI-2 synthesis), Lsr transporter system, LsrR regulator, QseC/QseB two-component system	Motility regulation, biofilm formation, acid stress resistance, virulence gene regulation, host interaction	Enteric and systemic infections	21
Streptococcus pneumoniae	Competence system (Com)	CSP (competence-stimulating peptide; linear peptide)	ComC (precursor), ComAB (processing/export), ComD (sensor kinase), ComE (response regulator)	Genetic competence, DNA uptake, biofilm formation, stress response, fratricide	Respiratory infections	22

4. Host cell reprogramming: Mechanistic foundations

Host–pathogen interactions are increasingly understood not merely as encounters between invading microorganisms and static host defenses, but as dynamic processes involving extensive remodeling of host cellular functions. Upon infection, host cells undergo coordinated changes in gene expression, signaling cascades, metabolic pathways, and chromatin architecture. While many of these responses are initially protective, pathogens frequently manipulate or redirect them to create permissive niches for survival and replication. This phenomenon is broadly termed **host cell reprogramming** and represents a central axis of infection biology and provides the mechanistic context for understanding quorum sensing–mediated host modulation.

Host cell reprogramming encompasses multiple interconnected layers: transcriptional remodeling, epigenetic modification, metabolic rewiring, immune polarization, and regulation of cell survival pathways. These layers do not operate independently; rather, they form integrated networks responsive to microbial signals, inflammatory mediators, and environmental cues.

4.1 Transcriptional reprogramming during infection

Host cells rapidly reconfigure their transcriptional landscape upon microbial invasion through activation of innate immune sensing pathways. Pattern recognition receptors (PRRs), including Toll-like receptors (TLRs), NOD-like receptors, and RIG-I–like receptors, detect conserved microbial motifs and initiate intracellular signaling cascades that drive coordinated gene expression programs.[23,24] These responses extend beyond simple defense mechanisms and represent an adaptive process integrating inflammation, metabolism, and cell fate decisions (**see** Fig. 4).

4.2 NF-κB signaling

NF-κB functions as a central regulator of infection-induced transcriptional responses. In unstimulated cells, NF-κB is sequestered in the cytoplasm by inhibitory IκB proteins. Activation of PRRs triggers phosphorylation and degradation of IκB, allowing NF-κB to translocate into the nucleus and initiate transcription of genes involved in inflammation and immunity.[25]

NF-κB–dependent transcription includes pro-inflammatory cytokines, chemokines, and molecules involved in leukocyte recruitment and survival. Importantly, the strength and duration of NF-κB activation

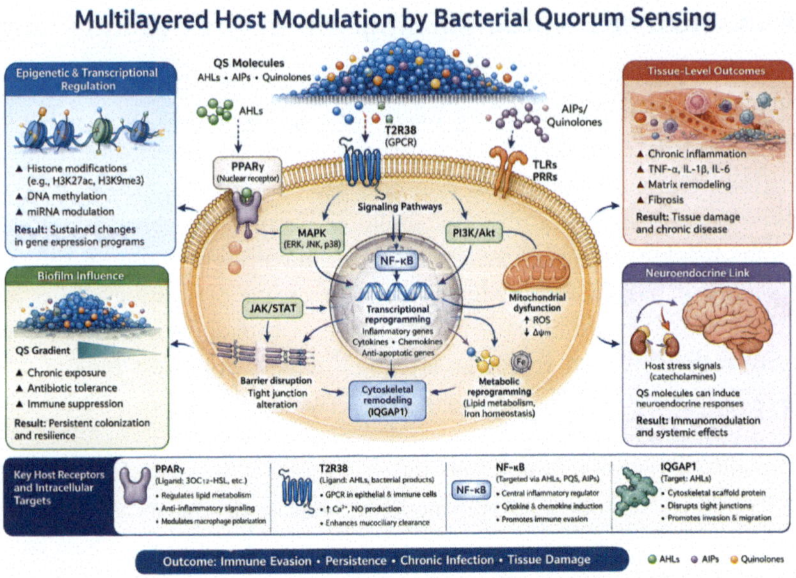

Fig. 4 **Mechanistic framework of quorum sensing–mediated host cell reprogramming.** Bacterial quorum sensing (QS) molecules (AHLs, AIPs, quinolones) interact with host cells via specific receptors and intracellular targets, including PPARγ, T2R38, and NF-κB. These interactions activate signaling pathways (MAPK, PI3K/Akt, JAK–STAT), leading to transcriptional, epigenetic, and metabolic reprogramming. Consequences include cytoskeletal remodeling, barrier disruption, immune modulation, and chronic inflammation, ultimately promoting persistence and disease progression.

determine the outcome of the response: transient activation supports pathogen clearance, whereas prolonged activation contributes to tissue damage and chronic inflammation.[26] Many pathogens target this pathway to either dampen immune responses or exploit inflammatory signaling to enhance their survival.

4.3 MAPK signaling cascades

Mitogen-activated protein kinase (MAPK) pathways—comprising ERK, JNK, and p38—are activated downstream of PRRs and cytokine receptors. These signaling modules regulate transcription factors such as AP-1 and coordinate cellular responses to stress and infection.[27]

MAPK signaling contributes to cytokine production, apoptosis, and tissue remodeling. Its role is highly context-dependent; for example, p38 activation is often associated with inflammatory gene expression, whereas ERK signaling may support cell survival. Pathogens frequently manipulate MAPK pathways to alter host responses.

4.4 Interferon-mediated responses

Interferon signaling represents a critical component of host defense, particularly in antiviral and intracellular bacterial infections. Activation of interferon receptors leads to JAK–STAT signaling and induction of interferon-stimulated genes (ISGs), which establish an antimicrobial state within cells.[28]

ISGs encode proteins that interfere with pathogen replication, enhance antigen presentation, and regulate immune cell activation. While essential for host defense, interferon responses must be tightly controlled, as excessive signaling can contribute to immunopathology and tissue injury.

4.5 Epigenetic regulation of transcription

Transcriptional responses during infection are reinforced by epigenetic mechanisms that control chromatin accessibility. Modifications such as histone acetylation and methylation regulate gene expression by altering chromatin structure.[29]

These changes enable both rapid gene activation and longer-term adaptations in innate immune cells. In some cases, prior exposure to pathogens results in enhanced responsiveness upon re-infection, a phenomenon described as trained immunity.

4.6 Metabolic integration

Transcriptional reprogramming is closely linked to cellular metabolism. Activated immune cells frequently shift toward aerobic glycolysis to support rapid ATP generation and biosynthesis.[30] This metabolic shift also influences signaling pathways and transcription factor activity.

Mitochondria contribute to immune responses through the production of reactive oxygen species (ROS), which participate in antimicrobial defense and signaling processes.

4.7 Systems-level coordination

The host response to infection is governed by a highly interconnected network of signaling pathways rather than isolated molecular events. Key regulatory axes—including NF-κB, MAPK, and interferon-mediated signaling—interact extensively through feedback loops and crosstalk mechanisms to coordinate transcriptional and functional outcomes.

This systems-level integration enables host cells to fine-tune the magnitude, duration, and specificity of immune responses. For instance, simultaneous activation of NF-κB and MAPK pathways can synergistically enhance pro-inflammatory gene expression, while interferon signaling introduces an additional layer of antimicrobial regulation and immune modulation.

Importantly, this coordination ensures that cellular responses remain proportionate to the nature and intensity of the infection. However, disruption of these regulatory networks—whether through pathogen-derived factors or host dysregulation—can lead to maladaptive outcomes, including chronic inflammation, immune suppression, or tissue damage.

Understanding host cell reprogramming therefore requires a network-based perspective in which signaling pathways are viewed as integrated modules rather than independent cascades. This framework is essential for interpreting how external signals, including those derived from bacterial quorum sensing systems, can reshape host cellular behavior at multiple regulatory levels. These integrated mechanisms are illustrated in Fig. 4.

5. Quorum sensing-mediated host modulation

Quorum sensing (QS) molecules exert direct effects on host cells through well-defined molecular targets rather than nonspecific interactions. These include nuclear receptors, G-protein–coupled receptors (GPCRs), scaffold proteins, and intracellular signaling intermediates that collectively mediate host cell reprogramming (Table 2) (Fig. 5).

5.1 Molecular targets of QS signals in host cells

Acyl-homoserine lactones (AHLs), particularly N-3-oxo-dodecanoyl-L-homoserine lactone (3OC12-HSL) produced by *Pseudomonas aeruginosa*, readily diffuse across mammalian cell membranes due to their amphipathic nature. Once internalized, 3OC12-HSL interacts with multiple intracellular targets:

- **Peroxisome proliferator-activated receptors (PPARs):** 3OC12-HSL functions as a ligand for PPARγ, modulating transcription of genes involved in lipid metabolism and inflammation. Activation of PPARγ leads to altered cytokine profiles and suppression of pro-inflammatory responses.
- **IQ motif–containing GTPase-activating protein 1 (IQGAP1):** AHL binding to IQGAP1 disrupts cytoskeletal organization and epithelial barrier integrity, contributing to increased permeability and tissue invasion.

Table 2 Host cellular pathways modulated by quorum sensing molecules.

Host pathway	QS molecule (s) implicated	Mechanistic effect	Functional outcome
NF-κB signaling	AHLs (e.g., 3OC12-HSL)	Altered IκB degradation; modulation of transcription factor activation	Modified cytokine production (TNF-α, IL-6, IL-8)
MAPK pathway (ERK, p38)	AHLs, QS-regulated toxins	Phosphorylation cascade activation	Inflammation, epithelial barrier disruption
JAK–STAT signaling	QS-regulated effectors	Altered interferon signaling	Immune modulation
Mitochondrial ROS signaling	AHLs, PQS	Mitochondrial membrane depolarization	Oxidative stress, apoptosis
Apoptosis pathways	AHLs	Caspase activation	Barrier compromise, dissemination
Autophagy	QS-regulated virulence factors	Altered autophagic flux	Intracellular survival advantage
Epigenetic modification	QS-regulated toxins	Histone acetylation/ methylation changes	Sustained inflammatory gene expression

- **NF-κB signaling pathway:** 3OC12-HSL modulates NF-κB activation through both inhibitory and stimulatory mechanisms depending on concentration and cellular context, influencing transcription of inflammatory mediators such as IL-6, IL-8, and TNF-α.
- **Mitochondrial pathways:** AHLs induce mitochondrial dysfunction, leading to reactive oxygen species (ROS) production and activation of apoptosis pathways via caspase signaling.

Bitter taste receptors (T2Rs), particularly T2R38, represent an additional class of host sensors responsive to bacterial QS molecules. These GPCRs, expressed in airway epithelial cells and immune cells, detect bacterial products including AHL-like molecules. Activation of T2R38 triggers:

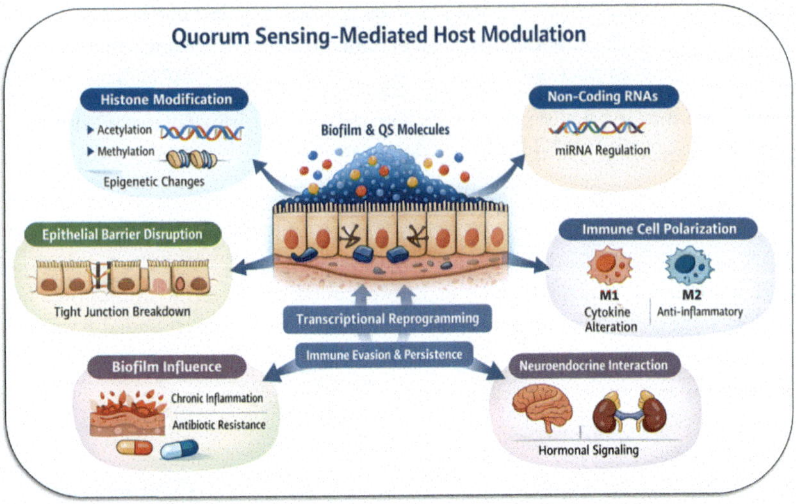

Fig. 5 Quorum sensing-mediated host modulation. Schematic illustrating how bacterial quorum sensing (QS) molecules influence host responses, including epigenetic regulation, non-coding RNA modulation, epithelial barrier disruption, immune cell polarization, biofilm-associated effects, and neuroendocrine interactions, ultimately promoting infection persistence.

- Intracellular calcium signaling
- Nitric oxide (NO) production
- Enhanced mucociliary clearance and antimicrobial responses

This pathway represents a direct link between QS detection and innate immune activation.

Quinolone-based signals, such as the Pseudomonas quinolone signal (PQS), exert host effects through:

- **Iron chelation and redox modulation**, altering intracellular iron availability and oxidative stress pathways
- **Modulation of MAPK and NF-κB signaling**, influencing inflammatory gene expression
- Potential interaction with lipid membrane domains, affecting membrane-associated signaling complexes

Autoinducing peptides (AIPs) from Gram-positive bacteria, although less permeable, interact with host cells primarily through:

- **Pattern recognition receptors (PRRs)** such as Toll-like receptors (TLRs)

- Indirect modulation of immune signaling cascades via cytokine induction
- Possible engagement with membrane-associated receptors, though these mechanisms remain less well defined compared to AHL systems

5.2 From molecular interaction to host reprogramming

These receptor-level interactions translate into coordinated host cell reprogramming through:

- Activation or suppression of transcription factors (e.g., NF-κB, PPARs)
- Epigenetic remodeling via histone modification pathways
- Metabolic rewiring through mitochondrial and lipid signaling
- Cytoskeletal reorganization and barrier disruption

Thus, quorum sensing molecules do not act as passive signals but as active modulators that hijack specific host molecular nodes, enabling pathogens to reshape cellular behavior in a controlled and context-dependent manner.

5.3 Non-coding RNA regulation

Non-coding RNAs (ncRNAs), particularly microRNAs (miRNAs), serve as critical regulators of gene expression at the post-transcriptional level. During infection, QS signaling can indirectly influence ncRNA expression through activation of host signaling pathways.[31]
QS-mediated modulation of ncRNAs may lead to:

- Down regulation of pro-inflammatory responses Certain miRNAs suppress the translation of cytokines and immune signaling molecules.
- Altered immune signaling networks miRNAs can target key transcription factors and signaling intermediates, reshaping host defense pathways.
- Regulation of apoptosis and cell survival Changes in ncRNA expression can influence host cell fate, potentially benefiting pathogen survival.

Additionally, some long non-coding RNAs (lncRNAs) may interact with chromatin-modifying complexes, linking ncRNA regulation with epigenetic control. This creates a multi-layered regulatory network through which QS signals fine-tune host responses.

5.4 QS and epithelial barrier dysfunction

Epithelial tissues represent a crucial physical and immunological barrier that prevents microbial invasion. QS-regulated virulence factors can compromise this barrier through coordinated disruption of cellular structures.[32]

Mechanisms include:

- Disruption of tight junction proteins QS-controlled factors can degrade or redistribute proteins such as occludin, claudins, and zonula occludens, weakening cell–cell adhesion.
- Cytoskeletal remodeling Alterations in actin filament organization lead to structural instability and loss of epithelial integrity.
- Increased paracellular permeability The breakdown of junctional complexes allows pathogens and toxins to pass between cells.

Barrier dysfunction facilitates bacterial invasion, immune cell infiltration, and amplification of inflammatory responses, ultimately contributing to tissue damage and disease progression.[32]

5.5 Immune cell polarization and cytokine landscape

QS signaling plays a significant role in shaping the immune microenvironment by influencing immune cell behavior and cytokine production.[33]

Macrophages, in particular, exhibit functional plasticity and can polarize into:

- M1 phenotype (pro-inflammatory)
- M2 phenotype (anti-inflammatory and tissue repair)

QS molecules can shift this balance by:

- Reducing pro-inflammatory cytokines such as TNF-α
- Modulating anti-inflammatory cytokines such as IL-10
- Altering chemokine gradients that control immune cell recruitment

This reprogramming often results in a less effective immune response, enabling pathogens to evade clearance and establish persistent infections.

5.6 Biofilm-derived QS gradients and microenvironmental reprogramming

Biofilms are structured microbial communities in which QS signaling plays a central role in coordinating collective behavior. Within biofilms, QS molecules form spatial and concentration gradients, exposing nearby host cells to continuous signaling.[34]

This sustained exposure leads to:

- Chronic low-grade inflammation
- Fibrotic tissue remodeling
- Impaired wound healing
- Reduced antibiotic penetration and efficacy

Furthermore, prolonged QS exposure may induce immune tolerance-like states, where immune cells become less responsive to stimuli. This contributes to the persistence and resilience of chronic infections.

5.7 Integration with host neuroendocrine signaling

An emerging aspect of QS research is its interaction with host neuroendocrine systems. Certain bacteria possess receptors capable of detecting host-derived catecholamines such as epinephrine and norepinephrine.[7]
This interaction enables:

- Sensing of host stress conditions
- Regulation of bacterial virulence gene expression
- Integration of microbial and host signaling networks

The bidirectional nature of this communication highlights QS as a mediator of inter-kingdom signaling, linking microbial behavior with host physiological states.

5.8 Systems biology perspective

A comprehensive understanding of QS-mediated host modulation requires integration of multi-omics approaches and computational modeling. Techniques such as transcriptomics, proteomics, and dual RNA sequencing have revealed coordinated changes in host and bacterial gene expression.[35]
Key insights include:

- Synchronized transcriptional responses between host and pathogen
- Complex regulatory networks governing infection dynamics
- Adaptive mechanisms that enhance infection stability

Mathematical models further demonstrate that QS-mediated interactions contribute to robustness and persistence of infections, particularly under fluctuating environmental conditions.

6. Biofilms, chronic infection, and tissue remodeling

Biofilm formation represents one of the most clinically significant outcomes of quorum sensing (QS) activity in pathogenic microorganisms. In contrast to planktonic growth, bacteria within biofilms adopt a sessile lifestyle characterized by multicellular organization, extracellular matrix production, and metabolic heterogeneity. This structural complexity provides enhanced resistance to antimicrobial agents and host immune defenses. Importantly, biofilm-associated infections are typically chronic, marked by persistent inflammation, continuous immune activation, and progressive tissue remodeling.[36,37]

QS signaling plays a central regulatory role in coordinating biofilm development and maintaining the host–pathogen interface. By synchronizing gene expression across bacterial populations, QS ensures efficient adaptation to environmental conditions and promotes long-term persistence within host tissues.[20,34]

6.1 Quorum sensing as a driver of biofilm development

Biofilm formation is a dynamic and regulated process involving initial attachment, microcolony formation, maturation, and eventual dispersion. QS systems coordinate gene expression at each stage, enabling bacteria to transition between planktonic and sessile states.[20]

6.1.1 Regulation of extracellular polymeric substance (EPS) production

A hallmark of biofilms is the extracellular polymeric substance (EPS) matrix, composed of polysaccharides, extracellular DNA (eDNA), proteins, and lipids. QS regulates genes involved in EPS synthesis and secretion, thereby controlling matrix composition and architecture.[36]

In Gram-negative bacteria, AHL-dependent QS systems regulate polysaccharide production and secretion pathways, while Gram-positive bacteria utilize peptide-based signaling to control matrix-associated proteins. The EPS matrix not only stabilizes biofilm structure but also functions as a diffusion barrier, limiting antimicrobial penetration and modulating nutrient gradients within the biofilm.[38]

6.1.2 Spatial organization and QS gradients

Within biofilms, limited diffusion and high cell density lead to the accumulation of QS molecules, creating spatial gradients of signaling compounds. These gradients drive heterogeneous activation of QS-regulated genes, resulting in functionally distinct bacterial subpopulations.[10]
Consequences of this heterogeneity include:

- Subpopulations with increased antibiotic tolerance
- Metabolic diversity (aerobic vs. anaerobic niches)
- Differential expression of virulence factors

Such complexity enhances the resilience of biofilms and contributes to their resistance to host immune clearance.

6.2 Immune evasion in biofilm-associated infections

Biofilm-associated bacteria employ multiple QS-regulated mechanisms to evade host immune defenses.

6.2.1 *Impaired phagocytosis*

The EPS matrix physically restricts access of phagocytic cells such as neutrophils and macrophages. Additionally, QS-regulated virulence factors can impair immune cell function by inducing apoptosis, inhibiting oxidative burst activity, and interfering with complement pathways.[34,39]

These mechanisms collectively reduce effective bacterial clearance and promote persistence within host tissues.

6.2.2 *Modulation of cytokine responses*

Chronic biofilm infections are characterized by sustained yet ineffective inflammatory responses. QS-regulated molecules can alter cytokine production, leading to dysregulated immune signaling.[40]

Typical alterations include:

- Persistent IL-6 and IL-8 secretion
- Dysregulated TNF-α production
- Imbalance between pro- and anti-inflammatory responses

This cytokine dysregulation contributes to tissue damage while failing to eradicate the pathogen.

6.3 Chronic inflammation and host tissue remodeling

Prolonged exposure to QS signals and biofilm-associated virulence factors drives progressive changes in host tissue structure.

6.3.1 *Epithelial barrier disruption*

QS-regulated enzymes and toxins degrade tight junction proteins such as occludin and claudins, compromising epithelial integrity. This increases permeability and facilitates bacterial invasion.[32]

Barrier dysfunction also enhances exposure of underlying tissues to inflammatory mediators, amplifying immune activation.

6.3.2 Fibrosis and extracellular matrix remodeling

Chronic inflammation stimulates fibroblast activation and extracellular matrix deposition. Key mediators such as transforming growth factor-β (TGF-β) and matrix metalloproteinases (MMPs) drive collagen accumulation and tissue remodeling.[41]

These processes result in tissue stiffening, structural distortion, and functional decline, particularly in chronic diseases such as cystic fibrosis and non-healing wounds.

6.3.3 Angiogenesis and hypoxia

Biofilm growth often generates hypoxic microenvironments due to high metabolic activity and limited oxygen diffusion. Stabilization of hypoxia-inducible factor-1α (HIF-1α) promotes angiogenesis, metabolic reprogramming, and inflammatory gene expression.[42]

QS-mediated bacterial metabolism may further reinforce hypoxic conditions, contributing to persistent inflammation and tissue remodeling.

6.4 Antibiotic tolerance and persistence

QS contributes to antibiotic tolerance through multiple mechanisms, including regulation of efflux pumps, induction of stress responses, formation of persister cells, and reduced metabolic activity within biofilms.[43]

These adaptations represent phenotypic plasticity rather than genetic resistance, allowing bacteria to survive antibiotic treatment. Consequently, infections often recur after therapy due to the persistence of biofilm-associated populations.

6.5 Host immune exhaustion and tolerance

Chronic biofilm infections can lead to immune exhaustion, characterized by reduced responsiveness of immune cells. Macrophages and T cells exposed to prolonged QS signaling may exhibit decreased cytokine production, altered metabolism, and impaired antimicrobial activity.[44]

This dysfunction reinforces infection persistence and limits effective immune clearance.

6.6 Clinical contexts of QS-driven biofilm pathology

QS-regulated biofilm infections are associated with numerous clinical conditions, including:

- Chronic lung infections (e.g., cystic fibrosis)
- Medical device-associated infections

- Chronic wounds
- Osteomyelitis
- Endocarditis

In these contexts, conventional antibiotic therapy is often insufficient, and treatment may require mechanical disruption or removal of infected tissues.[37]

6.7 Systems-level integration of chronic infection

From a systems biology perspective, biofilm-associated infections represent stable host–pathogen ecosystems. These systems are characterized by persistent QS signaling, sustained inflammatory feedback, metabolic interactions, and progressive tissue remodeling.[34]

This framework emphasizes that chronic infection is not merely a failure of immune defense but a dynamically maintained equilibrium shaped by microbial communication and host responses.

7. Therapeutic and translational strategies targeting quorum sensing

The recognition that quorum sensing (QS) governs virulence, biofilm maturation, immune modulation, and chronic infection persistence has catalyzed the development of anti-QS therapeutic strategies. Unlike conventional antibiotics, which target essential processes required for bacterial survival, QS-targeted approaches aim to attenuate pathogenicity without exerting direct bactericidal pressure. This anti-virulence paradigm is hypothesized to reduce the emergence of resistance while preserving commensal microbial communities.[45,46]

QS-targeted interventions can be broadly categorized into enzymatic quorum quenching, small-molecule inhibitors, natural compounds, biofilm-disrupting approaches, and host-directed therapies. Increasingly, systems and synthetic biology strategies are also being explored to disrupt microbial communication networks at multiple levels.

7.1 Quorum quenching enzymes

Quorum quenching refers to the enzymatic inactivation or degradation of QS signaling molecules, thereby preventing receptor activation and downstream transcriptional responses. This strategy directly targets the communication system rather than bacterial viability.

7.1.1 AHL lactonases and acylases

In Gram-negative bacteria, AHL-mediated QS systems can be disrupted by enzymes such as lactonases and acylases. Lactonases hydrolyze the homoserine lactone ring, while acylases cleave the amide bond linking the acyl chain to the lactone moiety, effectively inactivating the signal molecule.[47,48]

By reducing autoinducer concentrations below activation thresholds, these enzymes disrupt positive feedback loops essential for QS amplification. Experimental models have demonstrated that expression of lactonases can significantly reduce biofilm formation, attenuate virulence factor production, and improve host survival outcomes.[49]

Importantly, because quorum quenching does not inhibit bacterial growth directly, it aligns with the anti-virulence strategy and may impose reduced selective pressure for resistance development.

7.1.2 AI-2 disruption strategies

Autoinducer-2 (AI-2) represents a universal signaling molecule involved in interspecies communication. Targeting AI-2 signaling is more complex due to its integration with central metabolic pathways. Nevertheless, strategies aimed at blocking AI-2 transport systems or receptor binding have shown promise in polymicrobial infection models, where interspecies signaling contributes to pathogenic synergy.[21]

7.2 Small-molecule QS inhibitors (QSIs)

Small-molecule QS inhibitors (QSIs) represent one of the most actively investigated therapeutic classes. These compounds can interfere with QS signaling at multiple stages, including signal synthesis, receptor binding, and transcriptional activation.

7.2.1 Signal synthesis inhibitors

Inhibitors targeting signal synthase enzymes (e.g., LuxI homologs) can reduce the production of autoinducers. Structural analogs of S-adenosylmethionine (SAM) and acyl-acyl carrier protein (acyl-ACP) substrates have been explored as competitive inhibitors.[50]

However, challenges remain regarding specificity, potential off-target effects on host metabolism, and pharmacokinetic stability in vivo.

7.2.2 Receptor antagonists

Signal analogs that mimic native QS molecules can competitively bind to QS receptors without activating transcription. These antagonists effectively block signal perception and downstream gene expression.[4]

Key considerations for clinical development include binding affinity relative to endogenous signals, susceptibility to resistance via receptor mutations, and potential cross-reactivity with host signaling pathways. Despite these challenges, several synthetic QS antagonists have demonstrated efficacy in reducing virulence in preclinical infection models.

7.3 Natural compounds and phytochemicals

A growing body of evidence highlights the QS-inhibitory potential of natural compounds, particularly plant-derived phytochemicals such as flavonoids, phenolic acids, and essential oils.[51]

These compounds may act through multiple mechanisms, including:

- Interference with receptor–ligand interactions
- Inhibition of signal synthesis enzymes
- Modulation of membrane properties affecting signal diffusion

Their structural diversity and relative safety profiles make them attractive candidates for combinatorial therapies, particularly in the context of chronic infections.

7.4 Anti-biofilm strategies

Given the central role of QS in biofilm formation and maintenance, targeting QS pathways offers a promising approach to destabilize biofilm architecture.

7.4.1 Matrix-degrading agents

Enzymes that degrade biofilm matrix components—such as DNases, polysaccharide hydrolases, and proteases—can enhance antibiotic penetration and immune access.[52] When combined with QS inhibitors, these agents may synergistically disrupt biofilm integrity and reduce bacterial persistence.

7.4.2 Nanotechnology-based delivery systems

Nanoparticle-based delivery systems provide a means to enhance the efficacy of QS-targeted therapies. These systems can:

- Increase local drug concentration at infection sites
- Protect QS inhibitors from degradation
- Enable controlled and sustained release

Additionally, nanocarriers can be engineered to co-deliver antibiotics and QS inhibitors, thereby targeting both bacterial viability and communication pathways simultaneously.[53]

7.5 Host-directed therapies

An alternative therapeutic strategy involves targeting host pathways modulated by QS signaling rather than directly interfering with bacterial communication systems.
Potential approaches include:

- Modulation of NF-κB–mediated inflammatory responses
- Enhancement of autophagic clearance mechanisms
- Restoration of epithelial barrier function
- Correction of metabolic dysregulation

These strategies aim to reverse QS-induced host cell reprogramming, thereby limiting tissue damage and enhancing immune-mediated pathogen clearance.[54] A summary of current therapeutic strategies targeting quorum sensing pathways and host–pathogen interactions is provided in Table 3.

7.6 Resistance considerations and evolutionary constraints

Quorum sensing inhibitors (QSIs) have been widely proposed as anti-virulence strategies with a reduced likelihood of resistance development compared to conventional antibiotics. This assumption is based on the premise that QSIs do not directly affect bacterial viability, thereby exerting weaker selective pressure. However, emerging evidence suggests that this paradigm requires critical re-evaluation. Emerging evidence challenges the notion that quorum sensing inhibitors are evolutionarily robust. Mutations in efflux pumps (e.g., MexAB-OprM in P. aeruginosa) can reduce intracellular accumulation of QS inhibitors, conferring resistance. Additionally, social "cheater" dynamics may select for variants that bypass QS dependency, thereby maintaining virulence even under quorum quenching pressure.

7.6.1 Evolutionary pressure on anti-virulence strategies

- Although QSIs do not kill bacteria, they can still impose **fitness costs** by disrupting cooperative behaviors such as virulence factor production and biofilm formation. In infection contexts where these traits are essential for survival or transmission, inhibition of QS can Cheaters may gain short-term advantages but can destabilize population fitness

Table 3 Therapeutic strategies targeting quorum sensing and host–pathogen interactions.

Therapeutic class	Target	Mechanism of action	Representative examples	Functional outcome	References
Quorum quenching enzymes	Signal molecules (AHLs, AIPs)	Enzymatic degradation or modification of autoinducers (e.g., lactone hydrolysis, amide cleavage)	AHL lactonases (AiiA), acylases	Reduced signal accumulation, inhibition of QS activation	47,48
Signal synthesis inhibitors	QS signal synthase (e.g., LuxI homologs)	Inhibition of autoinducer biosynthetic pathways	SAM analogs, acyl-ACP analogs	Decreased production of QS signals	50
Receptor antagonists	QS receptors (e.g., LuxR-type, AgrC)	Competitive binding to receptors without activation	Halogenated furanones, AHL analogs	Blocked signal perception and gene activation	4,55
Signal analogs (agonist/antagonist modulators)	QS regulatory circuits	Modulation of receptor activity (agonistic or antagonistic)	Synthetic AHL mimics, peptide analogs	Dysregulation of QS signaling networks	45
Natural compounds (phytochemicals)	Multiple QS targets	Interference with signal synthesis, receptor binding, or membrane dynamics	Flavonoids, phenolic acids, essential oils	Attenuation of virulence and biofilm formation	51

(continued)

Table 3 Therapeutic strategies targeting quorum sensing and host–pathogen interactions. (*cont'd*)

Therapeutic class	Target	Mechanism of action	Representative examples	Functional outcome	References
Anti-biofilm agents	Biofilm matrix and QS-regulated pathways	Degradation of extracellular matrix components; disruption of biofilm integrity	DNases, dispersin B, proteases	Enhanced antibiotic penetration and biofilm dispersal	52
Nanotechnology-based delivery systems	Biofilm/QS microenvironment	Targeted delivery of QS inhibitors and antimicrobials	Liposomes, polymeric nanoparticles	Improved drug stability and localized efficacy	53
Host-directed therapies	Host signaling pathways	Modulation of host responses affected by QS (e.g., NF-κB, autophagy)	Anti-inflammatory agents, metabolic modulators	Restored immune function and reduced tissue damage	54
Synthetic biology approaches	QS regulatory genes/networks	Genetic or engineered disruption of QS circuits	CRISPR-based systems, engineered probiotics	Targeted suppression of QS signaling	56

- Under certain conditions, QS-proficient strains regain dominance, especially when virulence traits are required for host colonization
- QSI exposure may shift population structure, indirectly selecting for strains with restored or modified QS functionality create selective pressure favoring resistant variants.

7.6.2 Cheater dynamics and social evolution

Quorum sensing systems are inherently linked to **cooperative behaviors**, making them susceptible to exploitation by "cheater" populations that do not produce signals but benefit from communal outputs. While QSIs may initially promote the emergence of cheaters, long-term evolutionary dynamics are more complex:

Thus, social evolution does not eliminate resistance risk but instead reshapes it.

7.6.3 Efflux pumps and reduced intracellular accumulation

A growing body of evidence indicates that bacteria can develop resistance to QSIs through **upregulation of multidrug efflux pumps**, which reduce intracellular concentrations of inhibitory compounds. For example:

- Efflux systems such as MexAB-OprM in Pseudomonas aeruginosa can export QSIs
- This mechanism confers cross-resistance to structurally diverse inhibitors
- Efflux-mediated resistance can arise rapidly under selective pressure

This highlights that QSIs are not exempt from classical resistance mechanisms.

7.6.4 Receptor mutations and signal pathway rewiring

Mutations in QS regulatory proteins (e.g., LuxR-type receptors or AgrC/AgrA systems) can reduce inhibitor binding while maintaining responsiveness to native signals. Additionally:

- Bacteria may activate **alternative signaling pathways**
- Redundant or hierarchical QS systems (e.g., Las, Rhl, PQS in Pseudomonas) enable **compensatory regulation**
- Network plasticity allows maintenance of virulence even when one pathway is inhibited

Such adaptability underscores the resilience of QS networks.

7.6.5 Context-dependent resistance development

Importantly, resistance to QSIs is highly **context-dependent**:

- In acute infections, where virulence is essential, selective pressure may be strong
- In chronic biofilm-associated infections, spatial structure and heterogeneity may buffer selective effects
- Host immune interactions further influence evolutionary trajectories

These factors complicate predictions regarding resistance emergence.

7.6.6 Implications for therapeutic development

Collectively, these findings challenge the notion that anti-virulence therapies are inherently "resistance-proof." Instead, QSIs should be viewed as part of an evolutionarily dynamic therapeutic landscape.
Future strategies should consider:

- Combination therapies (QSIs + antibiotics)
- Targeting multiple QS pathways simultaneously
- Designing inhibitors less susceptible to efflux
- Incorporating evolutionary principles into drug design

A nuanced understanding of microbial social dynamics and adaptive mechanisms is essential for the successful deployment of quorum sensing–targeted therapies.

7.7 Clinical development and translational challenges

Despite promising preclinical data, the translation of QS-targeted therapies into clinical practice remains limited. Key challenges include:

- Optimization of pharmacokinetics and bioavailability
- Stability under physiological conditions
- Assessment of toxicity and off-target effects
- Regulatory hurdles for anti-virulence therapeutics

Furthermore, the diversity of QS systems across bacterial species complicates the development of broad-spectrum QS inhibitors. Precision approaches tailored to specific pathogens or infection contexts may therefore be required.

8. Future directions and emerging technologies

The expanding recognition that quorum sensing (QS) systems influence not only bacterial population dynamics but also host cellular reprogramming has opened new conceptual and technological frontiers. Future research will likely move beyond descriptive mechanistic studies toward integrative, systems-level approaches that combine microbiology, immunology, computational biology, and translational medicine. This section outlines key directions that may shape the next decade of research in QS-mediated host–pathogen interactions.

8.1 Multi-omics integration and systems biology

Traditional studies often analyze either bacterial or host responses in isolation. However, infection is fundamentally a bidirectional process. Emerging **multi-omics approaches**—including transcriptomics, proteomics, metabolomics, and epigenomics—enable simultaneous profiling of both host and pathogen.

8.1.1 Dual RNA sequencing

Dual RNA-seq technologies allow concurrent analysis of bacterial and host transcriptomes within infected tissues. This approach provides insight into:

- Temporal coordination between QS activation and host transcriptional responses
- Feedback loops between inflammatory signaling and bacterial regulatory networks
- Heterogeneity within infection microenvironments

Such data can identify synchronized regulatory nodes that may serve as therapeutic targets.

8.1.2 Spatial transcriptomics and single-cell analysis

Biofilm-associated infections exhibit strong spatial heterogeneity. Spatial transcriptomics and single-cell sequencing technologies now permit mapping of gene expression patterns across infected tissues.
These tools can reveal:

- Localized QS activation gradients
- Immune cell polarization states adjacent to biofilms
- Microenvironment-specific metabolic interactions

Understanding spatial organization will be critical for designing targeted interventions.

8.2 Structural biology and rational drug design

Advances in cryo-electron microscopy and computational modeling have accelerated structural characterization of QS receptors and regulatory complexes. High-resolution structural data enable:

- Rational design of receptor antagonists
- Optimization of binding affinity and specificity
- Identification of allosteric regulatory sites

Machine learning–assisted drug discovery may further accelerate development of next-generation QS inhibitors.

8.3 Synthetic biology and engineered microbiota

Synthetic biology approaches are increasingly applied to modulate microbial communication systems.

8.3.1 Engineered quorum quenching microbes

Genetically engineered commensal bacteria capable of degrading QS signals represent a promising avenue for microbiome-based therapy. Such organisms could:

- Disrupt pathogenic communication networks
- Reduce virulence without eliminating beneficial microbes
- Restore ecological balance in dysbiotic environments

8.3.2 Programmable gene circuits

CRISPR-based gene regulation systems offer potential for selectively silencing QS genes in pathogenic populations. Delivery methods remain a challenge, but targeted bacteriophage vectors may provide a solution.

8.4 Host-centric therapeutic innovation

Given the impact of QS on host signaling pathways, future strategies may increasingly focus on **host-directed therapies**.
Potential avenues include:

- Enhancing epithelial barrier resilience
- Modulating metabolic reprogramming to restore bactericidal states
- Targeting epigenetic regulators altered during chronic infection
- Correcting immune exhaustion phenotypes

Combining host-directed therapy with QS inhibition may yield synergistic benefits.

8.5 Microbiome context and ecological perspectives

QS signaling does not occur in isolation but within complex polymicrobial communities. Future studies must incorporate ecological frameworks to understand:

- Interspecies communication
- Competitive or cooperative signaling networks
- Microbiome-driven modulation of host immunity

Disruption of QS in one species may have unintended effects on microbial community dynamics. Therefore, ecological modeling will be essential.

8.6 Artificial intelligence and predictive modeling

Computational modeling and artificial intelligence (AI) tools are poised to transform QS research.
Applications include:

- Predictive modeling of QS network behavior
- Simulation of host–pathogen signaling interactions
- Drug screening through in silico docking
- Personalized infection risk assessment

AI-driven integration of multi-omics datasets may reveal emergent properties not apparent from reductionist approaches.

8.7 Translational and clinical challenges

Despite significant advances, key challenges remain:

- Interpatient variability in QS activation profiles
- Limited biomarkers for monitoring QS activity in vivo
- Complexity of polymicrobial infections
- Balancing anti-virulence strategies with microbiome preservation

Future clinical translation will require standardized assays, robust biomarkers, and interdisciplinary collaboration.

8.8 Conceptual shift: Infection as a communication network

Perhaps the most profound future direction is conceptual rather than technical. Viewing infection as a **communication-driven ecosystem** reframes pathogenesis as an emergent property of interacting regulatory networks.

In this framework:

- Bacterial QS systems
- Host signaling pathways
- Metabolic circuits
- Immune responses are components of a dynamic communication web. Disrupting pathogenic outcomes may therefore require network-level interventions rather than single-target strategies.

9. Conclusion

Quorum sensing (QS) has evolved from a conceptual framework describing bacterial population-density–dependent gene regulation into a central paradigm for understanding the complexity of host–pathogen interactions. What was once considered an intra-microbial communication system is now recognized as a sophisticated regulatory network capable of influencing not only bacterial collective behavior but also host cellular physiology. The growing body of evidence demonstrating that QS molecules directly and indirectly modulate host signaling pathways underscores the depth of cross-kingdom communication occurring during infection.

This chapter has outlined the molecular architecture of QS systems across Gram-negative and Gram-positive bacteria, emphasizing their hierarchical organization, signal diversity, and integration with global regulatory networks. In major pathogens, QS coordinates virulence expression, biofilm formation, metabolic adaptation, and stress responses—functions that are tightly linked to infection progression and persistence. Importantly, these systems do not operate in isolation; rather, they interface with host tissues through direct signaling interactions and through QS-regulated effector molecules.

Host cell reprogramming represents a complementary dimension of infection biology. Infection triggers coordinated changes in transcriptional programs, epigenetic landscapes, metabolic states, and cell survival pathways. While many of these responses are protective, pathogens have evolved mechanisms to reshape them to their advantage. QS molecules, particularly acyl-homoserine lactones and peptide autoinducers, contribute to this reprogramming by modulating key host pathways, including NF-κB, MAPK, JAK–STAT signaling, and mitochondrial function.

Through these mechanisms, bacterial communication systems influence inflammatory tone, epithelial barrier integrity, immune cell polarization, and apoptosis regulation.

The integration of QS signaling with biofilm development further amplifies its impact on host tissues. Biofilm-associated infections illustrate how sustained QS activity can create structured microenvironments characterized by immune evasion, metabolic heterogeneity, chronic inflammation, and progressive tissue remodeling. In such settings, infection becomes less an acute battle and more a stabilized ecological interaction maintained by persistent signaling networks.

Therapeutically, targeting QS presents an attractive alternative to traditional antimicrobial strategies. Anti-virulence approaches—such as quorum quenching enzymes, receptor antagonists, and biofilm-disrupting technologies—seek to attenuate pathogenic potential without necessarily compromising bacterial viability. While translational challenges remain, advances in structural biology, nanotechnology, synthetic biology, and computational modeling are expanding the toolkit for precise QS modulation. Moreover, host-directed therapies that reverse QS-mediated cellular reprogramming may provide complementary avenues for intervention.

Looking forward, the convergence of multi-omics technologies, spatial transcriptomics, artificial intelligence, and systems-level modeling will deepen our understanding of QS-driven host–pathogen communication. These integrative approaches are likely to reveal emergent properties of infection networks that cannot be captured through reductionist analyses. Importantly, they reinforce a conceptual shift: infection should be viewed not simply as microbial invasion, but as a dynamic communication ecosystem shaped by reciprocal signaling processes.

In summary, quorum sensing represents a central regulatory axis linking microbial collective behavior to host cellular reprogramming. Its influence extends from molecular signaling cascades to tissue-level pathology and clinical outcomes. Elucidating and therapeutically targeting this interface holds significant promise for addressing chronic infections, antibiotic resistance, and inflammatory tissue damage. As research continues to unravel the complexity of microbial communication networks, the integration of QS biology with host systems science will remain a critical frontier in infection biology and translational medicine.

References

1. Bassler BL. Small talk: cell-to-cell communication in bacteria. *Cell.* 2002;109(4):421–424.
2. Miller MB, Bassler BL. Quorum sensing in bacteria. *Annu. Rev. Microbiol.* 2001;55:165–199.
3. Papenfort K, Bassler BL. Quorum sensing signal–response systems. *Nat. Rev. Microbiol.* 2016;14(9):576–588.
4. Ng WL, Bassler BL. Bacterial quorum-sensing network architectures. *Annu. Rev. Genet.* 2009;43:197–222.
5. Nealson KH, Platt T, Hastings JW. Cellular control of the synthesis and activity of the bacterial luminescent system. *J. Bacteriol.* 1970;104(1):313–322.
6. Fuqua WC, Winans SC, Greenberg EP. Quorum sensing in bacteria: the LuxR-LuxI family of cell density-responsive transcriptional regulators. *J. Bacteriol.* 1994;176(2):269–275.
7. Hughes DT, Sperandio V. Inter-kingdom signalling: communication between bacteria and their hosts. *Nat. Rev. Microbiol.* 2008;6(2):111–120.
8. Medzhitov R, Schneider DS, Soares MP. Disease tolerance as a defense strategy. *Science.* 2012;335(6071):936–941.
9. Kendall MM, Sperandio V. What a dinner party! Mechanisms and functions of inter-kingdom signaling in host–pathogen associations. *Nat. Rev. Microbiol.* 2016;14(11):665–679.
10. Stewart PS, Franklin MJ. Physiological heterogeneity in biofilms. *Nat. Rev. Microbiol.* 2008;6(3):199–210.
11. Novick RP, Geisinger E. Quorum sensing in staphylococci. *Annu. Rev. Genet.* 2008;42:541–564.
12. Xavier KB, Bassler BL. LuxS quorum sensing: more than just a numbers game. *Curr. OpMicrobiol.* 2003;6(2):191–197.
13. Lee J, Zhang L. The hierarchy quorum sensing network in *Pseudomonas aeruginosa. Protein & Cell.* 2015;6(1):26–41.
14. Diggle SP, Cornelis P, Williams P, Cámara M. 4-Quinolone signalling in *Pseudomonas aeruginosa*: old molecules, new perspectives. *Int. J. Med. Microbiol.* 2006;296(2–3):83–91.
15. Diggle SP, Griffin AS, Campbell GS, West SA. Cooperation and conflict in quorum-sensing bacterial populations. *Nature.* 2007;450(7168):411–414.
16. Rutherford ST, Bassler BL. Bacterial quorum sensing: its role in virulence and possibilities for its control. *Cold Spring Harb Perspect. Med.* 2012;2(11):a012427.
17. Williams P, Cámara M. Quorum sensing and environmental adaptation in *Pseudomonas aeruginosa*: a tale of regulatory networks and multifunctional signal molecules. *Philos. Trans. R. Soc. B: Biol. Sci.* 2009;364(1524):1247–1257.
18. Rumbaugh KP, Griswold JA, Iglewski BH, Hamood AN. Contribution of quorum sensing to the virulence of Pseudomonas aeruginosa in burn wound infections. *Infect. Immun.* 1999;67(11):5854–5862.
19. Hammer BK, Bassler BL. Quorum sensing controls biofilm formation in *Vibrio cholerae. Mol. Microbiol.* 2003;50(1):101–104. https://doi.org/10.1046/j.1365-2958.2003.03688.x
20. Davies DG, et al. The involvement of cell-to-cell signals in biofilm development. *Science.* 1998;280(5361):295–298.
21. Pereira CS, Thompson JA, Xavier KB. AI-2-mediated signalling in bacteria. *FEMS Microbiol. Rev.* 2013;37(2):156–181.
22. Johnston C, Martin B, Fichant G, Polard P, Claverys JP. Bacterial transformation: distribution, shared mechanisms and divergent control. *Nat. Rev. Microbiol.* 2014;12(3):181–196.
23. Medzhito R. Recognition of microorganisms and activation of the immune response. *Nature.* 2007;449(7164):819–826.

24. Takeuchi O, Akira S. Pattern recognition receptors and inflammation. *Cell.* 2010;140(6):805–820.
25. Hayden MS, Ghosh S. Shared principles in NF-κB signaling. *Cell.* 2008;132(3):344–362.
26. Rahman MM, McFadden G. Modulation of NF-κB signalling by microbial pathogens. *Nat. Rev. Microbiol.* 2011;9:291–306.
27. Arthur JSC, Ley SC. Mitogen-activated protein kinases in innate immunity. *Nat. Rev. Microbiol.* 2013;13:679–692.
28. Ivashkiv LB, Donlin LT. Regulation of type I interferon responses. *Nat. Rev. Microbiol.* 2014;14:36–49.
29. Netea MG, Quintin J, van der Meer JWM. Trained immunity: a memory for innate host defense. *Cell Host & Microbe.* 2016;9(5):355–361.
30. O'Neill LAJ, Kishton RJ, Rathmell J. A guide to immunometabolism for immunologists. *Nat. Rev. Microbiol.* 2016;16:553–565.
31. O'Connell RM, Rao DS, Baltimore D. MicroRNA regulation of inflammatory responses. *Annu. Rev. Immunol.* 2012;30:295–312.
32. Turner JR. Intestinal mucosal barrier function in health and disease. *Nat. Rev. Immunol.* 2009;9(11):799–809.
33. Murray PJ, Wynn TA. Protective and pathogenic functions of macrophage subsets. *Nat. Rev. Immunol.* 2011;11(11):723–737.
34. Bjarnsholt T. The role of bacterial biofilms in chronic infections. *APMIS Supplementum.* 2013;136:1–51.
35. Westermann AJ, Gorski SA, Vogel J. Dual RNA-seq of pathogen and host. *Nat. Rev. Microbiol.* 2012;10(9):618–630.
36. Flemming HC, Wingender J. The biofilm matrix. *Nat. Rev. Microbiol.* 2010;8(9):623–633.
37. Costerton JW, Stewart PS, Greenberg EP. Bacterial biofilms: a common cause of persistent infections. *Science.* 1999;284(5418):1318–1322.
38. Flemming HC, et al. Biofilms: an emergent form of bacterial life. *Nat. Rev. Microbiol.* 2016;14(9):563–575.
39. Jesaitis AJ, et al. Compromised host defense on Pseudomonas aeruginosa biofilms. *J. Immunol.* 2003;171(8):4329–4339.
40. Høiby N, et al. Antibiotic resistance of bacterial biofilms. *Int. J. Antimicrobial Agents.* 2010;35(4):322–332.
41. Wynn TA. Cellular and molecular mechanisms of fibrosis. *J. Pathol.* 2008;214(2):199–210.
42. Semenza GL. Hypoxia-inducible factors in physiology and medicine. *Cell.* 2012;148(3):399–408.
43. Lewis K. Persister cells, dormancy and infectious disease. *Nat. Rev. Microbiol.* 2007;5(1):48–56.
44. Wherry EJ, Kurachi M. Molecular and cellular insights into T cell exhaustion. *Nat. Rev. Immunol.* 2015;15(8):486–499.
45. LaSarre B, Federle MJ. Exploiting quorum sensing to confuse bacterial pathogens. *Microbiol. Mol. Biol. Rev.* 2013;77(1):73–111.
46. Rasko DA, Sperandio V. Anti-virulence strategies to combat bacteria-mediated disease. *Nat. Rev. Drug Discov.* 2010;9(2):117–128.
47. Dong YH, Wang LH, Zhang LH. Quorum-quenching microbial infections. *Nature.* 2001;411:813–817.
48. Fetzner S. Quorum quenching enzymes. *J. Biotechnol.* 2015;201:2–14.
49. Hentzer M, Givskov M. Pharmacological inhibition of quorum sensing for the treatment of chronic bacterial infections. *J. Clin. Invest.* 2003;112(9):1300–1307.

50. Geske GD, O'Neill JC, Miller DM, Blackwell HE. Modulation of bacterial quorum sensing. *J. Am. Chem. Soc.* 2008;130(4):1096–1105.
51. Vattem DA, et al. Dietary phytochemicals as quorum sensing inhibitors. *J. Food Sci.* 2007;72(4):R50–R55.
52. Kaplan JB. Therapeutic potential of biofilm-dispersing enzymes. *Int. J. Artif Organs.* 2009;32(9):545–554.
53. Pelgrift RY, Friedman AJ. Nanotechnology in antimicrobial therapy. *Adv. Drug Delivery Rev.* 2013;65(13–14):1803–1815.
54. Ritchie JM, et al. Host-pathogen interactions and therapeutic targeting. *Cell Host & Microbe.* 2015;18(4):441–451.
55. Hentzer M, et al. Attenuation of Pseudomonas aeruginosa virulence by quorum sensing inhibitors. *EMBO J.* 2003;22(15):3803–3815.
56. Hwang IY, et al. Engineered probiotic systems for therapeutic applications. *Nat. Biotechnol.* 2017;35(11):1080–1087.

CHAPTER THREE

Mechanism of biofilm formation and host interactions

Parul Thapar[a,*], Moumita Dev[b], and Nagamani Bora[c]
[a]Department of Nutrition and Dietetics, School of Allied Health Sciences, Manav Rachna International Institute of Research and Studies (Deemed to be University), Faridabad, Haryana, India
[b]Department of Food and Nutrition, Swami Vivekananda University, Barrackpore, Kolkata, West Bengal, India
[c]Division of Microbiology, Brewing and Biotechnology, School of Biosciences, University of Nottingham, Sutton Bonington Campus, Leicestershire, United Kingdom
*Corresponding author. e-mail address: parul.thaparparul.thapar@gmail.com

Contents

Progress in Molecular Biology and Translational Science, Volume 224
ISSN 1877-1173, https://doi.org/10.1016/bs.pmbts.2026.07.016

Abstract

The prevalence of antimicrobial resistance in most of the microbial species, majority of human microbial infections and diseases occur due to the formation of biofilms in Gram- positive and Gram-negative bacteria. The species of bacteria include *Pseudomonas aeruginosa, Staphylococcus epidermidis, Escherichia coli, Klebsiella pneumoniae, Proteus mirabilis, Streptococcus viridans, Staphylococcus aureus, Enterococcus faecalis* etc. Not only bacteria but there exists some fungi and protists that may also form biofilms is the factors are favourable. The biofilms are constituted as 10 % microbial mass and 90 % water. 50–90 % of the biofilm matrix is composed of polysaccharides as the organic component. The biofilm makes a microbial species pathogenic in nature. The mechanism of biofilm formation involves the process of quorum sensing. It is a cell-to-cell communication process that controls the regulation of gene expression for the biosynthesis of molecules required for bacterial biofilms. Depending upon the type and the genes present, the mechanism of quorum sensing differs between species to species. In this chapter, different mechanisms involved in biofilm formation in species will be discussed in detail and their specific interactions with the host organisms.

1. Introduction

Biofilms are defined as communities of properly organised microorganisms attached to a substrate and embedded in a self-produced extracellular matrix (extracellular polymeric substances – EPS), which aids cell interactions.[1] Usually, biofilms are multicellular form of life in which microorganisms live, grow, and exchange materials, positioning themselves to exploit the surrounding resources or create environmental conditions favourable to biofilm growth. The chemical constituents of the extracellular matrix maintain biofilm architecture, as well as ensuring that cells interact with each other and with the substrate upon which they grow. In addition to maintaining biofilm architecture, the EPS forms a shelter that supports biofilm survival, protecting it from biotic (viruses, predators, and grazers) and abiotic (physical removal of cells, nutrient starvation, changes in temperature, salinity, pH, or turbulence) forces.[2] The microorganisms that form the biofilms include bacteria, fungi and protists. The examples of these microorganisms[3] are mentioned in the Table 1. Due to the formation of these biofilms, some species of Gram positive and Gram-negative bacteria are developing resistance to most of the antibiotics. This chapter has elaborated the mechanism of formation of biofilms and their interaction with the host organisms.

Table 1 Microorganisms forming biofilms.

Bacteria	Fungi	Protists
Acinetobacter baumanii	*Apophysomyces elegans*	*Cryptosporidium spp.*
Bacillus subtilis	*Aspergillus fumigatus*	*Cyclospora cayetanesis*
Campylobacter jejuni	*Blastoschizomyces capitatus*	*Toxoplasma gondii*
Clostridium difficile	*Candida albicans*	
Clostridium perfringens	*Candida parapsilosis*	
Enterococcus faecalis	*Coccidioides immitis*	
Escherichia coli	*Cryptococcus lorentii*	
Fusobacterium nucleatum	*Cryptococcus neoformans*	
Klebsiella pneumoniae	*Exophiala dermatitidis*	
Listeria monocytogenes	*Histoplasma capsulatum*	
Proteus mirabilis	*Lichtheimia corymbifera*	
Pseudomonas aeruginosa	*Malassezia pachydermatis*	
Salmonella spp.	*Trichospora asahii*	
Staphylococcus aureus	*Trichophyton rubrum*	
Staphylococcus epidermidis	*Trichophyton mentagrophytes*	
Streptococcus agalactiae	*Paracoccidioides brasiliensis*	
Vibrio cholerae	*Pneumocystis carinii* *Rhizomucor pusillus* *Rhizopus oryzae*	

2. Antimicrobial resistance in microorganisms

2.1 Prevalence of antimicrobial resistance in microorganisms involved

The occurrence of antimicrobial resistance in microorganisms has become one of the major health concerns. The prevalence of antimicrobial resistance is high in low-middle income countries that lack in access to medical facilities, poor water sanitation, limited medicines and diagnostics Many diseases caused due to certain microorganisms are being difficult to be treated. It is regarded as

Table 2 Antibiotic resistant species and diseases caused.

Antibiotic resistant species	Antibiotics	Infections caused	References
Salmonella	Fluoroquinolones, ceftriaxone and azithromycin	Intestinal inflammation, diarrhoea	8,9,16,17
Campylobacter jejuni	Ciprofloxacin, fluoroquinolones, nalidixic acid, tetracycline, ampicillin, erythromycin, gentamicin and enrofloxacin	Intestinal inflammation, diarrhoea	18
Helicobacter pylori	Amoxicillin, metronidazole, clarithromycin, tetracycline, levofloxacin,	Watery diarrhoea	8,19
Shigella dysenteriae, *Shigella flexneri* and *Shigella sonnei*	Ciprofloxacin, third- generation antibiotics- cephalosporin- ceftriaxone and azithromycin	Dysentery	10,20
Enterococcus faecalis	Vancomycin and beta-lactam antibiotics	Urinary tract infections, endocarditis, mastitis and bacteremia	21,22
Staphylococcus aureus	Methicillin, carbapenems, first-line drugs- co-trimoxazole	Nosocomial infections including skin rash and cuts; lung infections, blood stream infections; sepsis and even death	23

Staphylococcus haemolyticus	Last-line antibiotics glycopeptides	Noscomial infections	24–27
Escherichia coli	Third- generation cephalosporins, carbapenems, first-line drugs- ampicillin and co-trimoxazole; second line drugs-fluoroquinolones	Blood stream infections	Sustainable Development Goals SDGs AMR indicator28; GLASS methodology report,29
Klebsiella pneumoniae	carbapenems	Blood steam infections	GLASS methodology report,29
Mycobacterium tuberculosis	Rifampicin and other drugs	Pulmonary tuberculosis	4
Neisseria gonorrhoeae	Ciprofloxacin	Gonorrhoea	GLASS methodology report,29

silent pandemic posing threats to humans and ecosystems around the world.[4] It has been estimated that the increased abundance of antimicrobial resistance could lead to the loss of 3.8 % of World's annual GDP. In extreme cases, additional 24 million people may be forced towards extreme poverty, hunger and malnutrition if not controlled.[5]

The microorganisms that develop antimicrobial resistance are referred to as "superbugs. These may be the species of bacteria, fungi, virus and protozoa. The resistant bacterial species of resistant Gram-negative bacteria that are pathogenic, contagious and cause food borne illness[6,7] include the species of Salmonella, *Helicobacter pylori,*[8] *Campylobacter jejuni*[9]; *Shigella dysenteriae, Shigella flexneri*[10] and *Shigella sonnei.*[11] Gram-negative bacteria are found to show intrinsic resistance to multiple drugs.[12–14] Some Gram-positive bacteria have also been found to be multi-drug resistant including the species of *Enterococcus faecalis, Staphylococcus aureus,*[15] Methicillin resistant *Staphylococcus epidermidis* (MRSE) and multi-drug resistant (MDR) *Staphylococcus epidermidis.* The infections caused by these organisms and their resistance to certain antibiotics are mentioned in Table 2.

3. Causes of antimicrobial resistance

The formation of biofilms on the outer membranes of bacteria is one of the major hinderances for the effectiveness of certain antibiotics and their inappropriate use. It allows microbes to resist host immune responses and effect of antibiotics.[30] Both Gram-positive and Gram-negative bacteria can form biofilms, leading to an ongoing battle on various surfaces.

The infection caused due to antimicrobial resistance developed through "biofouling" of biofilms is called as biofilm infection. The process of bio-fouling through biofilms is a polysaccharide-encased multicellular bacterial aggregation adhering to various surfaces. Biofilm infection is recognized to be responsible for more than 65 % of infections, posing a huge threat to public health.[31] The microorganisms having biofilms possess different characteristics like collective cooperation, survival to antimicrobials and source capturing.[32]

3.1 Biofilms in microorganisms

Bacteria exist in two phases- planktonic phase (free-floating) and sessile phase (surface adherence). When a bacterium attaches to a surface (biotic or abiotic), there occurs modification in the genetic expression of maturation and production of exopolysaccharides (EPS). This EPS is also known as "slime" that exist as a part of biofilm.[33–35]

Biofilms are immobile complex structures that are formed by certain bacterial species (single or multiple), irreversibly attached to a surface and are surrounded by an extracellular polymeric substance produced by the bacteria. Depending upon their origin, biofilms can be considered as neutral or harmful. Neutral biofilms are the ones that are formed in natural environment whereas the ones that grow on open wounds and thereby cause infection are the harmful biofilms.[31]

3.2 Characteristics of biofilms

The complex structure of the biofilms can be observed by visualization through scanning and confocal microscopy. As observed, these are the complex population of cells that are surrounded by layers of exopolysaccharides (EPS). The layers consist of permeable water channels; uniform deposition of cells and slime. Once the biofilms are formed, they initiate the process of disease emergence through mechanisms like production of endotoxins; clusters of bacterial cells detachment; preventing the response from host immune system and formation of a protective barrier around the microbes, leading to formation of immune-resistant organisms.[31] The biofilms have viscoelastic property and rubbery texture. The detailed composition of biofilms[36–40] in different organisms is mentioned in the Table 3.

The polysaccharides are found as chain-like structure woven in a dense-mesh. The hydroxyl groups of the polysaccharides interact with each other through hydrogen bonds. This provides mechanical strength to the biofilms, thereby increasing their thickness upto 300 μm and adhering them to the surface.[41,42] The cations and anions form the cross bridge between the polymers and protect the biofilms to withstand extreme environmental conditions and any external shearing stress.[43]

Microbial biofilms can be sustained in high energy micro-environments and different types of climatic conditions, highlighting the key roles of biofilms in shaping the ecosystems in extremely arid areas.[1] Bacteria at different depths, ages, and locations within a structure tend to have distinct physiologic states. When concerning the physical and chemical environment, it is important to understand that due to the inherent metabolic activities occurring within these ecosystems, biofilms harbor complex nutrient gradients that significantly shape the multispecies communities present within them. These unique gradients are a direct result of the various biological processes and interactions taking place throughout the biofilm.[44]

Table 3 Composition of biofilms in micoorganisms.

Components	Gram positive bacteria	Gram negative bacteria	Coagulase-negative bacteria
Microbial mass	10 %	10 %	10 %
Water	90 %	90 %	90 %
Polysaccharides	50–90 %	50–90 %	50–90 %
Ions	Cations- Ca^{2+}, Mg^{2+}	anions	cations
Uronic acids	D-glucuronic acid, D-galacturonic acid, D-mannuronic acid and ketone-linked pyruvate	D-glucuronic acid, D-galacturonic acid, D-mannuronic acid and ketone-linked pyruvate	D-glucuronic acid, D-galacturonic acid, D-mannuronic acid and ketone-linked pyruvate
Teichoic acids and proteins including enzymes in EPS	absent	<1–2 %	<1–2 %
Extracellular DNA	<1–2 %	<1–2 %	<1–2 %
Examples	Streptococci spp.	Escherichia spp.	Staphylococci spp.

The microorganisms that are found to form biofilms include the species of *Staphylococcus epidermidis*, *Pseudomonas aeruginosa*, *Klebsiella pneumoniae*, *Escherichia coli*, *Proteus mirabilis*, *Streptococcus viridans*, *Staphylococcus aureus* and *Enterococcus faecalis*.[45]

3.3 Factors contributing biofilm formation

There are certain factors that contribute to formation of biofilms. These include- structure of the organism- pili and flagella, environmental factors- nutrition deficiency, oxygen levels, pH, temperature.

3.4 Structure of organism

The bacterial extracellular matrix (ECM) consisting of pili and flagella are the mostly commonly associated components of biofilms. These help the bacteria to adhere to the biotic or abiotic surfaces. The pili is also required for twitching motility of biofilms (spread over biofilm surface), example- *Pseudomonas aeruginosa* and *Acinetobacter baumanii*.[46] Not only this, ECM also contains polymers in the form extracellular polysaccharides (EPS), synthesized by bacteria, cyanobacteria, protists, fungi and microalgae. These polymers form a group of substances- polysaccharides, proteins, carbohydrates, lipids, extracellular DNA (eDNA) and acids.[47–55] EPS is found to maintain the biofilms integrity while eDNA usually is involved in exchange of genetic materials in between biofilms.[3]

3.5 Environmental factors

3.5.1 Nutritional deficiency

Low nutrient conditions create shear stress on microbial biofilms. This causes a layer of biofilms to rupture under extreme stress, one of the most important parts of biofilm mechanism (details in next section). The reduction of carbon (glycerol and glutamic acid) boosts the expression of EPS promoter, thereby increasing the matrix formation of biofilms.[56,57]

3.5.2 Oxygen levels

When there occurs decrease in the oxygen levels in the environment, it may lead to:

Hypoxia in the bacterial cells- Due to lack of oxygen, the bacterial respiration decreases that enables programmed cell death in bacteria. This causes cell lysis, that on the other hand stimulates the release of proteins and eDNA. These are responsible for the formation of EPS that triggers biofilm formation. Example- *Staphylococcus aureus*.[58]

Apoptosis- Within the layer of biofilms, bacterial cells are exposed to lower levels of oxygen, leading to cellular apoptosis. This enhances the release of intercellular substances into the matrix, thereby increasing the biofilms strength.[59,60]

Also, low levels of oxygen favour the maturation process of biofilm formation, as observed in *Pseudomonas aeruginosa* and *Staphylococcus aureus*.[61,62]

3.5.3 Temperature and pH

The formation of biofilms in different organisms is also favoured by varied temperature and pH ranges. Several types of bacterial species and strains undergo transition from planktonic/ sessile state to biofilm under different temperature and pH. The optimum temperature and pH required for the microorganisms for biofilm formation is mentioned in the Table 4.

Table 4 Optimum temperature and ph required for biofilm formation.

Species	**Optimum temperature (°C)**	**pH**	**References**
Salmonella spp. and *Pseudomonas aeruginosa*	30 °C	7.0	63
Listeria monocytogenes	4 °C–12 °C	7.5–8.0	64
Klebsiella pneumoniae Vibrio cholerae, Clostridium perfringens, Escherichia coli and Streptococcus agalactiae	37 °C	5.0–5.5	3,65–67
Aeromonas hydrophila	20 °C–25 °C	7.0–8.0	68
Escherichia coli MG1655	25 °C	7.4	66

3.6 Mechanism of biofilm formation

The biofilms formation is dependent on Reynolds number (Re). It is the number that helps in the prediction of patterns of flow of the fluid. When the Re number is low, it indicates laminar flow, whereas a high Re number- turbulent flow. The biofilms are formed when the flow is turbulent Re = >5000. A rough or smooth surface may be favourable for biofilms formation.[31] It is a sequential phenomenon, which also depends on a plethora of factors (physical, chemical, and biological) in its different phases: attachment, microcolony formation, maturation, and dispersion. During these phases, which may vary a little in their complex pattern, microbial biofilms usually form three-dimensional structures separated by porous water channels and cavities. These channels and cavities allow nutrient entrance and waste discharge. The polysaccharides that comprise the EPS form the skeleton of the biofilm, providing structural stability. The 3D architecture of a biofilm can change its thickness, shape, and structure through time, but this evolution does not necessarily result in progressive orders. In fact, biofilms are dynamic structures whose evolution depends on both biotic and abiotic factors acting on time scales ranging from seconds to years. Biofilm formation is an advantageous outcome for microbial survival and a crucial step in bacterial evolution.[69] The various steps involved in microbial biofilm formation are as follows[70] (Fig. 1A, B, C, D and E):

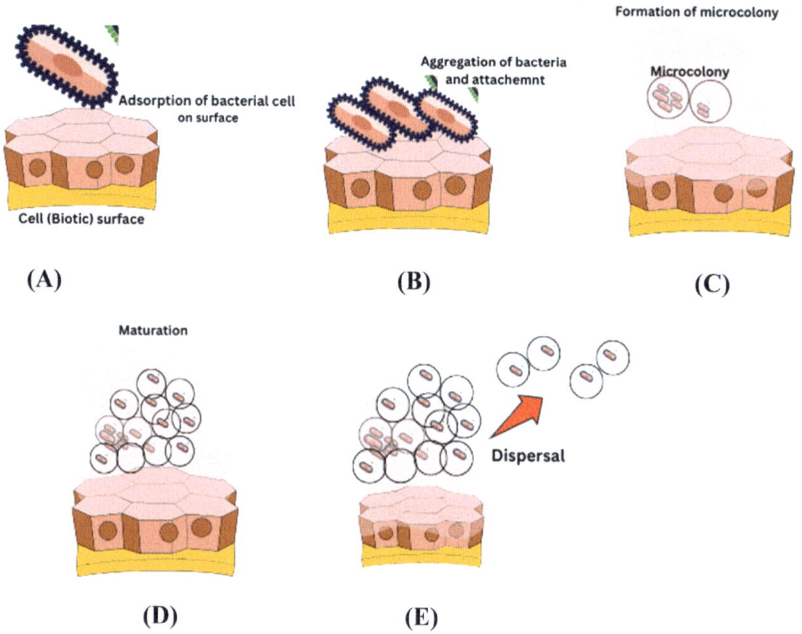

Fig. 1 **Mechanism of biofilm formation by a single sessile bacterium**. (A) Adsorption of the bacterial cell on the biotic surface (B) Release of EPS causing irreversible attachment and cellular aggregation leading to permanent adherence on the surface (C) Growth and multiplication of bacterium cells forming microcolonies that contain individual bacterial cells. This forms an external matrix of complex biomolecules (D) Maturation and multiplication of microcolonies (E) The microcolonies are detached. The bacterial colonies regain sessile state to further form biofilms. *Created on Canva.com.*

3.7 Adsorption

The bacterial cells may be free-floating or sessile that meets the surface either through convection, Brownian motion or sedimentation.[71] The microorganisms adsorb on the surface (biotic and abiotic) and forms an intersection with a media like water (if abiotic/biotic) or blood (if biotic). Once adsorbed, the microorganism initiates the attachment onto the surface.

3.8 Attachment (Initial)

Simple attachment occurs when microorganisms adhere on submerged surfaces in a reversible manner.[2] This stage is physiochemical by nature and is mainly driven by hydrophobic and electrostatic interactions and Van der

Waals forces. The bacterial flagella or pili attach to the surface through these forces. The flagella possess the property to repel the resistant forces that do not allow the bacteria to adhere.[72] In this context, environmental water turbulence and shear force are responsible for the so-called "encounter rates".[34,73–76]

3.9 Irreversible attachment

Due to the randomness as mentioned above, biofilm formation is intrinsically stochastic. When the cumulative effect of hydrophobic, electrostatic and Vander Waals forces counter the shear force, it provides mechanical strength to the bacteria and forms a monolayer that causes an irreversible adherence to the surface.[34,73–76] Once attached, microorganisms grow and reproduce leading to formation of aggregates.

3.10 Bacterial adhesion and aggregation

In this stage, the bacterial cells release exopolysaccharides (EPS). EPS interacts and binds to specific ligands present on the bacterial pili or fimbriae. It helps the bacteria to adhere to the surface tightly and irreversibly. The release of EPS stabilizes the biofilm structure and prevents the effect of antimicrobial agents. Example- The unique saccharides released from the species of *Pseudomonas aeruginosa* include alginate, Pel and Psl. Some lipids like surfactin, emulsan and viscosin found in EPS have surface active properties that helps in binding to heavy metals and colonize waxy surface in Rhodococcus spp.[77] Also, extracellular DNA (e-DNA) found in EPS plays an important role in bacterial aggregation and maintaining the biofilm integrity in the strains of *Listeria monocytogenes*.[78] Once adhered on to the surface, the bacterial cells release certain adhesins that allow the cells and other surface-bound organisms to adhere to one another and form aggregates. Once aggregated, in the presence of certain environmental factors like temperature and pressure, the bacterial cells transient from sessile to a planktonic state. Example- The species of *Staphylococcus epidermidis* releases a polysaccharide called PIA (Polysaccharide Intracellular Adhesin) that helps in cell-to-cell adhesion, thereby formation of biofilm.[79–85]

3.11 Formation of microcolony

This step is biological. Microcolony formation involves auto aggregation and more production of exopolysaccharides (EPS), enzymes, and quorum sensing (QS) molecules, contributing to a higher complexity and stability of the emerging biofilm. The bacterial cell division and multiplication takes place. During cellular aggregation, layers of cell clusters are formed on the surface. These clusters develop into microcolonies which are enveloped inside the EPS. During this process, intercellular signalling and quorum sensing (QS) takes place. In strains of *Pseudomonas aeruginosa*, rhamnolipids (a class of surfactants) have been found to initiate the formation of microcolony, shaping the biofilm and also in discharge of dispersal layer of biofilm.[86,87] The biofilm containing microbes consist of various micro-communities that coordinate among one another for the exchange of nutrients, metabolites and microbial waste elimination. The type of association in which the bacterial cells coordinate among one another and depend on each other for the exchange of metabolic products to fulfil their energy demands is called as syntropic relationship.[88,89]

3.12 Maturation

The maturation of the microcolonies activate the expression of biofilm-specific genes. Also, the EPS are released during the entire process of maturation that provides biofilm stability. The maturation occurs in two steps. Step 1 and Step 2.

Step 1- In this step, there are synthesis of autoinducer molecules- N-acylated homoserine lactone (AHL) by Gram negative bacteria[90] and phenol-soluble modulins by Gram positive bacteria including *Staph. aureus*,[91,92] coded through biofilm-specific genes. This leads to cell-to-cell contact (quorum sensing). The bacteria recognize the adjacent cells and forms clusters through bond formation effectively.[90]

Step 2- Formation of established colony- The microcolony expands in the size and their thickness increases to about 100 μm, thereby an established colony is formed.[90]

A biofilm consists of two layers: One layer is the foundational layer-The bacteria thrive in this layer. The second layer is the surface layer also called as dispersal zone. This layer disintegrates and spreads into surroundings.

3.13 Dispersion

After hours of growth and maturation, a three-dimensional structure emerges wherein aggregates of cells are embedded in a highly hydrated matrix, accumulating nutrients, signalling molecules and metals. The dispersal layer of the biofilm disseminates in the environment and spreads its infection, the process called as metastatic seeding. It is one of the most important steps in the biofilm growth and development. In medical, the dispersion phase of the biofilm is a major cause of the chronic infections. The dispersion occurs due to a programmed process associated with any nutrient deficiency or oxygen level (aerobic bacteria). During the deficiency, fatty acids like DSF (cis-11-methyl 1–2-dodecenoic acid) gets activated through the process of autophosphorylation. This causes activation of an enzyme- cyclic-diguanylate Guanosine monophosphate (c-di-GMP) phosphodiesterase. This enzyme metabolizes c-di-GMP, that leads to decomposition of the biofilms layer through shear pressure, which in turn dissolves the parts of EPS and causes of release of the dispersion layer. This layer may then infect the new surface (cell or body) creating new biofilms on different organs of the body.[34,73,93–95] In Gram positive bacteria, the phenol- soluble modulins also play a role in biofilm detachment and bacterial dispersal.[91]

During the laminar flow Re = >2000, reversible microbial attachment occurs to the surface like small device medical catheters like pocket catheters and early colonization is favoured due to minimal hydrodynamic forces.[96,97] The biofilm structure formed are usually patchy and uneven surface distribution. The cells were found to be heterogenous consisting of clusters separated by interstitial voids[98] in the strains of *Pseudomonas fluorescens* and *Bacillus cereus*.[99] In laminar flow, biofilms are mainly unstable but can be contagious.

3.14 Host interactions with biofilms

The microorganisms with biofilms evolve the property to attach to biotic and abiotic surfaces. The abiotic surfaces include all types of medical devices whereas biotic includes skin, bone, connective tissue, intestinal mucosa, airway and vascular endothelium.[100] These interactions enable them to various tissue associated chronic infections.[101] Once multilayered biofilms are formed on the medical devices, it results in device-related infections (DRIs) that increases the difficulty to treat the disease and often there are chances of a relapse. In *Acinetobacter baumanii*, a gene AdeRSedA encodes biofilm associated protein (Bap) called Bpf. This protein helps in cell-to-cell communication when adhered to abiotic catheters and enhances the adhesive capacity to bind to abiotic surfaces[102,103] like urinary catheters, venous catheters, tracheostomy tubes, mechanical ventilation, T tubes, hemodialysis and neurosurgical procedures promoting biofilm formation.[104] The flagellum *of Acrobacter butzleri* consists of characteristics that are involved in biofilm formation in abiotic devices, mainly medical devices made up of polystyrene and stainless steel. Within the mutant species of *A. butzleri*, the genes encoding for (p)ppGpp synthase enzyme regulates LuxS system. This system is involved in interspecies quorum sensing mechanism that synthesizes an autoinducer AI-2 signalling molecule.[105] This facilitates intracellular adhesion by polysaccharides of microbes and increase adherence of biofilms either outside the surface of catheter or inner lumen.[106] The species of MDR *Staphylococcus epidermidis* is most found to form a strong biofilm on the implanted devices within patients underwent surgeries for prosthetic heart valves, cardiac devices, prosthetic joints, central lines, contact and intraocular lens, urinary and intravascular catheters and intravenous drug use.[107–109] The mechanism of interaction of biofilms on biotic hosts may occur through these principles- 1. Detachment of biofilm cells or mass of cells- mainly during blood or urine infection/ emboli formation; 2. Interchange resistance plasmids within biofilms; 3. Endotoxin production by biofilm-associated by bacteria; 4. Resistance to host immune system. Biofilms that grow on open wounds of diabetic patients may cause chronic illness.[110] The exterior of deep wounds is found to be infected by biofilms of aerobic resistant species of *P. aeruginosa*, *S. aureus*, *S. epidermidis*, *E. coli*, *K. pneumoniae*, *E. faecalis*, *A. baumanii* and *S. pyogenes* whereas in the interior of the wound, anaerobic species of Clostridium spp. and *B. fragilis* biofilms have been reported.[111] Being multi-drug resistant, the treatment with antibiotics has no effect on patients to cure the infections. Some species of coagulase-negative Staphylococci (CoNS)- *Staphylococcus aureus* have shown to cause infections along with *S. epidermidis*.[25,112,113] It is found that in species of *S. aureus* and *S. epidermidis*, biofilms can differentiate the gene expressions through various virulence factors that may prevent phagocytosis by macrophages and neutrophils (Table 5). The virulence factpossessing the essential traits on microorganisms that are encoded by genes, possessing infectious mechanisms on the host cells.[114] In *Staphylococcus haemolyticus*, the biofilms are formed of proteins and DNA[115] whereas in *Streptococcus uberis*, the biofilms are composed of exopolysaccharides, water, proteins and DNA.[116] Type IV pili and flagellum in *Pseudomonas aeruginosa* are factors mainly responsible for adhesion, motility and organization of micro-colonies. During biofilm formation on host cells, quorum sensing by *P. aeruginosa* is responsible for intercellular communication. It also affects the immune response in the hosts as biofilms of *P. aeruginosa* affects its genetic expression. Similarly, virulence factors in Candida spp. have also shown gene regulated affects on immune system through biofilms (Table 5). Therefore, the biofilms persist and thrive in hospital environment, leading to nosocomial

Table 5 Virulence factors regulated by genes and their effects by host-biofilm interaction.

Strain	Virulence factor	Virulence gene	Host interactions and effects	References
S. aureus strain ST239-SCCmec III	Structural surface protein	*sasX*	Adherence to epithelial cells for biofilm formation. Enhances immune evasion mechanism that does not allow immune system to recognize the pathogens	118–120
S. aureus USA300 LAC	Leukocidins	*pvl, lukAB, hlgAB, hlgCB* and *lukED*	Release of leukocidins from biofilms that may destroy leukocytes.	121,122
S. epidermidis RP62A	Accumulation associated protein	*aap*	Adherence and association of strains on surface that prevents phagocytosis	123
P. aeruginosa	Quorum sensing proteins	*las* and *rhl*	Biofilm information inhibits the release of cytokines, thereby increase infection	124,125
P. aeruginosa	Enzymatic elastases LasA and LasB	*lasR* and *rhlR*	Facilitates biofilm infection in cystic fibrosis by degrading IgA and IgG antibodies	125
Toxins of *P. aeruginosa*	Enterotoxin A and B	*lasR* and *rhlR*	Cytokines repression in blood	126,127
	Pyocyanin and Pyoverdin	*lasR* and *rhlR*	Biofilm formation inhibits TNF-α and IL-1 in macrophages	128,129
Candida albicans and C. tropicalis	Transcriptional factors in hyphal morphogenesis	*ASH1, ACE2, EFG1, FLO8* and *NDT80*	Cellular intertwining in biofilms through hyphae that penetrate and rupture immune cells	130

and other infections.[117] Other embedded microbial inhabitants can either generate waste products that may be harnessed as nutrients, mitigate toxic by-products, protect biofilms from biofilm-targeting drugs, or engage in exploitive behaviour. It is essential to better understand these interactions and to incorporate them into approaches or models that are comprehensive and predictive. It is also critical to identify and exploit the metabolic products of one species in a biofilm that can be targeted to selectively kill other species. Understanding these interactions will also provide insights into designing new and better treatment options for biofilm-associated infections.

A growing emphasis has been placed on the importance of biofilm dissemination following the emergence of biofilm-induced infection.[131,132] For instance, small eddies effectively transport essential nutrients towards the centre of the biofilm structure, where the bulk of the biomass is primarily composed of chemolithotrophic bacteria that thrive on inorganic materials. Simultaneously, larger aggregates within the biofilm work to transport oxygen downwards, thereby favouring the growth and metabolic functions of nitrate- and sulphide-reducing bacteria, which are critical for nutrient cycling and maintaining ecological balance. In addition, the presence of intricate water channels contributes to the formation of a complex hydraulic architecture within the biofilm.

3.15 Mechanism

The unique arrangement of intricate water channels not only facilitates the efficient dispersion of nutrients but also ensures proper waste discharge and adequate aeration throughout the biofilm community. As a result, the hydraulic dynamics play a vital role in sustaining the diverse microbial life present within the structure. Moreover, the diffusion of ions, small molecules, and various macromolecules both into and out of the biofilm matrix occurs in a manner that can be described as Fickian. This diffusion leads to localized depletion of diffusible substrates and compounds, highlighting the crucial role of diffusion in influencing the kinetics of biofilm-associated biochemical reactions and the overall functionality of these complex communities.[133]

3.15.1 Importance

Interactions among bacteria of different species and types are important for the success, persistence, and dynamic behaviour of biofilms at the community scale.[134]

4. Conclusion

Biofilms represent a vehicle for genetic material transfer, seen as a mechanism of genetic diversification. The most important gene carriers, plasmids, do not regard the lineage of the species, and conjugative elements are known among bacteria of different phyla, including Gram-positives and Gram-negatives, occurring even between organisms from completely different environments. Flemming et al..[135,136] Biofilm formation provides certain advantages for bacteria in terms of survivability, persistency, and ultimate pathogenicity. However, researchers are still at a fundamental stage in elucidating the biological mechanism of biofilm formation and establishment, switching from planktonic form to the biofilm mode of growth. With an increasing awareness of the significance of biofilms in the medical field, there has been a steep rise in biofilm-related publications in the recent past.[137] Despite many efforts, a comprehensive and universally applicable biophysical and synthetic model of biofilm evolution is still absent. There is a vast majority of cellular behaviours and control mechanisms required for biofilm maturation and dispersion yet to be uncovered.

References

1. Luis Souza dos Santos A, Clara Milesi Galdino A, Pereira de Mello T, et al. What are the advantages of living in a community? A microbial biofilm perspective!. 2018. ncbi.nlm.nih.gov.
2. Mazza GM. The physics of biofilms - an introduction. *J Phys D: Appl Phys.* 2016;49(20):203001. https://doi.org/10.1088/0022-3727/49/20/203001
3. Samrot AV, Abubakar Mohamed A, Faradjeva E, et al. Mechanisms and impact of biofilms and targeting of biofilms using bioactive compounds—A review. *Medicina.* 2021;57:839. https://doi.org/10.3390/medicina57080839
4. Global tuberculosis report. World Health Organization. 2022 [website] https://www.who.int/teams/globaltuberculosis-programme/data#csv_files.
5. World Health Organization report. Hunger numbers stubbornly high for three consecutive years as global crises deepen: UN report. 2023 [website] https://www.who.int/news/item/24-07-2024-hunger-numbers-stubbornly-high-for-three-consecutive-years-as-global-crises-deepen–un-report.
6. Rossolini GM, Arena F, Pecile P, Pollini S. Update on the antibiotic resistance crisis. *Curr Opin Pharmacol.* 2014;18:56–60. https://doi.org/10.1016/j.coph.2014.09.006
7. Andersson DI, Hughes D. Antibiotic resistance and its cost: is it possible to reverse resistance? *Nat Rev Microbiol.* 2010;8:260–271. https://doi.org/10.1038/nrmicro2319
8. Islam MS, Sazzad HM, Satter SM, et al. Nipah virus transmission from bats to humans associated with drinking traditional liquor made from date palm sap, Bangladesh, 2011-2014. *Emerg Infect Dis.* 2016;22(4):664–670. https://doi.org/10.3201/eid2204.151747
9. Mukherjee P, Ramamurthy T, Mitra U, Mukhopadhyay AK. Emergence of high-level azithromycin resistance in campylobacter jejuni isolates from pediatric diarrhea patients in Kolkata, India. *Antimicrob Agents Chemother.* 2014;58(7):4248. https://doi.org/10.1128/AAC.02931-14

10. Nandy S, Mitra U, Rajendran K, Dutta P, Dutta S. Subtype prevalence, plasmid profiles and growing fluoroquinolone resistance in Shigella from Kolkata, India (2001–2007): a hospital-based study. *Trop Med Int Health*. 2010;15:1499–1507. https://doi.org/10.1111/j.1365-3156.2010.02656.x
11. Nandy S, Dutta S, Ghosh S, Ganai A, Rajahamsan J, Theodore RB, et al. Foodborne-associated Shigella sonnei, India, 2009 and 2010. *Emerg Infect Dis*. 2011;17(11):2072–2074. https://doi.org/10.3201/eid1711.110403
12. Roemhild R, Andersson DI. Mechanisms and therapeutic potential of collateral sensitivity to antibiotics. *PLoS Pathog*. 2021;17:e1009172. https://doi.org/10.1371/journal.ppat.1009172
13. Tyers M, Wright GD. Drug combinations: a strategy to extend the life of antibiotics in the 21st century. *Nat Rev Microbiol*. 2019;17:141–155. https://doi.org/10.1038/s41579-018-0141-x
14. Lehtinen S, Blanquart F, Lipsitch M, et al. On the evolutionary ecology of multidrug resistance in bacteria. *PLoS Pathog*. 2019;15:e1007763. https://doi.org/10.1371/journal.ppat.1007763
15. Sonbol IF, Ahmed AAA, Tarek EEB, Omnia MAF. Antimicrobial activity of bacteriocins produced by Enterococcus isolates recovered from Egyptian homemade dairy products against some foodborne pathogens. *Int Microbiol*. 2020. https://doi.org/10.1007/s10123-020-00127-z
16. Bager A, Helmuth T. Epidemiology of resistance to quinolones in *Salmonella*. *Vet Res*. 2001;32(3/4):285–290.
17. Klemm EJ, Shakoor S, Page AJ, et al. Emergence of an extensively drug-resistant Salmonella enterica serovar typhi clone harboring a promiscuous plasmid encoding resistance to fluoroquinolones and third-generation cephalosporins. *MBio*. 2018;9(1).
18. Portes BA, Panzenhagen P, dos Santos PMA, Junior CAC. Antibiotic resistance in *Campylobacter*: a systematic review of South American isolates. *Antibiotics (Basel)*. 2023;12(3):548. https://doi.org/10.3390/antibiotics12030548 9.
19. Boyanova L, Hadzhiyski P, Gergova R, Markovska R. Evolution of *Helicobacter pylori* resistance to antibiotics: a topic of increasing concern. *Antibio (Basel*. 2023;12(2):332. https://doi.org/10.3390/antibiotics12020332
20. Lefèvre S, Njamkepo E, Feldman S, et al. Rapid emergence of extensively drug-resistant *Shigella sonnei* in France. *Nat Commun*. 2023;14.
21. Mwikuma G, Kainga H, Kallu SA, Nakajima C, Suzuki Y, Hang'ombe BM. Determination of the prevalence and antimicrobial resistance of *Enterococcus faecalis* and *Enterococcus faecium* associated with poultry in four districts in Zambia. *Antibiotics*. 2023;12(4):65.
22. Guan L, Beig M, Wang L, et al. Global status of antimicrobial resistance in clinical *Enterococcus faecalis* isolates: systematic review and meta-analysis. *Ann Clin Microbiol Antimicro*. 2024;23(80):1–16.
23. Centre for Disease control and Prevention report. Staphylococcus aureus basics: disease resistance. 2025 [website] https://www.cdc.gov/staphylococcus-aureus/about/index.html.
24. Veach L, Pfaller M, Barrett M, Koontz F, Wenzel R. Vancomycin resistance in *Staphylococcus haemolyticus* causing colonization and bloodstream infection. *J Clin Microbiol*. 1990;28:2064–2068.
25. Becker K, Heilmann C, Peters G. Coagulase-negative staphylococci. *Clin Microbiol Rev*. 2014;27(4):870–926.
26. Kranjec C, Kristensen SS, Karolina TB, et al. A bacteriocin-based treatment option for *Staphylococcus haemolyticus* biofilms. *Sci Rep*. 2021;11:13909. https://doi.org/10.1038/s41598-021-93158-z

27. Froggatt JW, Johnston JL, Galetto DW, Archer GL. Antimicrobial resistance in nosocomial isolates of Staphylococcus haemolyticus. *Antimicrob Agents Chemother.* 1989;33(4):460–466. https://doi.org/10.1128/AAC.33.4.460
28. Sustainable Development Goals (SDGs) AMR indicator. *The Global Health Observatory*. Geneva: World Health Organization; 2021. https://www.who.int/data/gho/data/themes/topics/global-antimicrobial-resistancesurveillance-system-glass/sustainabledevelopment-goals-amr-indicator.
29. GLASS. *Methodology for Surveillance of National Antimicrobial Consumption.* Geneva: World Health Organization; 2020. https://www.who.int/publications/i/item/9789240012639.
30. Mah TF, O'Toole GA. Mechanisms of biofilm resistance to antimicrobial agents. *Trends Microbiol.* 2001;9:34–39. https://doi.org/10.1016/s0966-842x(00)01913-2
31. Sharma S, Mohler J, Mahajan SD, Schwartz SA, Bruggemann L, Aalinkeel R. Microbial biofilm: a review on formation, infection, antibiotic resistance, control measures, and innovative treatment. *Microorganisms.* 2023;11:1614. https://doi.org/10.3390/microorganisms11061614
32. Cochrane DM, Brown MR, Anwar H, Weller PH, Lam K, Costerton JW. Antibody response to Pseudomonas aeruginosa surface protein antigens in a rat model of chronic lung infection. *J Med Microbiol.* 1988;27:255–261.
33. Bjarnsholt T, Ciofu O, Molin S, Givskov M, Hoiby N. Applying insights from biofilm biology to drug development—Can a new approach be developed? *Nat Rev Drug Discov.* 2013;12:791–808.
34. Gupta P, Sarkar S, Das B, Bhattacharjee S, Tribedi P. Biofilm, pathogenesis and prevention–a journey to break the wall: a review. *Arch Microbiol.* 2016;198:1–15.
35. Irie Y, Borlee BR, O'Connor JR, et al. Self-produced exopolysaccharide is a signal that stimulates biofilm formation in Pseudomonas aeruginosa. *Proc Natl Acad Sci USA.* 2012;109:20632–20636.
36. Vandana, Das S. Structural and mechanical characterization of biofilm-associated bacterial polymer in the emulsification of petroleum hydrocarbon. *3 Biotech.* 2021;11:239.
37. Donlan RM. Biofilms: microbial life on surfaces. *Emerg Infect Dis.* 2002;8:881–890.
38. Fong JNC, Yildiz FH. Biofilm matrix proteins. *Microbiol Spectr.* 2015;3(2):1–10. https://doi.org/10.1128/microbiolspec.MB-0004-2014
39. Hussain M, Wilcox MH, White PJ. The slime of coagulase-negative staphylococci: biochemistry and relation to adherence. *FEMS Microbiol Rev.* 1993;10:191–207.
40. Okshevsky M, Meyer RL. The role of extracellular DNA in the establishment, maintenance and perpetuation of bacterial biofilms. *Crit Rev Microbiol.* 2015;41(3):341–352. https://doi.org/10.3109/1040841X.2013.841639
41. Singh S, Datta S, Narayanan KB, Rajnish KN. Bacterial exo-polysaccharides in biofilms: role in antimicrobial resistance and treatments. *J Genet Eng Biotechnol.* 2021;19:140.
42. Limoli DH, Jones CJ, Wozniak DJ. Bacterial extracellular polysaccharides in biofilm formation and function. *Microbiol Spectr.* 2015;3.
43. Chen MY, Lee DJ, Tay JH, Show KY. Staining of extracellular polymeric substances and cells in bioaggregates. *Appl Microbiol Biotechnol.* 2007;75:467–474.
44. Joshi RV, Gunawan C, Mann R. We are one: multispecies metabolism of a biofilm consortium and their treatment strategies. *Front Microbiol.* 2021 (frontiersin.org).
45. Chen M, Yu Q, Sun H. Novel strategies for the prevention and treatment of biofilm related infections. *Int J Mol Sci.* 2013;14:18488–18501.
46. Luo LM, Wu LJ, Xiao YL, et al. Enhancing pili assembly and biofilm formation in Acinetobacter baumannii ATCC19606 using non-native acyl-homoserine lactones. *BMC Microbiol.* 2015;15:62.

47. Roca C, Alves V, Freitas F, Reis M. Exopolysaccharides enriched in rare sugars: bacterial sources, production, and applications. *Front Microbiol.* 2015;6:288.
48. Voberková S, Hermanová S, Hrubanová K, Krzyžánek V. Biofilm formation and extracellular polymeric substances (EPS) production by Bacillus subtilis depending on nutritional conditions in the presence of polyester film. *Folia Microbiol.* 2015;61:91–100.
49. Czaczyk K, Myszka K. Biosynthesis of extracellular polymeric substances (EPS) and its role in microbial biofilm formation. *Pol J Env Stud.* 2007;16:799–806.
50. Hwang HJ, Kim SW, Xu CP, Choi JW, Yun JW. Morphological and rheological properties of the three different species of basidiomycetes Phiellinus in submerged cultures. *J Appl Microbiol.* 2004;96:1296–1305.
51. Jain R, Raghukumar S, Tharanathan R, Bhosle NB. Extracellular polysaccharide production by Thraustochytrid protists. *Mar Biotechnol.* 2005;7:184–192.
52. Li Y, Chen W, Liu Q, Xiang W, Zheng X, Qu J. Effect on composition and quantity of extracellular polymeric substances in an innovative hybrid membrane bioreactor under different sludge retention times. *Env Pollut Bioavailab.* 2019;31:103–111.
53. Marvasi M, Visscher PT, Martinez LC. Exopolymeric substances (EPS) from Bacillus subtilis: polymers and genes encoding their synthesis. *FEMS Microbiol Lett.* 2010;313:1–9.
54. Parikh A, Madamwar D. Partial characterization of extracellular polysaccharides from cyanobacteria. *Bioresour Technol.* 2006;97:1822–1827.
55. Pavlova K, Grigorova D. Production and properties of exopolysaccharide by Rhodotorula acheniorum MC. *Food Res Int.* 1999;32:473–477.
56. Zhang WB, Seminara A, Suaris M, Brenner MP, Weitz DA, Angelini TE. Nutrient depletion in Bacillus subtilis biofilms triggers matrix production. *N J Phys.* 2014;16:15028.
57. Petrova OE, Sauer K. Sticky situations: key components that control bacterial surface attachment. *J Bacteriol.* 2012;194:2413–2425.
58. Mashruwala AA, Guchte A, Boyd JM. Impaired respiration elicits SrrAB-dependent programmed cell lysis and biofilm formation in Staphylococcus aureus. *eLife.* 2017;6:e23845.
59. Thomas VC, Fey PD. Biofilms: take my breath away. *eLife.* 2017;6:e25739.
60. Ciofu O, Tolker-Nielsen T. Tolerance and resistance of Pseudomonas aeruginosa biofilms to antimicrobial agents-how P. aeruginosa can escape antibiotics. *Front Microbiol.* 2019;10:913.
61. Cramton SE, Ulrich M, Götz F, Döring G. Anaerobic conditions induce expression of polysaccharide intercellular adhesin in Staphylococcus aureus and Staphylococcus epidermidis. *Infect Immun.* 2001;69:4079–4085.
62. An S, Wu J, Zhang LH. Modulation of Pseudomonas aeruginosa biofilm dispersal by a cyclic-Di-GMP phosphodiesterase with a putative hypoxia-sensing domain. *Appl Env Microbiol.* 2010;76:8160–8173.
63. Speranza B, Sinigaglia M, Corbo MR. Effects of nutritional and environmental conditions on Salmonella sp. biofilm formation. *J Food Sci.* 2011;76:12–16.
64. Bonaventura GD, Piccolomini R, Paludi D, et al. Influence of temperature on biofilm formation by Listeria monocytogenes on various food-contact surfaces: relationship with motility and cell surface hydrophobicity. *J Appl Microbiol.* 2008;104:1552–1561.
65. Obana N, Nakamura K, Nomura N. A sporulation factor is involved in the morphological change of Clostridium perfringens biofilms in response to temperature. *J Bacteriol.* 2014;196:1540–1550.
66. Mathlouthi A, Pennacchietti E, De Biase D. Effect of temperature, pH and plasmids on in vitro biofilm formation in Escherichia coli. *Acta Nat.* 2018;10:129–132.

67. D'Urzo N, Martinelli M, Pezzicoli A, et al. Acidic pH strongly enhances in vitro biofilm formation by a subset of hypervirulent ST-17 Streptococcus agalactiae strains. *Appl Env Microbiol.* 2014;80:2176–2185.
68. Mizan FR, Jahid IK, Park SY, Myoung JJ. Effects of temperature on biofilm formation and quorum sensing of Aeromonas hydrophila. *Ital J Food Sci.* 2018;30:456–466.
69. Cheah YT, Chan DJC. Physiology of microalgal biofilm: a review on prediction of adhesion on substrates. *Bioeng.* 2021 (tandfonline.com).
70. Jo J, Price-Whelan A, Dietrich LEP. Gradients and consequences of heterogeneity in biofilms. *Nat. Rev. Microbiol.* 2022.
71. Palmer J, Flint S, Brooks J. Bacterial cell attachment, the beginning of a biofilm. *J Ind Microbiol Biotechnol.* 2007;34(9):577–588. https://doi.org/10.1007/s10295-007-0234-4
72. Zhao A, Sun J, Liu Y. Understanding bacterial biofilms: from definition to treatment strategies. *Front Cell Infect Microbiol.* 2023;13:1137947. https://doi.org/10.3389/fcimb.2023.1137947
73. Veerachamy S, Yarlagadda T, Manivasagam G, Yarlagadda PK. Bacterial adherence and biofilm formation on medical implants: a review. *Proc Inst Mech Eng H.* 2014;228:1083–1099.
74. von Eiff C, Jansen B, Kohnen W, Becker K. Infections associated with medical devices: pathogenesis, management and prophylaxis. *Drugs.* 2005;65:179–214.
75. Buttner H, Mack D, Rohde H. Structural basis of Staphylococcus epidermidis biofilm formation: mechanisms and molecular interactions. *Front Cell Infect Microbiol.* 2015;5:14.
76. Joo HS, Otto M. Molecular basis of in vivo biofilm formation by bacterial pathogens. *Chem Biol.* 2012;19:1503–1513.
77. Neu TR, Poralla K. An amphiphilic polysaccharide from an adhesive Rhodococcus strain. *FEMS Microbiol Lett.* 1988;49:389–392. https://doi.org/10.1016/j.jep.2012.08.045
78. Rather AM, Gupta K, Mandal M. Microbial biofilm: formation, architecture, antibiotic resistance, and control strategies. *Braz J Microbiology.* 2021;52:1701–1718. https://doi.org/10.1007/s42770-021-00624-x
79. Ziebuhr W, Krimmer V, Rachid S, Lossner I, Gotz F, Hacker J. A novel mechanism of phase variation of virulence in Staphylococcus epidermidis: evidence for control of the polysaccharide intercellular adhesin synthesis by alternating insertion and excision of the insertion sequence element IS256. *Mol Microbiol.* 1999;32:345–356.
80. Heilmann C, Gerke C, Perdreau-Remington F, Gotz F. Characterization of Tn917 insertion mutants of Staphylococcus epidermidis affected in biofilm formation. *Infect Immun.* 1996;64:277–282.
81. Heilmann C, Schweitzer O, Gerke C, Vanittanakom N, Mack D, Gotz F. Molecular basis of intercellular adhesion in the biofilm-forming Staphylococcus epidermidis. *Mol Microbiol.* 1996;20:1083–1091.
82. Leung JW, Liu YL, Desta T, Libby E, Inciardi JF, Lam K. Is there a synergistic effect between mixed bacterial infection in biofilm formation on biliary stents? *Gastrointest Endosc.* 1998;48:250–257.
83. Mack D, Fischer W, Krokotsch A, et al. The intercellular adhesin involved in biofilm accumulation of Staphylococcus epidermidis is a linear beta-1,6-linked glucosaminoglycan: purification and structural analysis. *J Bacteriol.* 1996;178:175–183.
84. Mack D, Nedelmann M, Krokotsch A, Schwarzkopf A, Heesemann J, Laufs R. Characterization of transposon mutants of biofilm-producing Staphylococcus epidermidis impaired in the accumulative phase of biofilm production: genetic identification of a hexosamine-containing polysaccharide intercellular adhesin. *Infect Immun.* 1994;62:3244–3253.
85. Merritt K, An YH. An YH, Friedman RJ, eds. *Factors Influencing Bacterial Adhesion.* Totowa, NJ, USA: Humana Press; 2000.

86. Boles BR, Thoendel M, Singh PK. Self-generated diversity produces "insurance effects" in biofilm communities. *Proc Natl Acad Sci USA*. 2004;101:16630–16635. https://doi.org/10.1073/pnas.0407460101
87. Kjelleberg S, Givskov M. *The Biofilm Mode of Life: Mechanisms and Adaptations*. First ed. Horizon Scientific Press; 2007.
88. Davey ME, O'Toole GA. Microbial biofilms: from ecology to molecular genetics. *Microbiol Mol Biol Rev*. 2000;64:847–867.
89. McKenney D, Hubner J, Muller E, Wang Y, Goldmann DA, Pier GB. The ica locus of Staphylococcus epidermidis encodes production of the capsular polysaccharide/adhesin. *Infect Immun*. 1998(66):4711–4720.
90. Gu H, Hou S, Yongyat C, De Tore S, Ren D. Patterned biofilm formation reveals a mechanism for structural heterogeneity in bacterial biofilms. *Langmuir*. 2013;29:11145–11153.
91. Otto M. Staphylococcal infections: mechanisms of biofilm maturation and detachment as critical determinants of pathogenicity. *Annu Rev Med*. 2013;64:175–188.
92. Nguyen HT, Nguyen TH, Otto M. The staphylococcal exopolysaccharide PIA–biosynthesis and role in biofilm formation, colonization, and infection. *Comput Struct Biotechnol J*. 2022;18:3324 34.
93. Oppenheimer-Shaanan Y, Steinberg N, Kolodkin-Gal I. Small molecules are natural triggers for the disassembly of biofilms. *Trends Microbiol*. 2013;21:594–601.
94. Grande R, Nistico L, Sambanthamoorthy K, et al. Temporal expression of agrB, cidA, and alsS in the early development of Staphylococcus aureus UAMS-1 biofilm formation and the structural role of extracellular DNA and carbohydrates. *Pathog Dis*. 2014;70:414–422.
95. Bjarnsholt T. The role of bacterial biofilms in chronic infections. *APMIS*. 2013;121(Suppl. 136):1–54.
96. Al-Sharify ZT, Al-Najjar SZ, Naser ZA, Alsherfy ZAI, Onyeaka H. The impact of fluid flow on microbial growth and distribution in food processing systems. *Foods*. 2025;14(3):401.
97. Alvarez-Ordóñez A, Coughlan LM, Briandet R, Cotter PD. Biofilms in food processing environments: challenges and opportunities. *Ann Rev Food Sci Technol*. 2019;10(1):173–195.
98. deBeer D, Stoodley P, Lewandowski Z. Measurement of local diffusion coefficients in biofilms by micro-injection and confocal microscopy. *Biotech Bioeng*. 1997;53:151–158.
99. Brading MG, Boyle G, Scott LMH. Biofilm formation in laminar flow using Pseudomonas fluorescens EX101. *J Ind Microbiol*. 1995;15:297–304.
100. Wi YM, Patel R. Understanding biofilms and novel approaches to the diagnosis, prevention, and treatment of medical device-associated infections. *Infect Dis Clin*. 2018;32(4):915–929.
101. Hall-Stoodley L, Costerton JW, Stoodley P. Bacterial biofilms: from the natural environment to infectious diseases. *Nat Rev Microbiol*. 2004;2(2):95–108.
102. Roy S, Chowdhury G, Mukhopadhyay AK, Dutta S, Basu S. Convergence of biofilm formation and antibiotic resistance in Acinetobacter baumannii infection. *Front Med*. 2022;9:793615.
103. Li Z, Ding Z, Liu Y, Jin X, Xie J, Li T. Phenotypic and genotypic characteristics of biofilm formation in clinical isolates of Acinetobacter baumannii. *Infect Drug Resist*. 2021:2613–2624.
104. Gedefie W, Demsis M, Ashagrie Y, Kassa M, Tesfaye M, Tilahun M. Acinetobacter baumannii biofilm formation and its role in disease pathogenesis: a review. *Infect Drug Resist*. 2021;1:3711–3719.
105. Salazar-Sánchez A, Baztarrika I, Alonso R, Fernández-Astorga A, Martínez-Ballesteros I, Martinez-Malaxetxebarria I. Arcobacter butzleri biofilms: insights into the genes beneath their formation. *Microorganisms*. 2022;10:1280.

106. Rupp ME, Ulphani JS, Fey PD, Mack D. Characterization of Staphylococcus epidermidis polysaccharide intercellular adhesin/hemagglutinin in the pathogenesis of intravascular catheter-associated infection in a rat model. *Infect Immun.* 1999;67:2656–2659.
107. Von Eiff C, Heilmann C, Peters G. New aspects in the molecular basis of polymer-associated infections due to staphylococci. *Eur J Clin Microbiol Infect Dis.* 1999;18:843–846.
108. Widerström M. Significance of *Staphylococcus epidermidis* in health care-associated infections, from contaminant to clinically relevant pathogen: this is a wake-up call!. *J Clin Microbiol.* 2016;54(7):1679–1681.
109. Cheung GY, Otto M. Understanding the significance of *Staphylococcus epidermidis* bacteremia in babies and children. *Curr Opin Infect Dis.* 2010;23(3):208.
110. Watters C, DeLeon K, Trivedi U, et al. Pseudomonas aeruginosa biofilms perturb wound resolution and antibiotic tolerance in diabetic mice. *Med Microbiol Immunol.* 2013;202:131–141.
111. Dowd SE, Sun Y, Secor PR, et al. Survey of bacterial diversity in chronic wounds using pyrosequencing, DGGE, and full ribosome shotgun sequencing. *BMC Microbiol.* 2008;8:43.
112. Gill SR, Fouts DE, Archer GL, Mongodin EF, DeBoy RT, Ravel J. Insights on evolution of virulence and resistance from the complete genome analysis of an early methicillin-resistant Staphylococcus aureus strain and a biofilm-producing methicillin-resistant Staphylococcus epidermidis strain. *J Bacteriol.* 2005;187(7):2426–2438.
113. Mirzaei R, Alikhani MY, Arciola CR, Sedighi I, Yousefimashouf R, Bagheri KP. Prevention, inhibition, and degradation effects of melittin alone and in combination with Vancomycin and rifampin against strong biofilm producer strains of methicillin-resistant *Staphylococcus epidermidis. Biomed Pharmacother.* 2022;147:112670.
114. Staniszewska M. Virulence factors in Candida species. *Curr Protein Pept Sci.* 2020;21(3):313–323. https://doi.org/10.2174/1389203720666190722152415
115. Fredheim EG, et al. Biofilm formation by *Staphylococcus haemolyticus. J Clin Microbiol.* 2009;47:1172–1180. https://doi.org/10.1128/JCM.01891- 08
116. Neely AN, Maley MP. Survival of enterococci and staphylococci on hospital fabrics and plastic. *J Clin Microbiol.* 2000;38:724–726.
117. Flemming H-C, Wingender J. The biofilm matrix. *Nat Rev Microbiol.* 2010;8:623–633. https://doi.org/10.1038/nrmicro2415
118. Bakthavatchalam YD, Triplicane DH, Munusamy E, Jennifer L, Veeraraghavan B. A distinct geographic variant of sasX in Methicillin- resistant Staphylococcus aureus ST239 and ST368 lineage from south India. *Microb Drug Resist.* 2019;25(3):413–420. https://doi.org/10.1089/mdr.2018.0292
119. De Backer S, Xavier BB, Vanjari L, et al. Remarkable geographical variations between India and Europe in carriage of the staphylococcal surface protein-encoding sasX/sesI and in the population structure of methicillin-resistant Staphylococcus aureus belonging to clonal complex. *Clin Microbiol Infect.* 2019;25(5):628.e1–628.e7. https://doi.org/10.1016/j.cmi.2018.07.024
120. Li M, Du X, Villaruz AE, et al. MRSA epidemic linked to a quickly spreading colonization and virulence determinant. *Nat Med.* 2012;18(5):816–819. https://doi.org/10.1038/nm.2692
121. Bhattacharya M, Berends ETM, Chan R, et al. Staphylococcus aureus biofilms release leukocidins to elicit extracellular trap formation and evade neutrophil-mediated killing. *Proc Natl Acad Sci.* 2018;115(28):7416–7421. https://doi.org/10.1073/pnas.1721949115
122. Golla R, Mishra B, Dang X, et al. Resistome of Staphylococcus aureus in response to human cathelicidin LL-37 and its engineered antimicrobial peptides. *ACS Infect Dis.* 2020;6(7):1866–1881. https://doi.org/10.1021/acsinfecdis.0c00112

123. Alabdullatif M, Ramirez-Arcos S. Biofilm-associated accumulation-associated protein (Aap): a contributing factor to the predominant growth of Staphylococcus epidermidis in platelet concentrates. *Vox Sang*. 2019;114(1):28–37. https://doi.org/10.1111/vox.12729
124. Alayande AB, Aung MM, Kim IS. Correlation between quorum sensing signal molecules and Pseudomonas aeruginosa's biofilm development and virulency. *Curr Microbiol*. 2018;75(7):787–793. https://doi.org/10.1007/s00284-018-1449-5
125. Luo J, Dong B, Wang K, et al. Baicalin inhibits biofilm formation, attenuates the quorum sensing-controlled virulence and enhances Pseudomonas aeruginosa clearance in a mouse peritoneal implant infection model. *PLoS One*. 2017;12(4):e0176883https://doi.org/10.1371/journal.pone.0176883
126. Jeske, A, Arce-Rodriguez A, Thöming JG, et al. Evolution of biofilm-adapted gene expression profiles in *lasR*-deficient clinical *Pseudomonas aeruginosa* isolates. *npj Biofilms Microbiomes*. 2022;8:6. https://doi.org/10.1038/s41522-022-00268-1
127. Li, H, Chuntian, Y, Enze Z, et al. Microbiologically influenced corrosion behavior of S32654 super austenitic stainless steel in the presence of marine Pseudomonas aeruginosa biofilm. *J. Mat. Sci. Technol*. 2017;33(12):1596–1603. https://doi.org/10.1016/j.jmst.2017.03.002
128. She P, Liu Y, Luo Z, et al. PA2146 gene knockout is associated with Pseudomonas aeruginosa pathogenicity in macrophage and host immune response. *Front Cell Infect Microbiol*. 2020;10. https://doi.org/10.3389/fcimb.2020.559803
129. Stellari F, Bergamini G, Sandri A, et al. In vivo imaging of the lung inflammatory response to Pseudomonas aeruginosa and its modulation by azithromycin. *J Transl Med*. 2015;13(1):251. https://doi.org/10.1186/s12967-015-0615-9
130. Villa S, Hamideh M, Weinstock A, et al. Transcriptional control of hyphal morphogenesis in Candida albicans. *FEMS Yeast Res*. 2020;20(1):5. https://doi.org/10.1093/femsyr/foaa005
131. Sadiq FA, Burmølle M, Heyndrickx M, et al. Community-wide changes reflecting bacterial interspecific interactions in multispecies biofilms. *Crit. Rev. Microbiol*. 2021;47(3):338–358.
132. Yao S, Hao L, Zhou R, Jin Y, Huang J, Wu C. Multispecies biofilms in fermentation: biofilm formation, microbial interactions, and communication. *Compr Rev food Sci food Saf*. 2022;21(4):3346–3375 (researchgate.net).
133. Creusot N, Tison-Rosebery J, Hubas C, et al. How metabolites and metabolism in aquatic biofilms reveal ecological responses to global change and their interactions. *Sci*. 2025;1:1–10.
134. Luo A, Wang F, Sun D, Liu X, Xin B. Formation, development, and cross-species interactions in biofilms. *Front Microbiol*. 2022 (frontiersin.org).
135. Flemming HC, van Hullebusch ED, Neu TR, et al. The biofilm matrix: multitasking in a shared space. *Nat Rev Microbiol*. 2023;21(2):70–86.
136. Pan M, Li H, Han X, et al. Effects of hydrodynamic conditions on the composition, spatiotemporal distribution of different extracellular polymeric substances and the architecture of biofilms. *Chemosphere*. 2022;307:135965 ([HTML]).
137. Pai L, Patil S, Liu S, Wen F. A growing battlefield in the war against biofilm-induced antimicrobial resistance: insights from reviews on antibiotic resistance. *Front. Cell Infect. Microbiol*. 2023;19(13):1327069. https://doi.org/10.3389/fcimb.2023.1327069

CHAPTER FOUR

Cytokine and chemokine dysregulation by biofilm components

Suman Das[a], Riyan Kaibarta[b], Soumyadip Pal[a], Debasis Mitra[c], Gustavo Santoyo[d], Sourav Chattaraj[e], Mousumi Ganguly[f], and Arindam Ganguly[a,*]

[a]Department of Microbiology, Bankura Sammilani College, Bankura, West Bengal, India
[b]Institute of Health Sciences, Presidency University, Kolkata, West Bengal. India
[c]Department of Microbiology, Graphic Era (Deemed to be University), Dehradun, Uttarakhand, India
[d]Genomic Diversity Laboratory, Institute of Biological and Chemical Research, Universidad Michoacana de San Nicolas de Hidalgo, Morelia, Mexico
[e]Centre for Industrial Biotechnology Research, School of Pharmaceutical Science, Siksha 'O' Anusandhan Deemed to be University, Bhubaneswar, Odisha, India
[f]Department of Nutrition, Bankura Sammilani College, Bankura, West Bengal, India
*Corresponding author. e-mail address: arindam_ganguly@yahoo.com

Contents

Abstract

Biofilms pose a major challenge in oral and systemic infections due to their capacity to evade host defences and modulate immune responses. Dysregulated cytokine and chemokine signaling induced by biofilm components plays a central role in disease progression, particularly in chronic conditions such as periodontitis, cystic fibrosis, and non-healing wounds. In periodontal disease, the shift from microbial homeostasis to dysbiosis is marked by enhanced pathobiont virulence, elevated inflammatory mediators including IL-6, IL-1β, TNF-α, and CXCL8, and a bias toward pro-inflammatory macrophage polarization. Localized aggressive periodontitis exhibits an exaggerated inflammatory response to bacterial surface molecules and biofilm-derived products, highlighting distinct mechanisms of immune activation. In systemic infections, biofilms

Progress in Molecular Biology and Translational Science, Volume 224
ISSN 1877-1173, https://doi.org/10.1016/bs.pmbts.2026.08.002

facilitate immune evasion through metabolic reprogramming of macrophages and neutrophils, sustaining chronic inflammation and tissue damage. Persistent biofilms in cystic fibrosis lungs, chronic wounds, and medical device-associated infections further illustrate the complex host-microbe interactions, where excessive neutrophil infiltration and prolonged cytokine release aggravate pathology. Moreover, biofilm constituents can suppress protective immune responses, creating an immunosuppressive microenvironment that favours bacterial persistence. This review summarizes current insights into biofilm-mediated immune modulation, emphasizing key cytokine and chemokine alterations across disease contexts and discussing emerging therapeutic strategies aimed at restoring immune homeostasis and improving clinical outcomes.

1. Introduction

Biofilm, which arises naturally from the formation of organism conglomeration, either as single or in multiple species accumulation, can comprise fungi, protozoa, and bacteria.[1] The total fixed carbon content comprises microorganisms, whose relative contribution to the total dry mass is somewhere less than about 10 %; with regard to matrix, it can account for over 90 % of the total. The matrix is the extracellular material produced largely by the organisms themselves, in which the biofilm cells are embedded. It is composed of a conglomerate of different types of biopolymers generally referred to as extracellular polymeric substances (EPS) that form the scaffold for the three-dimensional architecture of the biofilm. It is responsible for adhesion to biotic and abiotic surfaces and cohesion in the biofilm.[2–4] EPS have been called 'the dark matter of biofilms' because of the large range of matrix biopolymers and the difficulty in analyzing them.[5] EPS have been referred to as nutrients since some of the components are not degrading quickly, and thus require diverse types of enzymes to be entirely degraded. The matrix protects organisms from ultraviolet radiation, oxidizing or charged biocides, metallic cations, and desiccation.[6] Biofilm formation is developmental and begins with planktonic (free-living) organisms aggregating and/or transitioning to a surface-associated lifestyle, the last part being that cells disperse from the biofilm structure in a process referred to as dispersion to return to the single cell, planktonic mode of growth.[7] Biofilm formation consists of some specific processes, which include (a) initial attachment of microbes to a surface or each other, (b) microcolony formation, (c) maturation of the biofilm, and (d) biofilm dispersal.[8] The biofilm cells thus exhibit a character, which is specific from their planktonic counterparts. Innate resistance against the host immune defense and their increased stress tolerance include starvation, dehydration, antimicrobials, and immune evasion, for instance, by dysfunction through the release of bacterial toxins or byproducts and immune cell modulation or preventing phagocytosis (Fig. 1). It is also noteworthy that biofilms have been reported for their 10–1000-fold greater tolerance compared to their planktonic counterparts against various antibiotics Not surprisingly, hospitals have to contend with a number of nosocomial infections due to biofilm-forming bacterial pathogens that might compromise the wellbeing of any patient suffering from predispositions such as immune suppression or pre-existing diseases.[9,10]

Biofilms involve a very wide range of microbial infections occurring in the human body.[11,12] Biofilms on implanted medical devices pose the greatest risk associated with their tremendous antibiotic resistance, which then serves as a reservoir of bacteria that is constantly able to re-infect and may also trigger low-grade inflammation, which ultimately causes tissue damage and blockage of devices and general resistance to treatment. In the Gram-negative group, the predominant microorganisms related to nosocomial infection include *Pseudomonas aeruginosa, Klebsiella pneumoniae, Escherichia coli.* Among Gram-positives, they include Staphylococcus epidermidis, *Enterococcus faecalis*, and *Staphylococcus aureus.*[13] Biofilm formation is indeed found not only on medical devices like contact lenses, catheters, prostheses, heart valves, and pacemakers but also on various surfaces of the body, including skin and mucosal surfaces of the respiratory and digestive tracts.[10] In recent years, studies have shown that biofilms are capable of modulating the host immune response.[14] Cytokines involve nearly every aspect of inflammation and immunity, including cellular recruitment and activation, antigen presentation, innate immunity, and expression of adhesion molecules. Cytokines are typically small single-peptide proteins weighing 20 kD,[15] primarily derived from antigen presenting cells, and together with mononuclear phagocytic cells are particularly effective in promoting the infiltration of cellular and tissue damage characteristic of inflammation. This pathway may be one whereby processing of antigens takes place after uptake by antigen-presenting cells and after further processing and presentation to T-helper Lymphocytes for this class of cytokine production. Cytokines mediate cell-to-cell communication, which is distinct from, or may complement, contact-mediated cell-cell communication[16] and is produced as an outcome of immune insult to determine first whether immune response develops and second if the developed response will be cell-mediated, humoral, and cytolytic. Monocyte-derived cytokines

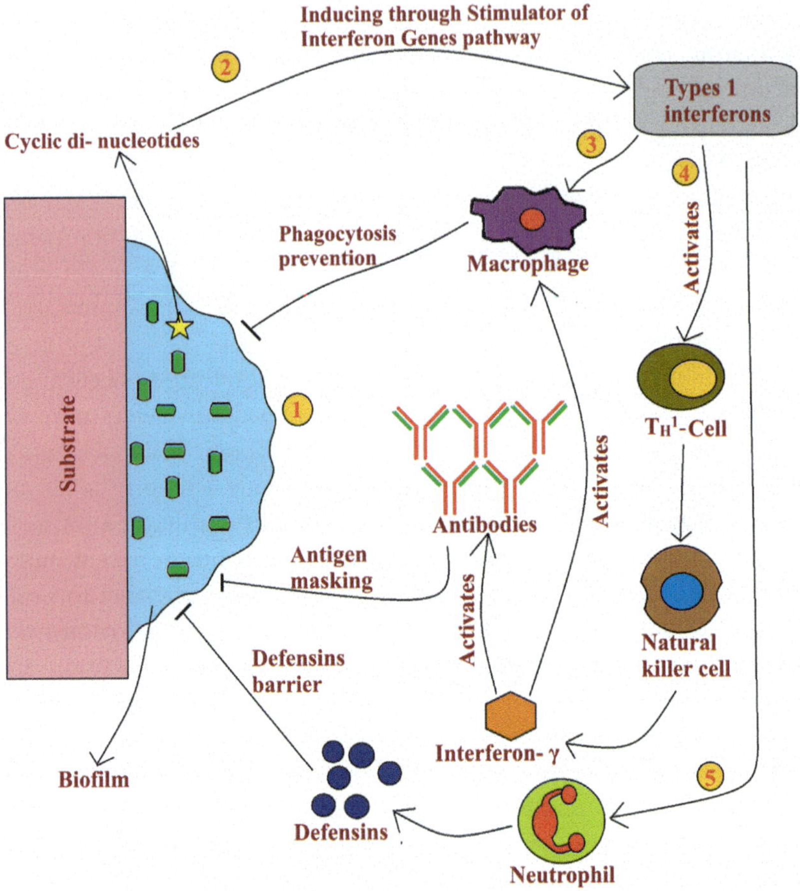

Fig. 1 Host-biofilm interaction. 1.The planktonic bacteria encounter a surface & use their flagella and other appendages to form a temporary contact with the surface. This is further enhanced & transforms into a permanent attachment when the bacteria secrete extracellular polymeric substances (EPS) consisting of proteins, polysaccharide, etc. which aid to attach them to the surface & form a biofilm; 2. The biofilms possess cyclic di-nucleotides, which act as an intracellular signalling molecules and are responsible for regulating the production of the EPS matrix, induces the activation of type-1 interferons through stimulations of interferon genes pathway; 3. The type-1 interferon activates the macrophage in response to the release of cyclic-di-nucleotides, but this macrophage is unable to phagocytose the biofilm because of the preventive measures of EPS matrix; 4. The type-1 interferon then activates the TH1 cells, which further activates the natural killer cells, that produce the interferon-γ. This interferon-γ then activates the antibodies. Which prove to be ineffective because of the antigen masking effect of the EPS matrix, which covers the surface antigen; 5. The type-1 interferons activate the neutrophils. Neutrophils contain defensins, known antimicrobial peptides that function like internal antibiotics. Defensins are ineffective due to the biofilm extracellular polymeric substances (EPS) which act as both physical and chemical shields against disruptors.

primarily include TNF and several IL species, including IL-1, IL-18, IL-8, IL-15, IL-6, IL-23, and IL-12. IL-8 is also categorized as the CXC chemokine CXCL8.[17] TNF (formerly TNF-α) is principally produced by macrophages and T lymphocytes. TNF has inflammatory, immunologic, and homeostatic effects like leading chemokinesis and activation of the respiratory burst in both macrophages and neutrophils; stimulating production of Prostaglandin E2 and collagenase.[18] There are many cytokine receptors that have common chains, such as the IL-2 receptor that has a common γ chain with receptors for IL-7, IL-9, and IL-15.[19]

Chemokines are small (8–12 kD) molecules that can induce chemotaxis of different types of cells, including eosinophils, monocytes, fibroblasts, neutrophils, and lymphocytes. They are classified as homeostatic chemokines that have very restricted expression patterns localized to one single tissue or organ, while inflammatory chemokines are produced by many nontissue-specific cell types at different locations. Many of the chemokine receptors are capable of binding to more than one ligand, allowing for an extensive overlap and redundancy of chemokine function.[17] Chemokines were described to be inflammatory, being produced on the site of infection or in response to a pro-inflammatory stimulus. The inflammatory chemokines, in conjunction with other signals, are capable of recruiting and activating leukocytes to initiate an immune response and kick-start wound healing.[20]

2. Biofilm structure and major components

The EPS of a biofilm is made up mostly of polysaccharides (Fig. 2), while a myriad of secondary metabolites can contribute to its composition, including proteins, nucleic acids, lipids, and more humic-type sub-stances.[21]

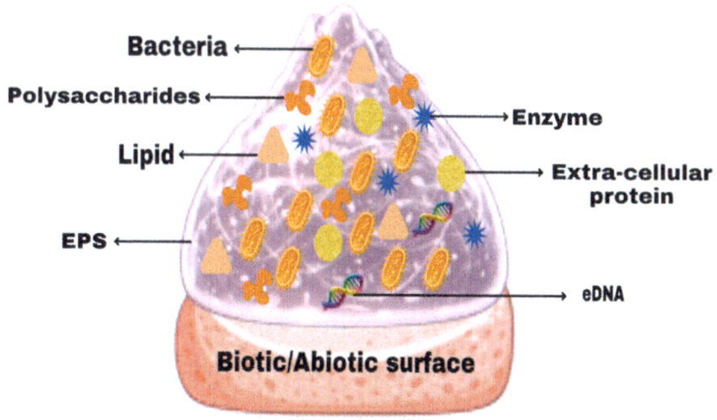

Fig. 2 Components of biofilm.

The dominant proteins and carbohydrates comprise anywhere from 75 to 90% of EPS composition.[22] The EPS systems interact with each other and form the matrix within which the microbial cells reside. Destruction of biofilms employing hydrolytic enzymes that target a range of EPS could further provide information regarding the composition of the biofilms and the importance of different matrix components in maintaining the integrity of the structure. Ultimately, the characterization of the matrix could be done by extraction and purification of EPS from the biofilm using physical and/or chemical treatments.[23]

The outer or extracellular portion of EPS is mostly composed of carbohydrates, such as carbon-containing neutral sugars (the majority being hexoses, and only a few pentoses), and uronic acids.[24] EPS can be divided into two major polysaccharide categories: homopolysaccharides and heteropolysaccharides. Experimental evidence indicated that homopolysaccharides consist of only one monosaccharide type, which is commonly D-fructose or D-glucose.[25] On the other hand, heteropolysaccharides contain two or more types of monosaccharides, the most common of these being L-rhamnose, D-galactose, and D-glucose in the EPS of *Bifidobacteria* and *Lactobacilli*.[26] The polysaccharide of *Bacillus subtilis* is associated with the formation of pellicle formation and colony surface structures. Cellulose also takes part in pellicle formation, matrix-fibriae interactions, and functions as a UV-protective compound. *Vibrio* polysaccharide acts on furtherance of environment biofilm and also in host infective cycle. The features and attributes of polysaccharides are very much data important in their physical and functional properties specially for looking into biofilm formation and biofilm stability.[27]

Extracellular proteins comprise many classes other than those mentioned above. They are cell surface adhesins, the protein subunit of pili and fimbriae, secreted extracellular proteins, proteins of outer membrane vesicles, and extracellular enzyme.[28] Structural proteins mainly interact with the formation of the polysaccharide matrix of the floc and bacterial surfaces to the environment outside the floc. The lectin-like proteins formed between activated sludge give a floc formation.[29] Biofilm Associated Protein (BAP) represents a protein that has a direct relationship with the biofilm matrix and has been characterized mainly from biofilms of *Staphylococci*. Biofilm associated proteins possess an adhesion and development property for biofilms, but they are also important virulence factors.[28] There are two types of organisms, namely prokaryote and eukaryote, in which N-glycosylated proteins are found. Glycosylation of protein takes place by covalent bonding of sugar to proteins. Functions include defense, communication, and integrity of the organism.[30]

Extracellular DNA (eDNA) has been identified as a crucial component of the extracellular matrix.[31] The appearance of eDNA is correlated with the releasing of the competence signaling peptides, which in turn promote the horizontal gene transfer in the EPS matrix in Streptococcus mutans.[24] All the overall functions of extracellular DNA in maintenance of biofilms protection, gene transfer, and adhesion were summarized in a comprehensive overview.[32] Apart from eDNA, the presence of extracellular RNA in Escherichia coli was also found to be associating with outer membrane vehicles but also exists as free molecules.[33]

Lipids are also found within the biofilm matrix.[34] Serratia marcescens secretes lipids externalized by their surface-active action. The other lipopolysaccharides are essential for the attaching of *Thiobacillus ferrooxidans* to pyrite surfaces.[35,36] The criteria Biosurfactants produced by microorganisms at the air-water interface of surface water create, and they obviously play a significant role in influencing surface tension and, thus, in the gas exchange the oceans have with the atmosphere.[37] Antibacterial and antifungal activities as well as significance of bacterial attachment and detachment from oil droplets are attaining their importance through biosurfactants.[38] Interestingly, rhamnolipids, surfactants as well, were discovered in EPS matrix of Pseudomonas aeruginosa.[39] Hydrophobic compounds were also described for other bacterial species. An example is *Staphylococcus aureus*, where a surfactant peptide was identified as biofilm structuring agent, which is also involved in the process of detachment. Emulsan is a polyanionic heteropus polysaccharide, a bio-emulsifier produced by *Acinetobacter calcoaceticus*. Due to its properties, it was patented as a measure to clean tankers from shipping oil.[40]

3. Cytokines and chemokines

Cytokines can be classified (Table 1) into different functional categories. These include lymphocyte growth factor, pro-inflammatory or anti-inflammatory effects, and polarization of the immune response to an antigen.[55]

3.1 Pro-inflammatory cytokines

Proinflammatory cytokines are mainly produced upon activation of macrophages and are involved in the up-regulation of inflammatory responses. There are evidences that some of these proinflammatory cytokines as IL-1β, IL-6, and TNF-α play important roles in pathological pain.[56]

Table 1 Key characteristics and functional classification of chemokines and cytokines.

Classification	Molecules	Primary cellular sources	Receptors	Major biological function	Significant effect	References
Pro-inflammatory cytokines	TNF-α/TNF- β	Mononuclear phagocytes (TNF-α), mast cells, neutrophils, endothelial cells, lymphocytes (TNF- β), activated lymphocytes	TNFR1, TNFR2	Regulates apoptosis, SAPK pathways, NF- κB activation	Activates neutrophils, increases adhesion molecules (ICAM-1, VCAM-1, and E- selectin), and fosters anticancer immunity	17,41
	IL-6	Endothelial cells, B and D cells, hepatocytes, mononuclear phagocytes, fibroblasts	IL-6R (gp130 complex)	Promote T-cell proliferation, activates microglia/ astrocytes, induces acute phase proteins	Crucial role in the way neurons respond to damage; produces an IL-1 receptor antagonist	42–44
	IL-1β	Monocytes, macrophages, non-immune cells	IL-1R	Induce hyperalgesia, mediate inflammation, CNS glial activation	Increased upon nerve damage causes hypersensitivity to pain	45,46
Anti-inflammatory cytokines	IL-10	Monocytes, Regulatory T cells (Tr1), B cells	IL-10R1/IL-10R2	Supresses IL-2, IFN-γ synthesis, inhabits ICAM-1and MHC- II expression, inhibits MMP production, activates JAK-1/ TYK2-STAT3 pathway	Immunosuppressive; inhibits metalloproteinases (MMPs)	47,48

	TGF-β (β1, β2, β3)	Monocytes, fibroblasts, chondrocytes, osteocytes, platelets, T cells	TGF- βR (heterotetrameric serine/threonine kinase receptors)	Regulates immune suppression, fibrosis, wound healing, inhibits B-cell immunoglobulin secretion and NK cell cytotoxicity, promotes apoptotic immune tolerance	Immunosuppressive condition following apoptosis; promotes tissue healing	49,50
Chemokines (CC family)	MCP-1 (CCL2)	Fibroblast, macrophages, endothelial cells, injured tissues, monocytes; induced by TNF-α, IFN-γ, IL−1, VEGF	CCR2	Strong chemotactic factor for monocytes/macrophages	Increased ulceration, diabetic problems, and renal inflammation	51
Chemokines (CXC family)	IL-8 (CXCL8)	Macrophages, monocytes, non-immune cells	NON-CXCR1, CXCR2	Potent neutrophil chemoattractant/activator	MAPK signaling is controlled by TNF-α and IL-1	52,53
Chemokines (CC family)	RANTES (CCL5)	T lymphocytes, platelets, macrophages	CCR1, CCR3, CCR4, CCR5	Recruitment of immune cells to inflammation/infection sites	HIV-1 (M-tropic strains) natural inhibitor; paradoxical viral enhancement at elevated levels	54

Tumor Necrosis Factor (TNF) consists of two homologous proteins mainly secreted from mononuclear phagocytes (TNF-α), also known as cachectin, and lymphocytes (TNF-β), also called lymphotoxin-α.[18,56] As earlier mentioned, TNF-α can also be released by endothelial cells, activated lymphocytes, mast cells, neutrophils, and mast cells.[17] TNF acts through two cell surface receptors, TNFR1 and TNFR2 that diverge in signaling pathways to regulate apoptotic pathways, NF kB activation of inflammation and activate stress-activated protein kinases (SAPKs). Both TNF-α receptors are present in neurons and glia.[41] TNF obviously has an impact on multiple distinct signaling pathways through 2 cell surface receptors, TNFR1 and TNFR2, responsible for the regulation of apoptotic pathways, and the NFkB activation of inflammation as well as mediating the activation of stress-activated protein kinase (SAPK) pathways. The presence of TNF-α receptors in neurons as well as in glial cells has been documented.[41] By inducing antitumor immunity by acting directly on cancer cells with a cytotoxic action and by promoting the antitumor immune responses. TNF works on the same path with endothelial cells: it induces adhesion molecules: intracellular adhesion molecule (ICAM-1), vascular cell adhesion molecule (VCAM-1), and E-selectin; thus the egress of granulocytes into inflammatory loci is permitted as part of the process. TNF strongly stimulates neutrophils.[57]

As per,[58] mononuclear phagocytes are considered the major source of interleukin 6 while IL-6 can also be produced by endothelial cells, B and T cells, hepatocytes, bone marrow cells, and fibroblasts. In men, this interleukin seems to act centrally on the neuronal reaction to nerve injury. In many instances, IL-6 also acts to activate microglia and astrocytes as well as on the regulation of expression of neuropeptides from neurons.[59,60] Interleukin-6 mediates activation, proliferation, and differentiation effects on T-lymphocytes. In this regard, IL-6 is the most potent inducer of hepatocyte synthesis of acute-phase proteins. IL-6 further induces Interleukin-1 receptor antagonist synthesis.[17]

IL-1β is produced by macrophages, monocytes and some non-immune cells. They release IL-1β during inflammation, invasion, cell injury and infection.[61] After IL-1β hyperalgesia is produced, it is possible to do intraperitoneal, intraplantar, and intracerebroventricular injections. After peripheral nerve crush injury and trauma, IL-1β expression is increased in microglia and astrocytes in the central nervous system (CNS).[62,63]

3.2 Anti-inflammatory cytokines

Major representatives of anti-inflammatory cytokines in the development of pathogenesis include Interleukin-10 and Transforming growth factor-β (TGF-β).

Interleukin-10 (IL-10) is a cytokine structural variant of the interferons, occurring predominantly as a homodimer in which each monomer comprises a single polypeptide chain of 160 residues creating a structure of 6 inter-linked alpha helices.[64–66] In humans, IL-10 is primarily secreted from Type 1 Regulatory T cells, while it also shows production from monocytes and B cells. IL-10 inhibits production of IL-2 and IFN-γ synthesis by TH1 lymphocytes. Additionally, it suppresses the expression of ICAM-1, MHC class II molecules on monocytes, B7, and CD23.[67,68] Binding to the IL-10R receptor initiates the action of IL-10. The IL-10 receptor (IL-10R) is a heterodimer consisting of the subunits IL-10R1 and IL-10R2. During the first step of receptor activation, IL-10 binds to the IL-10R1 subunit, which undergoes a conformational change that allows for the subsequent binding of the IL-10-IL-10R1 complex to the IL-10R2 subunit.[69] At this point, a cascade of intercellular signal transmission is started. This leads to the activation of the tyrosine kinases, JAK1 and TYK2. This, in turn, sends the cellular signal further by transcriptional activator, STAT3. Then, on binding to STAT-binding elements in the cells' nuclei, this molecule stimulates synthesis of IL-10 dependent products.[70] IL-10 inhibits the production of some of the members of the Matrix Metalloproteinase (MMPs) family of metalloproteinases.[71]

It is considered a regulatory molecule TGF-β or transforming growth factor-Beta, and it influences the differentiation, proliferation or migration of cells, survival, thus modulating a fusion of biological processes: wound healing, carcinogenesis, development, immune responses, and fibrosis.[50] The TGF-β superfamily includes bone morphogenetic proteins, activins, and growth differentiation factors.[72] TGF-β is primarily secreted by chondrocytes, osteocytes, fibroblasts, a few T cells, monocytes and platelets. A T-helper cell that produces TGF-β has been proposed to form a distinct phenotype-called T repressor (Tr1) or T-helper type 3 (TH3) cell. TGF-β is produced as an inactive precursor and is activated through proteolytic cleavage. In addition, it inhibits immunoglobulin secretion by B lymphocytes and cytotoxicity of mononuclear phagocytes and NK cells.[17] Analogs of TGF-β homologous isoforms in mammals: TGF-β1, -β2, and -β3, are encoded by different genes. The predominant isoform expressed is TGF-β1 in the immune system, but all three isoforms have similar properties in vitro.[73] TGF-β produced by apoptotic T cells creates an immunosuppressive milieu and accounts for the absence of inflammation and autoimmunity consequent to apoptotic cell death.[74] The remnant of a homodimer of this newly synthesized protein, known as latency associated protein (LAP), that is then noncovalently associated with a homodimer of the mature TGF-β. The combined term for this complex is latent TGF-β or small latent complex (SLC). The SLC may be secreted unaccompanied or in association with LTBP as a large latent complex (LLC).[75] TGF-β actually acts as a fibrotic agent which would enhance healing during the resolution of immune responses and other biologically distinct processes of wound healing, including bone damage, punch, incisional and excisional procedures, and ulcers.[76]

Chemokine molecules exert their activity by means of interacting with members of the 7-transmembrane G-protein-coupled receptor superfamily. Homeostatic chemokines tend to be expressed in a given tissue or organ, while, on the other hand, inflammatory chemokines can be produced by several cell types in many sites.[17]

Monocyte chemoattractant protein-1 (MCP-1; which is also generally termed as CCL2) belongs to and is a strong chemotactic stimulant for monocytes in the family of CC chemokines. MCP-1 is now considered to be identical to the gene now referred to as JE, which is expressed in response to platelet-derived growth factors in mouse fibroblasts.[77] This protein could also mediate two quite different host-dependent responses: cellular immune reactions and responses to tissue injury.[78] As predicted, the N-terminal hydrophobic signal of this precursor molecule for MCP-1 comprises 23 amino acids. The aforementioned mature protein is after cleavage of the signal peptide, a polypeptide consisting of 76 amino acids which produces a single polypeptide chain.[79] The gene for MCP-1 locates itself at chromosome 17 (chr17, q11.2) which can be induced through several mediation factors including tumor necrosis factor alpha, interferon gamma, IL-1, IL-4, platelet-derived growth factor, lipopolysaccharide from bacteria, and vascular endothelial growth factor.[80] MCP-1 is a secreted protein which attracts blood monocytes originating from the tissues, or tissue macrophages, via its cell surface receptor CCR2.[81] MCP-1 is produced in injured kidneys in response to many pro-inflammatory stimuli and its expression could predictably be found in renal diseases characterized by prominent inflammation.[82,83] MCP-1 exerts its effect by attaching to a G-protein-coupled receptor on the surface of the leukocytes it seeks to activate and move.[84] Monocyte chemotactic protein-1 is an important factor in diabetes insulin, for instance, resistance and complications such as retinopathy and diabetic nephropathy. Among those with type 1 and type 2 DM, there is a significant increase in the circulating level of MCP-1.[79] MCP-1 is early in the inflammatory response and is upregulated by TNF-a, modulating further leukocyte recruitment, thus making MCP-1 an important molecule in ulceration.[85]

Interleukin-8 (IL-8), also known as CXCL8, is regarded as an inflammatory CXC chemokine. The transcription of the gene IL 8 codes the sequenced protein domains, which have a length of 99 amino acids and are much shorter than the naturally processed proteins, which would yield a signal-active protein of either 77 amino acids in nonimmune cells or 72 amino acids in monocytes and macrophages.[53] IL-8 was then discovered as potent activators and chemoattractants for neutrophils, while most of the CC chemokines, such as monocyte chemotactic protein-1.[86] IL-8 utilizes CXCR1 and CXCR2 for binding and inducing nearly identical biological functions.[87] Expression of IL-8 has been shown to be primarily regulated by AP-induced transcriptional activity and/or through NF-ĸB, although the IL-8 gene promoter has additionally been characterized by other hormone response elements and NF-IL-6 consensus sites.[88] IL-8 is about a number of factors within a wide variety of cell responding to various proinflammatory stimuli, including cytokines, reactive oxidants, and viruses. The most strongly active proinflammatory cytokines in the majority of cell types were tumor necrosis factor-a and IL-1. This appears to be exactly what proinflammatory cytokines do to activate IL-8 expression.[89] Furthermore, the IL-8 signaling pathway also controls the cellular activities of MAPK signaling circuits, which consist of various serine/threonine kinases colocalized with scaffolding proteins within close proximity to cell surface receptors.[53]

Human RANTES is a chemokine whose proinflammatory activity promotes accumulation and activation of cells in chronic inflammatory settings.[90] RANTES is a small 68-amino acid protein belonging to the rapidly growing chemokine family.[91] The leukocyte movement initiated by RANTES is through binding to specific G protein-coupled receptors, namely CCR1, CCR3, CCR4, and CCR5.[90] This chemotactic agent, or RANTES, has thus now shed some direct denominative nomenclature within immunology, CCL5, with which it works from the recruitment of cells to inflammation and infection sites: natural killer cells, dendritic cells, basophils, T-lymphocytes, mast cells, and eosinophils.[92] Perhaps, it is one of the most important possible modulators for HIV-1 infection. It is the strongest endogenous natural chemokine inhibitor against M-tropic HIV-1 infections,[93] even though, with increased concentrations, it can act rather as a stimulator to increase the viral infection.[94]

4. Therapeutic strategies targeting immune dysregulation

Biofilm associated infections represent a foremost therapeutic challenge because of their ability to persist despite antimicrobial treatment and induce prolonged immune dysregulation. While antibiotics are the mainstay of management, their actions against biofilms, due to their decreased penetration, the different metabolically altered states of the bacteria, and the observable tolerance mechanisms possessed by bacteria, have been greatly reduced. Furthermore, the action of antibiotic therapy in itself does not neutralize the chronic inflammatory environment sustained by the dysregulated cytokine and chemokine cell signaling that propels tissue destruction and disease progression.[40] This gap has created a growing interest in therapeutic strategies that harbored immune dysregulation in the host along with microbial control. One possible strategy is to directly modulate proinflammatory cytokines that remain elevated during biofilm-associated infection processes. Cytokines like TNF-α, IL-1β, and IL-6 are predominant drivers for chronic inflammation in periodontal disease, lung pathology in cystic fibrosis, and wounds in chronic inflammation. The therapeutic blockade of these mediators has seen some success in a number of inflammatory disorders being explored lately related to biofilm. For example, the blockade of TNF-α and/or IL-1 stimulation has been shown to protect tissues from inflammation and provide better clinical outcomes in microbial dysbiosis-associated chronic inflammatory conditions.[95] However, judicious consideration should be taken, as unbridled suppression of cytokine activity may result in poor host defences and allow microbial persistence. Besides cytokines, the chemokine signalling pathway is another important focus of targeting recruitment and activation of immune cells. Most chronic biofilm infections are the ones that present hyper neutrophilia, as they involve continuous release of excess amounts of chemokines, the most of which are CXCL8 and corresponding ligands. Neutrophils are necessary for antimicrobial defence, but their prolonged presence is associated with collateral tissue damage due to the release of proteases, reactive oxygen species, and neutrophil extracellular traps. For example, therapeutic strategies that may target these chemokine receptor axes like CXCL8-CXCR1/2 would limit the more destructive kinds of inflammation while keeping in balance the fine tuning of immune processes. That is why modulation and not total inhibition of chemokine signalling seems to be the more relevant alternative for the mass restoration to immune balance. In the recent years, host-directed immunomodulatory therapies aimed at restoring immune homeostasis have emerged as promising alternatives to pathogen-focused treatments. Biofilms actively host-immune-responsive, skewing macrophage polarization to a pro-inflammatory or dysfunction phenotype that sustains cytokine release without effecting clearance. Strategies that will push pro-resolving macrophage phenotypes may mitigate this chronic inflammation ongoing in tissue repair. There lies an attraction to the larger principles of immune resolution biology in that these strategies aim to restore balance and not simply immune suppression.[96] There have been recent advances aimed at immunometabolic reprogramming in biofilm induced immune dysregulation. Biofilm derived signals may alter the metabolic state of macrophages and neutrophils and regulate cytokine production, cell survival, and functional capacity. Targeting metabolic pathways, such as glycolysis, oxidative phosphorylation, and lipid metabolism, would mainly provide indirect but effective immune modulation of inflammatory responses.[97] Such approaches would be expected to reduce cytokine and chemokine production but avoid the risk associated with direct blockade. Next promising strategy is to combine immune modulators with biofilm disruptive therapies. The biofilm matrix, consisting of extracellular polymeric substances, protects bacteria from immune recognition as well as antimicrobial agents. Degrading the biofilm matrix or interfering with quorum-sensing communication enhances access for overarching immunity while reducing the immunosuppressive properties of mature biofilms. Such approaches may reduce bacterial load while correcting immune dysregulation when used with immune-targeted therapies.[40,98] In particular, antimicrobial peptides are gaining considerable attention due to their double role in biofilm disruption and modulation of immune responses. The idea of trained immunity also has potential new therapeutic opportunities related to infections involving biofilms. Reprogramming innate immune cells to respond better but more contained through successive microbial exposures may aid in tolerance against immune sensing through continuous biofilms. Underlying trained immunity is the manipulation of epigenetic and metabolic pathways that would further enhance microbial clearance but limit excessive cytokine release.[97] Although these still largely experimental strategies show great promise for chronic infections characterized by immune exhaustion. Tailored and precision oriented therapeutic interventions are crucial for the future. They very much differ depending on host genetics, microbial composition, and disease context: biofilm-related immune responses. In this respect, profiling patient-specific cytokine and chemokine signatures may inform immune-targeted therapies specific to each one's inflammatory profiles. The identification of immune biomarkers for severity of disease, type of response to treatment, and presence or not of recurrence risk may help inform therapeutic decisions and possibly improve clinical outcomes.[96] Thus far, however, the advances have not solved a number of practical problems in bringing immune-targeted therapies into standard clinical practice. The chief issue has undeniably been about balancing immune suppression against an effective antimicrobial defense, particularly concerning the protection of immunocompromised patients. Understandably, long term safety, potential off-target effects, and the prospect of secondary infections need to be evaluated cautiously. Beyond all of that, most of the present evidence comes from experimental models, which strengthens the need for well-designed clinical trials to provide assessments of efficacy in biofilm-driven diseases. Therapeutic strategies aimed at resolving immune dysregulation are expected to bring a major paradigm shift in the treatment of biofilm-associated infections. Such a shift is created by the understanding that chronic inflammation and subsequent immune imbalance are two central factors in disease persistence rather than exclusive efforts to rid the body of bacteria. In this regard, the incorporation of immune modulation into biofilm-disruptive and antimicrobial treatment will build a more holistic approach to addressing the multifaceted nature of biofilm-associated pathology and ultimately improving long-term treatment success.

5. Future perspective

Yet, with considerable progress made in biofilm immunology, there are still major gaps in our understanding as to how complex biofilm communities orchestrate cytokines and chemokines dysregulation across time. Future work in this area would ideally adopt a systems approach-one combining transcriptomics, proteomics, metabolomics, and single-cell immune profiling-to define biofilm-specific immune signatures.[40,96] These approaches promise high value in distinguishing cytokine patterns that are relevant for early biofilm establishment versus mature chronic biofilms. Characterization of biofilm-derived extracellular polymeric substances, quorum-sensing molecules, and metabolic byproducts through molecular characterization will be one of the crucial future studies in immunomodulating biofilms. Most of these can activate pro-inflammatory mediators like IL-1β, IL-6, TNF-α, and CXCL8 but can also promote immune tolerance or exhaustion when stimulated chronically.[98,99] How these biofilms fine-tune the inflammatory-immunosuppressed balance will greatly influence the understanding of chronicity in the disease during periodontitis, cystic fibrosis, and persistent wound infections. Emerging evidence supports that innately immunometabolically reprogrammed cells have been targeted as one of the most important mechanisms underlying biofilm-induced cytokine dysregulation. Biofilm has been shown to alter the macrophage regarding polarization and neutrophil survival by reshaping cellular metabolic pathways that keep on-maintaining chronic inflammation and injury to tissues.[97] Targeting these metabolic checkpoints constitutes a promising strategy for modulation of the cytokine and chemokine responses without the broad suppression of protective immunity. In terms of pharmaceuticals, it is likely that treatments of the future would require combination strategies targeting both biofilm structure and host immune response. The biofilm disruptive approaches, quorum sensing inhibitors, matrix degrading enzymes, and antimicrobial peptides would likely be the most efficient in combination with host-directed therapies addressing normalization of aberrant cytokine and chemokine signaling.[40,99] Manipulation of chemokine receptors or macrophage differentiation pathways has emerged as a promising avenue for restoring immune homeostasis in chronic diseases related to biofilms. Finally, translational research should invest much effort in identifying biofilm-specific cytokines and chemokines as biomarkers for diagnosis and prognosis. By connecting the immune and microbiome along with clinical data, it can lead to bringing individualized treatment modalities according to the inflammatory and microbial profiles of a patient.[96] The advancement of these multi-disciplines into productive avenues will contribute dramatically to the development of new therapies, which not only eradicate biofilms but also recover immune dysregulation and promote subsequent durable tissue repair.

6. Conclusion

Biofilms serve as active and dynamic immunologically relevant microbial communities, which have significant implications on cytokine and chemokine signaling on the host side. By interacting with immune cells, biofilms progressively alter inflammatory pathways, skew leukocyte recruitment, and disturb effective microbial clearance by interfering with such processes continuously. Such processes possess the capability of triggering chronic inflammation and are important players to progressive tissue injury in ailments such as device-associated infections, cystic fibrosis, periodontal disease and chronic wounds. Chronic production of pro-inflammatory cytokines coupled with excessive chemokine-mediated neutrophil influx results in a self-renewing inflammatory environment that augments disease development. Simultaneously, biofilms cohort the immune escape by inhibiting protective immune responses, skewing macrophage polarization, and interfering with well-orchestrated innate and adaptive immunities. Co-existence of increased inflammation and diminished immune resolution conveys the intricate nature of biofilm-host relationships. Thus, treatment of biofilm-related diseases goes beyond microbial clearance. Such interventions will most probably produce lasting therapeutic effects by breaking up the biofilm structure while rebalancing the dysfunctional cytokine- and chemokine-prompted networks with which they might have interfered. Understanding these immune mechanisms more deeply will guide the design of targeted interventions aimed at limiting tissue destruction, restoring immune homeostasis, and enhancing long-term clinical management of pathology, particularly with regard to chronic infections driven by biofilms.

References

1. Ganguly A, Konar R, Karuppiah P, et al. Sustainable valorization of food waste through bacterial biofilm: a pathway towards environmental safety and economic viability. In: Sarkar T, Lahiri D, Nag M, Bhattacharya D, eds. *Biofilm Applications to Revolutionize Food Technology*. Cham: Springer; 2025. https://doi.org/10.1007/978-3-031-85205-3_13.
2. Donlan RM, Costerton JW. Biofilms: survival mechanisms of clinically relevant microorganisms. *Clin. Microbiol. Rev*. 2002;15(2):167–193.
3. Ganguly A, Kundu R, Kaibarta R, et al. The immune-modulating effects of extracellular polysaccharides: mechanistic insights and functional food potential. *Extracellular Polysaccharides*. Cham, Switzerland: Springer Nature; 2025:361–388.
4. Kaibarta R, Das S, Kundu R, Rath S, Chattaraj S, Ganguly A. Biodegradation of plastics: a microbial solution for a sustainable future. In: Thatoi H, ed. *Research and Innovations in Industrial and Marine Biotechnology: A Circular Economy 2024*. Taylor & Francis Group; 2026. https://doi.org/10.1201/9781003759003-20.

5. Flemming HC, Neu TR, Wozniak DJ. The EPS matrix: the "house of biofilm cells". *J. Bacteriol.* 2007;189(22):7945–7947.
6. Flemming HC, Wingender J. The biofilm matrix. *Nat. Rev. Microbiol.* 2010;8(9):623–633.
7. Rumbaugh KP, Sauer K. Biofilm dispersion. *Nat. Rev. Microbiol.* 2020;18(10):571–586.
8. Tolker-Nielsen T. Biofilm development. *Microb Biofilms.* 2015:51–66.
9. Davies D. Understanding biofilm resistance to antibacterial agents. *Nat. Rev. Drug Discov.* 2003;2(2):114–122.
10. Schulze A, Mitterer F, Pombo JP, Schild S. Biofilms by bacterial human pathogens: clinical relevance-development, composition and regulation-therapeutical strategies. *Microb. Cell.* 2021;8(2):28.
11. Chattaraj S, Ganguly M, Chattaraj M, Mitra D, Kaibarta R, Ganguly A. Integrated approach for sustainable management and control of Kyasanur forest disease in alignment with sustainable development goals. *Virology.* 2026;614:110749. https://doi.org/10.1016/j.virol.2025.110749
12. Ganguly A, Mahapatra S, Ray S, et al. The rising threat of Nipah virus: a highly contagious and deadly zoonotic pathogen. *Virol. J.* 2025;22:139. https://doi.org/10.1186/s12985-025-02728-4 (Article).
13. Donlan RM. Biofilms and device-associated infections. *Emerg. Infect. Dis.* 2001;7(2):277.
14. Gheitasi R, Weiss D, Müller MM, et al. Staphylococcus aureus biofilm-associated component PNAG stimulates the secretion of the immunomodulatory chemokine CXCL10 via dectin-1 signaling. *Commun. Biol.* 2025;8(1):1136.
15. Akira S, Hirano T, Taga T, Kishimoto T. Biology of multifunctional cytokines: IL 6 and related molecules (IL 1 and TNF). *FASEB J.* 1990;4(11):2860–2867.
16. Peters M. Actions of cytokines on the immune response and viral interactions: an overview. *Hepatology.* 1996;23(4):909–916.
17. Borish LC, Steinke JW. 2. Cytokines and chemokines. *J Allergy Clin Immunol.* 2003;111(2):S460–S475.
18. Beutler B, Cerami A. The biology of cachectin/TNF–a primary mediator of the host response. *Annu Rev Immunol.* 1989;7(1):625–655.
19. Stahl N, Yancopoulos GD. The alphas, betas, and kinases of cytokine receptor complexes. *Cell.* 1993;74(4):587–590.
20. Moser B, Loetscher P. Lymphocyte traffic control by chemokines. *Nat Immunol.* 2001;2(2):123–128.
21. Ganguly A, Kaibarta R, Das S, et al. Biodegradation and valorisation of plastic based food packets: a microbial solution for sustainability and circular economy. *Discov Sustain.* 2025;6:447. https://doi.org/10.1007/s43621-025-01257-y Article.
22. Marvasi M, Visscher PT, Casillas Martinez L. Exopolymeric substances (EPS) from Bacillus subtilis: polymers and genes encoding their synthesis. *FEMS Microbiol. Lett.* 2010;313(1):1–9.
23. Di Martino P. Extracellular polymeric substances, a key element in understanding biofilm phenotype. *AIMS Microbiol.* 2018;4(2):274.
24. Czaczyk K, Myszka K. Biosynthesis of extracellular polymeric substances (EPS) and its role in microbial biofilm formation. *Pol J Environ Stud.* 2007;16(6).
25. Dertli E, Colquhoun IJ, Côté GL, Le Gall G, Narbad A. Structural analysis of the α-D-glucan produced by the sourdough isolate Lactobacillus brevis E25. *Food Chem.* 2018;242:45–52.
26. Badel S, Bernardi T, Michaud PJBA. New perspectives for Lactobacilli exopolysaccharides. *Biotechnol Adv.* 2011;29(1):54–66.
27. Vu B, Chen M, Crawford RJ, Ivanova EP. Bacterial extracellular polysaccharides involved in biofilm formation. *Molecules.* 2009;14(7):2535–2554.
28. Fong JN, Yildiz FH. Biofilm matrix proteins. *Microb Biofilms.* 2015:201–222.

29. Park C, Novak JT. Characterization of lectins and bacterial adhesins in activated sludge flocs. *Water Environ Res.* 2009;81(8):755–764.
30. Hug I, Feldman MF. Analogies and homologies in lipopolysaccharide and glycoprotein biosynthesis in bacteria. *Glycobiology.* 2011;21(2):138–151.
31. Neu TR, Lawrence JR. Extracellular polymeric substances in microbial biofilms. *Microbial Glycobiology.* Academic Press; 2010:733–758.
32. Okshevsky M, Meyer RL. The role of extracellular DNA in the establishment, maintenance and perpetuation of bacterial biofilms. *Crit Rev Microbiol.* 2015;41(3):341–352.
33. Ghosal A, Upadhyaya BB, Fritz JV, et al. The extracellular RNA complement of Escherichia coli. *Microbiologyopen.* 2015;4(2):252–266.
34. Conrad A, Kontro M, Keinänen MM, et al. Fatty acids of lipid fractions in extracellular polymeric substances of activated sludge flocs. *Lipids.* 2003;38(10):1093–1105.
35. Matsuyama T, Nakagawa Y. Surface-active exolipids: analysis of absolute chemical structures and biological functions. *J MIcrobiol. Methods.* 1996;25(2):165–175.
36. Sand W, Gehrke T. Extracellular polymeric substances mediate bioleaching/biocorrosion via interfacial processes involving iron (III) ions and acidophilic bacteria. *Res Microbiol.* 2006;157(1):49–56.
37. Leck C, Bigg EK. Biogenic particles in the surface microlayer and overlaying atmosphere in the central Arctic Ocean during summer. *Tellus B: Chem Phys Meteorol.* 2005;57(4):305–316.
38. Ron EZ, Rosenberg E. Natural roles of biosurfactants: minireview. *Environ Microbiol.* 2001;3(4):229–236.
39. Davey ME, Caiazza NC, O'Toole GA. Rhamnolipid surfactant production affects biofilm architecture in Pseudomonas aeruginosa PAO1. *J Bacteriol.* 2003;185(3):1027–1036.
40. Flemming HC, Wingender J, Szewzyk U, Steinberg P, Rice SA, Kjelleberg S. Biofilms: an emergent form of bacterial life. *Nat Rev Microbiol.* 2016;14(9):563–575.
41. Boka G, Anglade P, Wallach D, Javoy-Agid F, Agid Y, Hirsch EC. Immunocytochemical analysis of tumor necrosis factor and its receptors in Parkinson's disease. *Neurosci Lett.* 1994;172(1-2):151–154.
42. Erta M, Quintana A, Hidalgo J. Interleukin-6, a major cytokine in the central nervous system. *Int J Biol Sci.* 2012;8(9):1254.
43. Erta Cañabate M. Role of astrocytic IL-6 and IL-6R in normal physiology and neuroinflammation. Universitat Autònoma de Barcelona 2014.
44. Schetters ST, Gomez-Nicola D, Garcia-Vallejo JJ, Van Kooyk Y. Neuroinflammation: microglia and T cells get ready to tango. *Front Immunol.* 2018;8:1905.
45. Mailhot B, Christin M, Tessandier N, et al. Neuronal interleukin-1 receptors mediate pain in chronic inflammatory diseases. *J Exp Med.* 2020;217(9):e20191430.
46. Guimarães RM, Aníbal-Silva CE, Davoli-Ferreira M, et al. Neuron-associated macrophage proliferation in the sensory ganglia is associated with peripheral nerve injury-induced neuropathic pain involving CX3CR1 signaling. *Elife.* 2023;12:e78515. https://doi.org/10.7554/eLife.78515
47. de Waal Malefyt R, Abrams J, Bennett B, Figdor CG, de Vries JE. Interleukin 10(IL-10) inhibits cytokine synthesis by human monocytes: an autoregulatory role of IL-10 produced by monocytes. *J Exp Med.* 1991;174(5):1209–1220. https://doi.org/10.1084/jem.174.5.1209 PMID: 1940799; PMCID: PMC2119001.
48. Kubo M, Motomura Y. Transcriptional regulation of the anti-inflammatory cytokine IL-10 in acquired immune cells. *Front Immunol.* 2012;3:275.
49. Chen W, Wahl SM. TGF-beta: receptors, signaling pathways and autoimmunity. *Curr Dir Autoimmun.* 2002;5:62–91.
50. Blobe GC, Schiemann WP, Lodish HF. Role of transforming growth factor β in human disease. *N Engl J Med.* 2000;342(18):1350–1358.

51. Yoshimura T. The chemokine MCP-1 (CCL2) in the host interaction with cancer: a foe or ally? *Cell Mol Immunol.* 2018;15(4):335–345. https://doi.org/10.1038/cmi.2017.135 Epub 2018 Jan 29. PMID: 29375123; PMCID: PMC6052833.
52. Cambier S, Gouwy M, Proost P. The chemokines CXCL8 and CXCL12: molecular and functional properties, role in disease and efforts towards pharmacological intervention. *Cell & Mol Immunol.* 2023;20(3):217–251.
53. Waugh DJ, Wilson C. The interleukin-8 pathway in cancer. *Clin Cancer Res.* 2008;14(21):6735–6741.
54. Silva MJA, Marinho RL, Rodrigues YC, et al. Molecular role of HIV-1 human receptors (CCL5–CCR5 Axis) in neuroAIDS: A systematic review. *Microorganisms.* 2024;12(4):782. https://doi.org/10.3390/microorganisms12040782
55. Dinarello CA. Historical insights into cytokines. *Eur J Immunol.* 2007;37(S1):S34–S45.
56. Zhang JM, An J. Cytokines, inflammation, and pain. *Int Anesthesiol. Clin.* 2007;45(2):27–37.
57. Tracey KJ, Fong Y, Hesse DG, et al. Anti-cachectin/TNF monoclonal antibodies prevent septic shock during lethal bacteraemia. *Nature.* 1987;330(6149):662–664.
58. Akira S, Taga T, Kishimoto T. Interleukin-6 in biology and medicine. *Adv Immunol.* 1993;54:1–78.
59. Hirota H, Kiyama H, Kishimoto T, Taga T. Accelerated nerve regeneration in mice by upregulated expression of interleukin (IL) 6 and IL-6 receptor after trauma. *J Exp Med.* 1996;183(6):2627–2634.
60. Klein MA, Möller JC, Jones LL, Bluethmann H, Kreutzberg GW, Raivich G. Impaired neuroglial activation in interleukin-6 deficient mice. *Glia.* 1997;19(3):227–233.
61. Copray JCVM, Mantingh I, Brouwer N, et al. Expr interleukin-1 beta rat Dors root ganglia. *J Neuroimmunol.* 2001;118(2):203–211.
62. Watkins LR, Wiertelak EP, Goehler LE, Smith KP, Martin D, Maier SF. Characterization of cytokine-induced hyperalgesia. *Brain Res.* 1994;654(1):15–26.
63. Yan HQ, Banos MA, Herregodts P, Hooghe R, Hooghe-Peters EL. Expression of interleukin (IL)-1β, IL-6 and their respective receptors in the normal rat brain and after injury. *Eur J Immunol.* 1992;22(11):2963–2971.
64. Tan JC, Indelicato SR, Narula SK, Zavodny PJ, Chou CC. Characterization of interleukin-10 receptors on human and mouse cells. *J Biol Chem.* 1993;268(28):21053–21059.
65. Vieira PD, de Waal-Malefyt R, Dang MN, et al. Isolation and expression of human cytokine synthesis inhibitory factor cDNA clones: homology to Epstein-Barr virus open reading frame BCRFI. *Proc Natl Acad Sci.* 1991;88(4):1172–1176.
66. Zdanov A, Schalk-Hihi C, Gustchina A, Tsang M, Weatherbee J, Wlodawer A. Crystal structure of interleukin-10 reveals the functional dimer with an unexpected topological similarity to interferon γ. *Structure.* 1995;3(6):591–601.
67. Del Prete G, De Carli M, Almerigogna F, Giudizi MG, Biagiotti R, Romagnani S. Human IL-10 is produced by both type 1 helper (Th1) and type 2 helper (Th2) T cell clones and inhibits their antigen-specific proliferation and cytokine production. *J Immunol. (Baltimore, Md: 1950).* 1993;150(2):353–360.
68. Ding L, Linsley PS, Huang LY, Germain RN, Shevach EM. IL-10 inhibits macrophage costimulatory activity by selectively inhibiting the up-regulation of B7 expression. *J Immunol.* 1993;151(3):1224–1234.
69. Kotenko SV, Krause CD, Izotova LS, Pollack BP, Wu W, Pestka S. Identification and functional characterization of a second chain of the interleukin-10 receptor complex. *EMBO J.* 1997.
70. Finbloom DS, Winestock KD. IL-10 induces the tyrosine phosphorylation of tyk2 and Jak1 and the differential assembly of STAT1 alpha and STAT3 complexes in human T cells and monocytes. *J Immunol. (Baltimore, Md: 1950).* 1995;155(3):1079–1090.
71. Wang Y, Lou S. Direct protective effect of interleukin-10 on articular chondrocytes in vitro. *Chin Med J.* 2001;114(07):723–725.

72. Chang H, Brown CW, Matzuk MM. Genetic analysis of the mammalian transforming growth factor-β superfamily. *Endocr Rev.* 2002;23(6):787–823.
73. Govinden R, Bhoola KD. Genealogy, expression, and cellular function of transforming growth factor-β. *Pharmacol. Ther.* 2003;98(2):257–265.
74. Chen W, Frank ME, Jin W, Wahl SM. TGF-β released by apoptotic T cells contributes to an immunosuppressive milieu. *Immunity.* 2001;14(6):715–725.
75. Annes JP, Munger JS, Rifkin DB. Making sense of latent TGFβ activation. *J Cell Sci.* 2003;116(2):217–224.
76. Li MO, Wan YY, Sanjabi S, Robertson AKL, Flavell RA. Transforming growth factor-β regulation of immune responses. *Annu Rev Immunol.* 2006;24(1):99–146.
77. Cochran BH, Reffel AC, Stiles CD. Molecular cloning of gene sequences regulated by platelet-derived growth factor. *Cell.* 1983;33(3):939–947.
78. Leonard EJ, Yoshimura T. Human monocyte chemoattractant protein-1 (MCP-1). *Immunol. Today.* 1990;11:97–101.
79. Singh S, Anshita D, Ravichandiran V. MCP-1: function, regulation, and involvement in disease. *Int Immunopharmacol.* 2021;101:107598.
80. Sheikine Y, Hansson G. Chemokines and atherosclerosis. *Ann Med.* 2004;36(2):98–118.
81. Boring L, Gosling J, Chensue SW, et al. Impaired monocyte migration and reduced type 1 (Th1) cytokine responses in CC chemokine receptor 2 knockout mice. *J Clin Investig.* 1997;100(10):2552–2561.
82. Rovin BH, Rumancik M, Tan L, Dickerson J. Glomerular expression of monocyte chemoattractant protein-1 in experimental and human glomerulonephritis. *Lab. Invest J Techni. Methods Pathol.* 1994;71(4):536–542.
83. Rovin BH, Yoshiumura T, Tan L. Cytokine-induced production of monocyte chemoattractant protein-1 by cultured human mesangial cells. *J Immunol.* 1992;148(7):2148–2153.
84. Yadav A, Saini V, Arora S. MCP-1: chemoattractant with a role beyond immunity: a review. *Clinica Chim Acta.* 2010;411(21-22):1570–1579.
85. Watanabe T, Higuchi K, Hamaguchi M, et al. Monocyte chemotactic protein-1 regulates leukocyte recruitment during gastric ulcer recurrence induced by tumor necrosis factor-α. *Am J Physiol-Gastrointest. Liver Physiol.* 2004;287(4):G919–G928.
86. Cross AK, Richardson V, Ali SA, Palmer I, Taub DD, Rees RC. Migration responses of human monocytic cell lines to α-and β-chemokines. *Cytokine.* 1997;9(7):521–528.
87. Remick DG. Interleukin-8. *Crit Care Med.* 2005;33(12):S466–S467.
88. Brat DJ, Bellail AC, Van Meir EG. The role of interleukin-8 and its receptors in gliomagenesis and tumoral angiogenesis. *Neuro-Oncology.* 2005;7(2):122–133.
89. Brasier AR, Jamaluddin M, Casola A, Duan W, Shen Q, Garofalo RP. A promoter recruitment mechanism for tumor necrosis factor-α-induced interleukin-8 transcription in type II pulmonary epithelial cells: dependence on nuclear abundance of Rel A, NF-κB1, and c-Rel transcription factors. *J Biol Chem.* 1998;273(6):3551–3561.
90. Schall TJ. Biology of the RANTES/SIS cytokine family. *Cytokine.* 1991;3(3):165–183.
91. Zlotnik A, Yoshie O. Chemokines: a new classification system and their role in immunity. *Immunity.* 2000;12(2):121–127.
92. Appay V, Rowland-Jones SL. RANTES: a versatile and controversial chemokine. *Trends Immunol.* 2001;22(2):83–87.
93. Cocchi F, DeVico AL, Garzino-Demo A, Arya SK, Gallo RC, Lusso P. Identification of RANTES, MIP-1α, and MIP-1β as the major HIV-suppressive factors produced by CD8+ T cells. *Science.* 1995;270(5243):1811–1815.
94. Gordon CJ, Muesing MA, Proudfoot AE, Power CA, Moore JP, Trkola A. Enhancement of human immunodeficiency virus type 1 infection by the CC-chemokine RANTES is independent of the mechanism of virus-cell fusion. *J virology.* 1999;73(1):684–694.

95. Hajishengallis G. Periodontitis: from microbial immune subversion to systemic inflammation. *Nat Rev Immunol.* 2015;15(1):30–44.
96. Belkaid Y, Harrison OJ. Homeostatic immunity and the microbiota. *Immunity.* 2017;46(4):562–576.
97. Netea MG, Domínguez-Andrés J, Barreiro LB, et al. Defining trained immunity and its role in health and disease. *Nat Rev Immunol.* 2020;20(6):375–388.
98. Hanke ML, Kielian T. Deciphering mechanisms of staphylococcal biofilm evasion of host immunity. *Front Cell Infect Microbiol.* 2012;2:62.
99. Hajishengallis G, Lamont RJ. Breaking bad: manipulation of the host response by P orphyromonas gingivalis. *Eur J Immunol.* 2014;44(2):328–338.

Further reading

Limoli DH, Jones CJ, Wozniak DJ. Bacterial extracellular polysaccharides in biofilm formation and function. *Microb Biofilms.* 2015:223–247.

CHAPTER FIVE

Biofilm-driven chronic inflammation and tumorigenesis

Shreyas Joshi[a,1], Shreya Pandey[b,1], Kartik Adake[c,d,*], and Kedar Ahire[d]
[a]Tata Institute of Fundamental Research, Hyderabad, India
[b]Department of Zoology, Daulat Ram College, University of Delhi, New Delhi, India
[c]Department of Zoology, St. Xavier's College (Empowered Autonomous Institute), Mumbai, Maharashtra, India
[d]Department of Zoology, Savitribai Phule Pune University, Pune, Maharashtra, India
*Corresponding author: e-mail address: kartik.adake@xaviers.edu

Contents

[1] These authors contributed equally to this work

Progress in Molecular Biology and Translational Science, Volume 224
ISSN 1877-1173, https://doi.org/10.1016/bs.pmbts.2026.07.001

Abstract

Biofilms of pathogenic bacteria are traditionally defined as microcolonies encased in an endogenously produced matrix upon host tissue. However, realistically, it is a highly heterogenous microenvironment composed of the bacterial cells and their extracellular polymeric substances (EPS), and host necrotic tissue, fibrotic tissue, fibrosis-associated cells and their extracellular matrix. This diverse milieu makes biofilms a potent immune evasion strategy. Its immune non-clearance can lead to relapse, potentially resulting in a persistent dysfunctional immune response; chronic inflammation. Characterised by repeated proinflammatory signalling, chronic inflammation may lead to localised or systemic instability by elevating mutagenesis and promoting anti-apoptotic pathways in host cells. This combined with the biofilm microenvironment's relative isolation from the host immune system significantly promotes tumourigenesis. The tumour microenvironment develops in continuum with the biofilm microenvironment as an immunosuppressive vascularised tissue. This tripartite pathophysiology of biofilm-induced chronic inflammation and tumourigenesis presents these clinical conditions, traditionally studied independently of each other, as an intricate web of interdependent disorder. Such interventions may include biofilm-disrupting agents, inflammatory pathway inhibitors targeting NF-κB/STAT3 signalling, and immune checkpoint-directed therapies aimed at dismantling the tumour-protective inflammatory niche.

1. Introduction

1.1 Biofilms in Inflammation — an overview

Prokaryotes, despite their single-celled nature, show incredible diversity and ubiquitous dominance, overcoming multiple environmental challenges. This global success can be attributed to rapid reproduction, metabolic diversity, and perhaps most notably, biofilm formation.[1], pp. 674–676.

Biofilms are three-dimensional usually polymicrobial colonies which form selectively permeable microenvironments on abiotic, as well as biotic surfaces.[2] This remarkably enhances resistances to environmental stressors, including antibiotics, in participating cells,[3,4] particularly because of the physiological and biochemical mediation by the self-produced extracellular polymeric substances (EPS) in the biofilm matrix.[5] Biofilm formation is an integral part of bacterial pathogenesis as the biofilm matrix also aids in evading the non-specific and specific, cellular and humoral components of the immune system.[6]

Our immune system has adapted to recognising these very EPS of the biofilm matrix as potent pathogen-associated molecular patterns (PAMPs), along with other bacterial components like cell wall lipopolysaccharides, to promote inflammation at the site of colonisation. Phagocytes like macrophages and dendritic cells (DC) identify and bind these EPS using Toll-like receptors (TLRs), lectins, and other such pattern recognising receptors (PRRs) to trigger downstream immune responses. This includes upregulation of proinflammatory cytokines like TNFα, IL-6, and IL-1β; discussed further.

1.2 Acute and chronic inflammation

Inflammation, characterised by *rubor*, *calor*, *dolor*, *tumor*, and *functio laesa*, is the hallmark of initial immune response. Acute inflammation is necessary for an effective immune response as inflammation creates a suitable environment for optimal functioning of immunocytes, their components, and products.

The entire idea of biofilms, the formation of a secluded microenvironment for safety and replication, functions as an immune evasion strategy innately present in pathogenic bacteria. The Cross Reactive Antigen (CRA) are a part of the extracellular components secreted by fibroblasts, which, on exceeding its "ON" threshold,[7] cause an intensified inflammatory response, contributing to high cross reactivity, thus perpetuating persistent immune response, facilitating non-clearance of bacterial infection. This also increases the chances of relapse, eventually leading to chronic inflammation (Fig. 1).

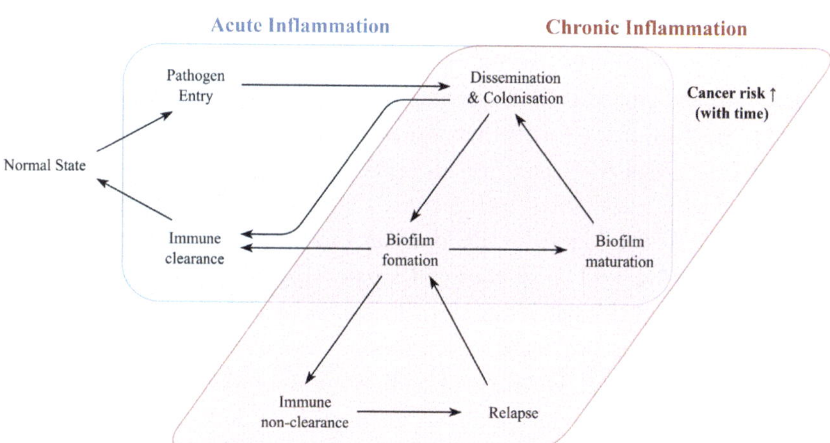

Fig. 1 Biofilm-driven chronic inflammation and tumorigenesis *(Made using draw.io)*.

Chronic inflammation is the persistent, dysregulated immune response resulting primarily due to non-clearance of immunogens and/or immune complexes. Immunogens and immune complexes are prime activators of proinflammatory signalling pathways, and their non-clearance leads to repeated inflammatory responses, drastically extending the inflammation bout.

2. Proinflammatory signalling

Recognition of PAMPs by phagocytic immunocytes like macrophages, dendritic cells, and neutrophils is usually the major source of proinflammatory response in bacterial infections. Bacterial lipopolysaccharide (LPS) binding to cell surface Toll-like receptor 4 (TLR4) of macrophages is a well studied example of the same, wherein classically activated macrophages (M1 subpopulation) are characterised by production of various proinflammatory cytokines; including TNFα, IL-6, and IL-1β.[8] Unlike planktonic bacterial exposure, biofilm-associated EPS creates sustained multivalent receptor clustering that prolongs TLR4 dimerisation and delays receptor internalisation. This altered ligand presentation amplifies NF-κB persistence and promotes chronic cytokine release rather than transient acute activation.

2.1 TNF-α signalling

LPS-TLR4 binding triggers downstream activation of NF-κB (Nuclear factor κB) in M0 and M1 macrophages. LPS binds the leucine-rich motif of TLR4 with help of LBP (LPS binding protein) and CD14 molecules, which induces dimerisation of TLR4. Mal (MyD88 adaptor-like) protein binds the TIR (Toll/IL-1 receptor) domain of the now dimerised TLR4, which bridges it to the MyD88 adaptors, which oligomerise, forming a scaffold "myddosome" for further signal transduction.[9]

IRAK1-IRAK4 (IL-1 receptor associated kinase 1–4 complex) is recruited to the myddosome, which in turn recruits TRAF-6 (TNF receptor associated factor 6). TRAF6 bridges the myddosome to the TAB-2 (TAK-1 binding protein 2) – TAK-1 (TGF-β activated kinase 1). TAK1 phosphorylates the IKK (inhibitor of κB kinase) complex. Phosphorylated IKK complex, in turn phosphorylates IκBα (inhibitor of κB alpha). Unphosphorylated IκBα binds and sequesters NF-κB; whereas, phosphorylation of IκBα induces a conformational change, liberating the NF-κB peptide. NF-κB is a transcription factor which translocates to the nucleus to upregulate the expression of proinflammatory cytokines such as TNF-α, IL-6, and IL-1β.[8,9]

TNF-α is synthesised as a transmembrane 26 kDa trimer (tmTNF-α), which is later cleaved by the membrane-bound protease TACE (TNF-α converting enzyme) to liberate water-soluble 17 kDa trimer (sTNF-α). sTNF-α functions as an autocrine, paracrine, as well as an endocrine proinflammatory signalling molecule by activating various immune cells, also including macrophages.[10,11]

sTNF-α binds the various cell surface TNFR (TNF-α receptors) isotypes on different immunocytes. In macrophages, for example, sTNF-α binds the extracellular N-terminal TNF-binding domains of the transmembrane TNFR1, inducing its trimerisation. TRADD (TNFR1-associated death domain protein) binds the now trimerised cytoplasmic C-terminal death domain of TNFR1, where it recruits TRAF (TNFR associate factor), IAP (inhibitor of apoptosis pathway), and RIP (receptor interacting protein). The TRADD-TRAF-IAP-RIP complex forms the TNFR1 adaptor complex. IAP also possesses U3 ligase (ubiquitin ligase) activity, catalysing autopolyubiquitination and also the polyubiquitination of RIP. Polyubiquitinated IAP inhibits apoptotic pathways by inactivating caspases; whereas polyubiquitinated RIP facilitates the binding of IKK complex.[10,12]

As mentioned previously, IKK phosphorylates IκBα, thereby liberating the NF-κB peptide, which translocates to the nucleus to upregulate the expression of proinflammatory cytokines, like IL-6 and IL-1β, and TNF-α itself; creating a positive feedback loop. It also upregulates angiogenesis-promoting factors like β-catenin and VEGF (Vascular Endothelial Growth Factor).

Additionally, NF-κB upregulates the production of apoptosis inhibitors. This is vital for cell survival as the upstream signalling pathway for NF-κB is linked with that of caspase-8 mediated apoptosis.

2.2 IL-6 signalling

As discussed above, IL-6 can be induced by various ligand-receptor activations on immune cells; including PAMP-TLR binding and TNF-TNFR binding. In fact, most cytokine production pathways show autoinduction and redundant signalling to enhance the control over inflammation.

IL-6 are single-chain 20 kDa peptides released as potent autocrine, paracrine, and endocrine proinflammatory cytokines. IL-6 binds the cell-surface ligand-binding domain of the transmembrane IL-6 receptor, inducing its dimerisation. The cytoplasmic signal-transducing domains of the receptor are associated with JAK (Janus kinase; previously just another kinase) proteins, which get activated by transphosphorylation due to proximity upon dimerisation of the receptors.[10,13]

Phosphorylated JAK proteins phosphorylate the signal-transducing domain of the IL-6 receptor at the conserved tyrosine residue (Tyr759). These phosphotyrosine residues serve as docking sites for STAT3 (signal transducer and activator of transcription 3). JAK phosphorylates STAT3, which undergoes dimerisation. STAT3 dimer is a transcription factor which translocates to the nucleus to upregulate the expression of acute phase peptides like CRP (C-reactive proteins) in hepatocytes, angiogenesis-promoting factors like β-catenin and VEGF in endothelial cells, and proinflammatory cytokines such as IL-1β, and IL-6 itself in immunocytes; creating a positive feedback loop.[10,13]

Similarly to NF-κB, STAT3 also upregulates the production of apoptosis inhibitors. This is vital for cell survival as the upstream signalling pathway for STAT3 is linked with that of caspase-3 mediated apoptosis.

2.3 IL-1β signalling

As mentioned earlier, IL-1β is also upregulated by NF-κB and STAT3. More accurately, these pathways promote the synthesis of a precursor pro-IL-1β (31 kDa), which undergoes caspase-1 proteolysis-dependent processing to produce bioactive IL-1β (17.5 kDa). This mature IL-1β is secreted by a rather unconventional pathway where the N-terminal pore-forming domain of GSDMD (gasdermin D), also proteolytically activated by caspase-1, allows efflux of the IL-1β peptide.[14,15]

IL-1β are also single-chain peptides released as potent autocrine, paracrine, and endocrine proinflammatory cytokines. IL-1β binds the cell-surface ligand-binding domain of the transmembrane IL-1R (IL-1 receptor). IL-1β binding-activated IL-1R recruits the IL-1RacP (IL-1R accessory protein), forming a heterodimeric transmembrane receptor complex. The cytoplasmic domain of this complex forms a TIR-like platform, facilitating myddosome formation (refer subsection 3.2.1. TNF-α signalling).[14,15] As discussed earlier, further downstream signalling activates NF-κB.

IL-1β also upregulates VEGF and β-catenin, and cell adhesion molecules on vascular endothelial cells to promote leukocyte extravasation. More importantly, IL-1β is a potent pyrogen, able to cross the blood-brain barrier. In chronic inflammation, overexpression of IL-1β can play a significant role in onset and progression of neuro-associated disorders like encephalitis and glioblastoma.[16] Together, these programs embed an IL-1β→NF-κB→IL-6/STAT3 feed-forward logic across tumour, myeloid, endothelial, and fibroblast compartments that stabilizes immune evasion while accelerating progression, which is discussed further in 3.6.3.

3. Oxidative stress in chronic inflammation

Elevated generation of ROS (reactive oxygen species) and RNS (reactive nitrogen species) are a ubiquitous characteristic feature in inflammation. The immune system uses these cytotoxic chemicals precisely against pathogens. In chronic inflammation, however, shows accumulation of these species, which now non-specifically react with and degrade self proteins and DNA.

3.1 ROS/RNS generation

ROS and RNS are generated by various channels. The largest contributor perhaps remains activated phagocytes, which engulf, ingest, and degrade antigens by various pathways. These degradation pathways actively involve utilisation of ROS and RNS for digestion of the antigen. One such pathway, which solely relies on ROS/RNS for antigen degradation, is the "respiratory burst" pathway. Phagosome membrane-bound NOX (NADPH oxidase) complexes use NADPH as reducing agents to convert molecular oxygen (O_2) to superoxide radical ($O_2^{-\cdot}$). Superoxide radical in itself is a potent ROS, but is preferentially converted to hydrogen peroxide (H_2O_2) or peroxynitrate ($ONOO^-$), which can further produce a diverse array of ROS and RNS (Punt et al.[17], pp. 315–317).

The H_2O_2 shunt is modulated by the cuproenzyme SOD (superoxide dismutase), which plays a dual role. SOD-Cu^{2+} oxidises the superoxide radical to regenerate molecular oxygen, ensuring the avoidance of hypoxic shock; while the now reduced SOD-Cu^+ reduces another superoxide radical to the highly unstable peroxide ion (O_2^{2-}), which rapidly accepts two protons from the phagosome lumen to produce H_2O_2.[18] H_2O_2 itself, again, is a potent ROS, but it can further be converted to hypohalous acids (HOX), most commonly hypochlorous acid (HOCl), a well known antiseptic agent, by the activity of MPO (myeloperoxidase), which condenses the H_2O_2 molecule with free halide ions (X^-); here, chloride ions (Cl^-), which are abundant in the cytoplasm, and by extension, the phagosome lumen (Punt et al.[17], pp. 315–317).

Alternatively, superoxide radicals can be used to generate RNS. $O_2^{-\cdot}$ spontaneously reacts with nitrous oxide (NO), synthesised by oxidative deimination of L-arginine to L-ornithine by NOS (nitrous oxide synthase). NO can also form other nitrogen oxides like nitrates (NO_2^-) and nitrites (NO_3^-) (Punt et al.[17], pp. 315–317).

3.2 Protein and nucleic acid damage

ROS and RNS have high redox potential and can as a result induce unwanted covalent modifications in proteins and nucleic acids. NO and $ONOO^-$, for instance, induce 3-nitration of tyrosine residues, causing conformational change in proteins and possible loss of function. This nitrotyrosine serves as a biomarker for various cancers.[19]

RNS are also known to inflict S-nitrosylation of sulphydryl (–SH) groups in cysteine residues and enzyme cofactors to generate S-nitrosothiol (–SNO) cysteine. S-nitrosylation is an important protein regulator, perhaps mediated by SNO-CoA (S-nitrosothiol coenzyme A)[20] however, its uncontrolled incidence by RNS prove to be detrimental to the cell. S-nitrosylation of Haemoglobin, specifically at Cysβ93, is although significant in physiological transport of NO, may facilitate systemic dissemination of nitrosative stress signals, extending localized biofilm-induced inflammatory injury into distal tissue dysfunction that predisposes systemic mutagenic environments.[21]

Sulphydryl group modifications are not restricted to RNS induction. ROS can also oxidise sulphydryl groups to sulphenic acid (–SOH), sulphinic acid (–SOOH), and/or sulphonic acid ($-S(O)_2OH$)·H_2O_2 can also cleave disulphide bonds in proteins to produce two free Cys–SOH. This, again, leads to a conformational change in proteins and possible loss of function. On the flip side, this same ROS reducing potential of the cysteine sulphydryl group makes free cysteine a strong endogenous antioxidant.[22]

In nucleic acids, guanidine residues take the brunt of the blow. Major lesion forming mutations involve base modifications to 8-oxoguanosine and 8-nitroguanosine. ROS can also create ssDNA nicks and, rarely, dsDNA breaks in both genomic and mitochondrial DNA; demanding error-prone repair mechanisms like NHEJ (non-homologous end joining). Damaged dysfunctional mitochondria are another site of ROS generation by electron "leakage" from the electron transport system (ETS), particularly the ETS complex IV and the F_o-F_1 ATPase complex. This ROS-induced ROS-release (RIRR) leads to a positive feedback loop of ROS efflux.[23]

3.3 Akt pathway and Telomerase dysfunction

3.3.1 AKT pathway and carcinogenesis

Accumulation of ROS also upregulate Akt, a viral-origin (AKT-8 Retrovirus) anti-apoptotic proto-oncogene first isolated from AKR mouse thymoma line, hence the name "AKT". Akt proto-oncogene is responsive to environmental stressors, especially the Reactive Oxygen

Species, (which have been discussed in detail in the previous section). In normal physiology, AKT acts as a central interpreter of extracellular cues to balance cell survival, growth, and metabolism. Upstream, growth factors, hormones, and cytokines stimulate receptor tyrosine kinases or, in some contexts, GPCRs (G-Protein Coupled Receptors). These receptors recruit and activate class I PI3Ks (p110α/β/δ/γ), which convert PIP2 into PIP3 at the inner leaflet of the plasma membrane. PIP3 then serves as a docking platform for the PH domain–containing proteins, most notably PDK1 and AKT. Membrane residency enables PDK1 to phosphorylate AKT on Thr308 in the activation loop, while mTORC2 sets Ser473 in the hydrophobic motif; both events are required for full catalytic activation. Several lipid phosphatases curb this input: PTEN removes the 3′ phosphate from PIP3 to regenerate PIP2, while SHIP and INPP4 enzymes reshape the PIP lipid landscape to tune signal intensity and duration. AKT activation is initiated at membranes but signals into both cytosolic and nuclear compartments.

Once active, AKT coordinates a broad array of downstream programs. It promotes survival by antagonizing pro-apoptotic machinery and by sustaining mitochondrial function. It drives anabolism through mTORC1, which is relieved from inhibition when AKT phosphorylates and suppresses the TSC1/2 complex; mTORC1 then enhances protein synthesis via S6K and 4E-BP1 and supports nucleotide and lipid biosynthesis. AKT increases glucose uptake and glycolytic flux by mobilizing glucose transporters and regulating key enzymes, while inhibiting GSK3 to favor glycogen synthesis and stabilize labile transcription factors such as c-Myc and β-catenin. In the nucleus, AKT phosphorylation of FOXO transcription factors promotes their export and degradation, dampening stress-response and cell-cycle inhibitory programs. AKT also facilitates NF-κB signaling via IKK, supports HIF-1α and other translation-dependent programs through mTORC1, and restrains autophagy by inhibiting ULK1. Negative feedback circuits, particularly S6K-mediated phosphorylation of IRS proteins, temper upstream receptor signaling to prevent runaway activation.

Tumor cells frequently co-opt the same circuitry but remove the brakes and amplify the drivers. Upstream, this occurs through activating mutations or amplification of receptor tyrosine kinases, oncogenic RAS, activating mutations in PIK3CA, amplification of PIK3CB in select contexts, hotspot mutations in AKT itself (such as E17K in the PH domain), and, very commonly, loss or silencing of PTEN. These lesions elevate PIP3, prolong membrane residency of AKT, and increase nuclear and cytosolic AKT outputs. mTORC1/2 can also be hyperactivated by lesions in their regulators, reinforcing the pathway. Chronic inflammatory signaling in the tumor microenvironment supplies additional tonic input through cytokines and chemokines that engage RTKs and GPCRs, further feeding the PI3K–AKT axis.

Downstream, sustained AKT rewires cell physiology to favor malignant traits. Persistent mTORC1 activity drives biomass accumulation and cell-cycle progression, while feedback loops are often bypassed, producing a high-flux anabolic state. Inhibition of GSK3 stabilizes β-catenin and c-Myc, boosting transcriptional programs that support proliferation and, in many settings, telomerase expression. Nuclear AKT silences FOXO-dependent anti-proliferative and pro-apoptotic genes, aiding survival under stress and therapy. The pathway enhances DNA repair capacity, supports redox homeostasis, and promotes epithelial–mesenchymal plasticity and invasion through cross-talk with NF-κB and STAT3. Metabolically, AKT enforces aerobic glycolysis and lipogenesis, supplying membranes and signaling lipids for rapid growth. These outputs together create resilience to nutrient stress, hypoxia, and cytotoxic therapies, and they foster immune evasion by reshaping cytokine production and antigen presentation. Importantly, the same feedback circuits that maintain homeostasis in normal cells are often blunted in tumors, so AKT signaling remains chronically elevated and self-reinforcing.

4. Telomerase activity and its dysfunction in tumourigenesis

Human telomerase is a ribonucleoprotein enzyme that solves the end-replication problem by extending telomeres. Its core consists of the catalytic reverse transcriptase (hTERT) and the RNA template (hTR/TERC). Biogenesis requires the H/ACA RNP assembly factors (dyskerin, NOP10, NHP2, GAR1) and TCAB1 for trafficking to Cajal bodies. At chromosome ends, recruitment and processivity are governed by the shelterin complex, particularly TPP1's TEL patch interaction with hTERT and POT1 binding to single-stranded telomeric DNA. In healthy humans, telomerase is tightly restricted: it is robust in the germline, early embryonic development, and specific stem/progenitor compartments but repressed in most somatic tissues. This developmental repression enforces progressive telomere shortening with cell divisions, eventually triggering a DNA damage response and replicative senescence when telomeres become critically short.

Physiologic regulation of hTERT expression integrates developmental and immune signals. The promoter is under control of transcription factors such as SP1, E-box–binding factors, ETS family members, and context-dependent inputs from NF-κB, STATs, NFAT, and β-catenin/TCF. Higher-order chromatin structure and telomere position effects contribute to repression in somatic cells. Post-transcriptionally, extensive alternative splicing of hTERT limits the pool of catalytically competent transcripts, and multiple microRNAs target the mRNA. Post-translationally, phosphorylation and chaperone interactions modulate hTERT stability and nuclear import, while SUMOylation and ubiquitination, fine-tune activity and turnover of the cell. Beyond telomere synthesis, hTERT can influence transcriptional networks and mitochondrial function, helping to manage oxidative stress and support tissue repair, but these non-telomeric roles are normally constrained by low expression, in tumors.[24]

Most cancers bypass the normal telomere clock by reactivating telomerase. A dominant route is genetic deregulation of the TERT promoter through two recurrent hotspot mutations (commonly referred to as C228T and C250T), which create de novo binding motifs for ETS transcription factors. These changes shift the promoter from a repressed to an active chromatin state, recruit epigenetic co-activators, and alter higher-order chromatin architecture, thereby installing a robust "on" state. Other genomic mechanisms include TERT amplification and structural rearrangements that hijack strong enhancers. Epigenetic remodeling—such as focal promoter methylation that blocks repressor binding—can further elevate expression. Alternative splicing patterns shift toward catalytically competent hTERT isoforms, while promoter-proximal and distal enhancer elements become more accessible to oncogenic transcription factors.

Tumors also exploit signaling inputs that load onto this permissive promoter architecture. Chronic inflammatory pathways activate NF-κB and STAT3; immune activation pathways converge on NFAT; and developmental cues stabilize β-catenin/TCF activity. These inputs enhance hTERT transcription atop the genetic "floor," while kinase signaling influences hTERT post-translationally to promote nuclear localization and stability. The downstream consequences are twofold. First, sustained telomerase activity maintains telomere length above crisis thresholds, enabling limitless replicative potential and preventing catastrophic end-to-end fusions. Second, non-telomeric functions of hTERT reinforce malignant behavior: cooperation with β-catenin and other transcriptional programs supports stemness and proliferation; effects in mitochondria dampen reactive oxygen species and apoptosis; and modulation of inflammatory and angiogenic pathways reshapes the tumor microenvironment to favor growth and dissemination. In some contexts, telomerase activity also feeds back into survival signaling circuits, further stabilizing oncogenic states established by PI3K–AKT and related pathways.

Together, these views underscore a common theme: in health, AKT and telomerase are tightly gated to match tissue needs, with multilayered negative controls and feedback to maintain balance. In cancer, those gates are pried open genetically and epigenetically, and chronic microenvironmental inputs keep the systems running. The result is sustained anabolism and survival signaling coupled to durable telomere maintenance and pro-tumor ecosystem remodeling—two interlocking engines that power malignant progression.

5. Wnt signalling and epithelial-mesenchymal transition

5.1 Overview of Wnt signalling pathway

One of the key pathways involved in the organogenesis at the embryonic stage of development and regulation of cellular processes such as cell proliferation, cell differentiation, and apoptosis is the Wnt signalling pathway. In healthy adult individuals, it plays an important role in tissue regeneration, renewal of stem cells, as well as maintaining the overall homeostasis of organs. This signalling takes place by the activation of the three pathways: Wnt/β-catenin Pathway (Canonical Pathway), Wnt/Ca2+ Pathway and the Wnt/PNP Pathway. Although these pathways work in mutual regulation, the Wnt/β-catenin Pathway is the most versatile and one of the key modulators that facilitate epithelial–mesenchymal transition (EMT), leading to eventual mediation of metastasis and tumour progression in cancer. Fusobacterium nucleatum adhesin FadA binds E-cadherin, directly activating β-catenin signalling and inducing oncogenic transcription in colorectal epithelium.

5.2 Wnt/β-catenin pathway

Wnt/β-catenin Pathway is a tightly regulated and an evolutionarily conserved pathway, which has primarily the following segments: an extracellular segment, consisting of proteins like Wnt glycoprotein ligands that activate the pathway, via the paracrine signalling. The Wnts are modified into lipid bound forms, which is mediated by an enzyme "Porcupine" (PORCN), in the endoplasmic reticulum. Following this modification, a group of proteins of p24 family, including TMED4/eclair, TMED5/opossum and TMED2/CHOp24 transport the modified Wnts from the ER to the Golgi apparatus, essentially facilitating their secretion. The membrane segment consists of Wnt receptor complex, including "Frizzled" protein (FZD), LRP–5/6 co-receptor. GSK3 Phosphorylated LRP6/5 acts as a binding site for AXIN, which is a negative regulator of the pathway and acts as a tumour suppressor, present in the cytoplasm.[25]

The AXIN provides a binding site for the N-terminal RGS domain which interacts with the components of the destruction complex, namely, the APC proteins. APC and AXIN synergistically promote kinase activity, thus leading to sequential phosphorylation of β-catenin, which is further degraded by E3 ubiquitin ligase, along with proteasomal degradation, leading maintenance of low levels of β-catenin in healthy cells. In aberrant the cysteine-rich domain (CRD) of the transmembrane FZD protein binds with Wnt ligands, upon which the DVL (Dishevelled) scaffolding protein are recruited, that have a PDZ domain which binds to the Lys-Thr-X-X-X-Trp (KTxxxW) motif, present at the cytoplasmic tail of FZD. This inhibits the phosphorylation of LRP5/6 by GSK3, and the further ubiquitination due to the disassembly of the destruction complex. Thus higher levels of β-catenin gets accumulated in the cytoplasm and is translocated to the nucleus leading to nuclear stabilisation of β-catenin, thus, interacts with the T-cell factors (TCF)/Lymphoid enhancer factor, leading to activation of target genes, thus increasing cellular activity such as cell proliferation and cell migration, which are hallmarks of cancer.[26]

5.3 EMT–Wnt/β-catenin– axis in tumorigenesis and cancer metastasis

In recent studies, the Epithelial Mesenchymal Transition (EMT) and its regulating factors have emerged to provide new insights in carcinogenesis. Interestingly enough, EMT is not specific to a certain type of tumour, but is regulated by multiple factors that cause tumourigenesis and cause metastasis. EMT is a process that facilitates the transition of epithelial cells to the mesenchymal cells during embryonic development. But in adults, its reactivation is a hallmark of several cancers, especially those of the gastrointestinal tract, such as gastric cancer, and colo-rectal cancer, the microbial causative agent for which happens to be Helicobacter pylori.[27] Although partially reactivated, it is responsible for tumourigenic activities like cancer cell motility, cancer progression in highly inflamed tissues. H. pylori biofilms provide persistent epithelial exposure to CagA and VacA virulence factors. CagA-mediated AKT/GSK3β inhibition stabilizes β-catenin, while sustained inflammatory NF-κB activation reinforces transcription of EMT drivers such as SNAIL and ZEB1, collectively reawakening developmental EMT programs in adult gastric mucosa.

The formation of mesenchymal cells from epithelial cells takes place by the degradation of the basement membrane and the consequent biochemical reactions which lead to the loss of cell adherence and invasiveness, increased migration and elevated resistance to apoptosis. This is initiated by several factors such as activation of transcription factors responsible for the expression to cell-surface specific proteins, rearrangement of the cytoskeletal elements, increased production in the enzymes that are responsible for degradation of the extracellular matrix.[28]

EMT can be divided into three types: Type 1 EMT – associated with implantation, embryogenesis and organogenesis; Type 2 EMT – associated with tissue regeneration and organ fibrosis meanwhile, Type 3 EMT – associated with cancer progression and metastasis.

One of the hallmarks of cancer is the EMT mediated de-differentiation of epithelial cells. The EMT pathway is initiated by the TGF-β cytokine and is also believed to be one of the key mediators of tumourigenesis and metastasis of cancer cells.

EMT via β-catenin is not a solitary axis; several pathways converge to sharpen or sustain it. TGF-β/SMAD cooperates with β-catenin at promoters/enhancers of EMT genes, with CBP/p300 acting as shared co-activators; AKT phosphorylates and inhibits GSK3β, stabilizing both β-catenin and SNAIL; NF-κB amplifies SNAIL/ZEB expression and links inflammatory inputs to EMT; hypoxia stabilizes HIF-1α, which upregulates select Wnt ligands and receptors and can co-occupy promoters with β-catenin; Notch and Hedgehog can reinforce partial EMT states; and Hippo pathway effectors YAP/TAZ physically partner with β-catenin/TCF and share target repertoires that promote motility and stemness. Non-canonical Wnt branches also intersect: Wnt5a/Wnt11 via Frizzled and ROR1/2 can activate planar cell polarity (RhoA/Rac1/JNK) or Ca2+ /PKC signaling to remodel the cytoskeleton and, context-dependently, either antagonize or facilitate canonical β-catenin–driven EMT.

Post-translational control adds further texture. Beyond the CK1α→GSK3β degron, β-catenin is phosphorylated at Tyr residues (e.g., by Src), loosening its binding to E-cadherin and favoring signaling; acetylation and O-GlcNAcylation modulate its stability and transcriptional potency; and nuclear export/import is regulated by phosphorylation-dependent interactions with 14-3-3 proteins and by APC's nuclear shuttling role. Conversely, GSK3β phosphorylates SNAIL to trigger β-TrCP-mediated degradation, so upstream kinases that restrain GSK3β (AKT, Wnt-on) stabilize both the initiator (β-catenin) and the EMT effectors (SNAIL), locking the program in. Epigenetically, EZH2 (PRC2) collaborates with SNAIL/ZEB to silence epithelial genes, while microRNA circuits—most notably the miR-200 family and miR-34—form double-negative feedback loops with ZEB and SNAIL that stabilize EMT or epithelial states.

6. Angiogenesis and immune evasion

Angiogenesis, the growth of new blood vessels from pre-existing vasculature, is a mechanism exploited by tumors to sustain themselves and enable metastasis. In cancers, hypoxia and oncogenic signalling converge on the hypoxia-inducible factors HIF-1α and HIF-2α, and the VEGF family to initiate and maintain vessel sprouting. Yet the tumor vasculature that emerges is structurally and functionally abnormal: haywire architecture, excessive permeability, increased interstitial pressure, that lead to poor function of such vasculature. All of these features directly contribute to the antitumour immunity, impairing dendritic cell (DC) maturation and antigen presentation, blunting endothelial adhesion molecule function required for lymphocyte trafficking, and favoring recruitment and polarisation of immunosuppressive myeloid and regulatory subsets.

6.1 VEGF/HIF: Signalling pathway

Hypoxia is a defining property of solid tumors. Under normoxia, prolyl hydroxylases mark HIF-α subunits for VHL-mediated ubiquitination and proteasomal degradation. Reduced oxygen inhibits these hydroxylases, stabilizing HIF-1α and HIF-2α, which dimerise with HIF-1β to transactivate VEGFA, ANGPTL4, CXCL12, glycolytic enzymes, and matrix-remodeling genes that collectively adapt tissues to hypoxia and promote neovascularization.[29] VEGF-A primarily signals through VEGFR2 on endothelial cells, with VEGFR1 contributing in endothelial and myeloid compartments. VEGFR2 activation triggers several core cascades. PI3K–AKT–mTOR supports endothelial survival and nitric oxide (NO) production, MAPK–ERK drives proliferation and motility, and PLCγ–PKC–RAF and calcium-dependent pathways regulate cytoskeletal dynamics and permeability. These signals are tightly integrated with other vascular guidance systems. VEGF-induced Dll4 in endothelial tip cells activates Notch in neighbouring stalk cells to restrain excessive sprouting, refining branching patterns; disrupting Dll4 can cause hyper-sprouting, yet functionally inefficient vasculature. The Angiopoietin–Tie axis governs vessel stability, where ANG1 stabilizes via Tie2, whereas ANG2 destabilises pericyte–endothelial contacts and sensitizes vessels to VEGF. PDGF-BB and TGF-β further regulate pericyte recruitment and basement membrane assembly; when dysregulated, vessels become thin-walled, leaky, and poorly supported.

These structural abnormalities create a vicious cycle. Poor perfusion maintains hypoxia and acidosis elevates interstitial fluid pressure (IFP), which further compromises perfusion and drug delivery. Hypoxia and acidosis impair antigen processing and presentation by tumor cells and DCs, upregulate inhibitory checkpoints, and select for glycolytic phenotypes that compete with T cells for glucose and other nutrients. Continuous VEGF exposure induces endothelial "anergy," characterized by diminished or improperly clustered ICAM-1/VCAM-1/E-selectin, which impairs leukocyte adhesion and diapedesis. Along with immunosuppressive cytokines, VEGF can drive expression of Fas ligand on tumor endothelium, selectively triggering apoptosis in infiltrating effector CD8+ T cells while sparing Tregs, establishing an endothelial barrier that filters antitumor immunity.[30]

6.2 Immunosuppressive cell recruitment

Angiogenic signalling and hypoxia remodel leukocyte composition and function through convergent chemokine, growth factor, and metabolic pathways.

Myeloid-derived suppressor cells expand under sustained VEGF exposure, which skews hematopoiesis toward immature myeloid fates and interferes with DC development by dampening NF-κB activation in precursors. Within tumors, HIF-1α drives MDSC glycolysis and expression of arginase-1 and iNOS, depleting L-arginine and generating NO and peroxynitrite that blunt T-cell receptor signalling and proliferation. MDSCs also upregulate PD-L1 and secrete IL-10 and TGF-β, creating immunosuppressive cytokine milieus. Chemokine pathways such as CXCL12/CXCR4 and CCL2/CCR2, and growth factors like G-CSF and GM-CSF, facilitate systemic mobilization and tumor homing.

Regulatory T (Treg) cells accumulate through mechanisms linked to hypoxia and VEGF. VEGF–Neuropilin-1 interactions can enhance Treg trafficking and retention in tumours,[31] and endothelial FasL expression induced by VEGF, IL-10, and PGE2 selectively eliminates effector CD8+ T cells while sparing Tregs, skewing the repertoire. Tregs reinforce suppression through metabolic and checkpoint pathways, including hydrolysis of ATP to adenosine via CD39/CD73 and expression of CTLA-4 and PD-1. Adenosinergic signalling (A2A receptor) suppresses cytotoxic activity and cytokine secretion by effector T cells and NK cells, with especially potent effects under hypoxia.

Tumour-associated macrophages are recruited via CSF1–CSF1R and CCL2–CCR2 and polarised toward M2-like states by hypoxia, lactic acidosis, and TGF-β.[32] These macrophages secrete VEGF, MMP9, and ARG1, remodel the ECM, and support angiogenesis and invasion, metastasising the tumour activity. ANG2 can activate Tie2-expressing macrophages, further destabilizing vessels and promoting immune suppression. TAMs also present inhibitory ligands and produce IL-10, reinforcing dysfunctional T-cell states.

Cancer-associated fibroblasts are activated by PDGF and TGF-β and shape the immune landscape through chemokines and matrix. CAF-derived CXCL12 forms chemokine "exclusion" zones that retain T cells at the peritumoural stroma, preventing deep infiltration.[33] TGF-β from CAFs enforces T-cell exclusion and restrains Th1 differentiation and cytotoxic function. CAFs deposit hyaluronan and collagen that elevate IFP and compress vessels, reinforcing hypoxia and thus VEGF transcription, creating positive feedback. (https://doi.org/10.1186/s12943-023-01860-5).

Endothelial anergy compounds these effects. Persistent VEGF signalling reduces or mislocalizes ICAM-1/VCAM-1/E-selectin on tumour endothelium and dampens responses to inflammatory cytokines, throttling lymphocyte entry despite systemic antitumour immunity. In parallel, VEGF impairs DC maturation and antigen-presenting capacity, reducing IL-12 and co-stimulatory molecules and increasing apoptosis under certain maturation contexts; fewer functional DCs translate into impaired priming and restimulation of effector T cells.

6.3 Biofilm matrix barrier effect

Biofilm matrix barrier effect works at two levels in tumourigenesis and cancer proliferation:

6.3.1 Tumor-associated microbial biofilms

In colorectal cancer, invasive polymicrobial biofilms are frequently observed on tumors and even on adjacent mucosa. Their EPS and extracellular DNA (eDNA) matrices impose diffusion barriers that create hypoxic microdomains, reshape metabolite and reactive oxygen species gradients, and concentrate proteases and growth factors. Functionally, biofilm-positive mucosal homogenates can transfer tumorigenic potential as seen in mouse models, accompanied by Th17 signaling and myeloid recruitment. Cooperative pathogens such as enterotoxigenic *Bacteroides fragilis* and *pks*$^{+}$ *Escherichia coli* disrupt E-cadherin, activate NF-κB/STAT3 pathways,[34] and introduce genotoxic stress, plausibly reinforcing HIF stabilization and VEGF production in adjacent epithelium and stroma. While explicit tissue-level amounts of HIF-1/2α and VEGF attribution to biofilms remain limitedly explored, the structural and inflammatory context aligns with sustained angiogenic and immunosuppressive programming. In addition, oral anaerobes like Fusobacterium can induce neutrophil extracellular traps (NETs) that are pro-angiogenic and can impair cytotoxic surveillance, while microbial pattern-recognition signals modulate endothelial activation and myeloid polarization.

6.3.2 Biofilm analogies in the tumor ECM

The desmoplastic ECM having complexes such as dense collagen I, hyaluronan, fibrillar glycoproteins, and enzymatic crosslinking—acts as a biofilm-analogous barrier in many tumors. Elevated IFP, vessel compression, and reduced convection and diffusion preserve hypoxia and acidosis, sustaining HIF signaling and VEGF production. Hypoxic matrices promote endothelial pruning, which reduces pericyte coverage, and blunt leukocyte trafficking. CAF-derived CXCL12 and TGF-β establish chemokine gradients and immune-excluded niches, while lactate and adenosine accumulation favors M2 polarization and tolerogenic DCs. Matrix-bound VEGF and ANG2 are released by proteases, that fuels disordered vascular sprouting and reinforces abnormal vascular phenotype.[35]

7. Temporal progression of biofilm-induced tumourigenesis

7.1 Bacterial infection — the foundation stone of tumour microenvironment

7.1.1 Mechanobiology of biofilm-associated fibrosis

Progressive collagen deposition and matrix crosslinking within chronic biofilm-associated fibrosis increase local tissue stiffness. This mechanical resistance activates integrin–FAK signalling and nuclear translocation of

YAP/TAZ transcriptional co-activators, promoting epithelial plasticity, survival signalling, and epithelial–mesenchymal transition (EMT). Such biomechanical signalling complements inflammatory cytokine-driven transcriptional reprogramming and may physically shield transformed niches from immune surveillance and therapeutic penetration.

7.1.2 Metabolic competition and acidic niche formation

Dense biofilm matrices obstruct oxygen and glucose diffusion, producing steep nutrient gradients. Combined glycolytic activity of infiltrating tumour and bacterial cells increases lactate accumulation, acidifying the micro-environment. This acidosis suppresses cytotoxic T-cell receptor signalling, promotes M2 macrophage polarization, stabilizes HIF-1α, and reinforces angiogenic adaptation.

Bacterial pathogens enter the body via different characteristic transmission routes, disseminate and colonise the tissue at the target site. This colonisation process can be delineated into four major phases; (1) initial adherence, (2) irreversible attachment, (3) EPS production and biofilm formation, and (4) biofilm maturation and dispersal.

Most invading pathogens are actually cleared out by the immune system during the entry and dissemination steps, especially in healthy individuals. In rare cases, particularly those with a high load of pathogens, or in immunocompromised individuals, bacterial colonisation is more often than not successful. During irreversible attachment, bacteria can secrete certain cytotoxins to locally induce necrosis for acquisition of nutrients and modify the surface to enhance attachment.

Sometimes, bacteria specifically produce tumourigenic toxins to enhance survival by exploiting the anti-immune properties of the tumour cells. For example: cag$^+$ Helicobacter pylori, a pathogen of the gastric lining, was classified as a group 1 carcinogen by IARC (International Agency for Research on Cancer), WHO (World Health Organisation) in 1994. The CagPAI in these strains of H. pylori code for a type VI secretion system (T4SS) which delivers the VacA bacteriotoxin and the CagA oncoprotein and into gastric MALT (mucosa-associated lymphoid tissue) cells, leading to the development of non-Hodgkin's gastric lymphoma.

Necrotic tissue undergoes fibrosis, the process often nucleated and promoted by bacterial EPS; downregulating normal wound healing pathways. Pathogen-associated fibrotic tissue is characterised by accumulation of exogenous (bacterial-origin) and endogenous (host-origin) EPS. While exogenous EPS, functioning as PAMPs, trigger proinflammatory pathways in responding immunocytes (subsection 3.2. Proinflammatory signalling), endogenous EPS mask the bacterial antigens, hindering immune clearance. This non-eradication of pathogens consequently leads to relapse and reformation of the biofilm. Resulting repeated triggering of proinflammatory responses due to immune non-clearance is the characteristic of chronic inflammation. The biofilm-protected dormant bacterial persister state mirrors cancer stem cell quiescence. Both populations exhibit metabolic suppression, therapeutic tolerance, and relapse potential. Biofilm-associated tumour niches may therefore function as protective sanctuaries for dormant cancer stem cells, shielding them from chemotherapy and facilitating recurrence.

7.1.3 Chronic inflammation — the scaffolding for tumourigenesis

In chronic inflammation has been through and through linked with tumourigenesis. Firstly, it accelerates and promotes fibrosis, strengthening the tumour microenvironment. Further, in chronic inflammation, abnormal rates of macrophage polarisation to M1 (inflammatory phase) macrophages, magnifying the feedback loops of proinflammatory cytokines.

This results in formation and accumulation of excess ROS and RNS (RONS), which directly modifies and damages proteins and nucleic acids, leading to higher rates of mutagenesis. Extensive DNA damage, particularly in actively proliferating and dividing cells, depends on error-prone repair mechanisms like NHEJ, translesion synthesis etc. RONS mediated dysregulation of maintenance proteins like p53 and telomerase heavily contributes to substantial genomic instability, a requisite for tumourigenesis.

The proinflammatory cytokine pathways; more specifically those of TNFα and IL-6, are deeply intertwined with apoptotic cascade pathways, and thus upregulate apoptosis-suppressing genes to prevent cell death. Accumulation of these cytokines and persistence of the inflammatory environment create apoptosis resistance cells. Cytokines like TNFα, IL-1β, IL-6, and TGF-β promote angiogenesis by VEGF upregulation, another hallmark of tumourigenesis.

Gradually, chronic inflammation causes systemic dysfunction of the immune cells, allowing recruitment of immunosuppressive cells which facilitate protection and immune evasion of the tumour.

7.1.4 Tumourigenesis — the construction of a cancer

What begins as a single cell that ignores the ‘stop signals’ of the cell cycle due to a random mutation in its DNA, eventually culminating into an invasive and aggressive malignancy, upon bypassing several pathways and checkpoints, leads to the formation of a tumour, also known as a 'neoplasm'.

The process of formation of a tumour is called tumourigenesis, which is an intricate amalgamation of disrupted pathways, spontaneous and random errors, chronic inflammation, and epigenetic stressors.

Haywire mechanisms in tumour cells initiate cytokine and chemokine fields that recruit and reprogramme stromal and immune compartments; in return, those compartments build fibrotic, hypoxic, and metabolically skewed niches that stabilize immune evasion, angiogenesis, and invasive behaviour. Threading these events is a resilient inflammatory control circuit—IL-1β primed and activated via innate sensors, NF-κB as the transcriptional integrator, and STAT3 driven by IL-6 family cytokines—that functions across tumour, myeloid, endothelial, and fibroblast lineages to maintain survival signaling, suppress anti-tumour immunity, and consolidate the physical scaffolding of disease. In what follows, each component of this "construction project" is developed in depth and tied back to this core circuitry, that sustains the tumour and furthers its progression.[36]

Tumour-associated macrophages are recruited by chemokines and growth factors and then polarized by local cytokines into phenotypes that support angiogenesis, immune suppression, and metastasis. CCL2 and CSF1 draw monocytes into the tumour, where IL-10, TGF-β, and metabolic cues bias them toward M2-like, tissue-remodelling states.[37] TAMs secrete VEGF and other mediators that sprout aberrant vasculature; they dampen T-cell function via IL-10, TGF-β, and checkpoint ligand expression; further deploying proteases and chemokines that degrade matrix, facilitating intravasation, and helping prepare pre-metastatic niches. NF-κB and STAT3 signalling within TAMs integrates IL-1/TLR inputs and IL-6/IL-10 family signals to lock in these programmes, creating reciprocal reinforcement with tumour and fibroblast outputs.

Cancer-associated fibroblasts act as architects of both stroma and immune tone. Activated by paracrine cues such as TGF-β and PDGF, CAFs assume matrix-producing, contractile, and cytokine-secreting phenotypes that restructure tissues and shape immune cell access.By depositing and remodelling ECM, CAFs increase stiffness and guide invasion; by releasing IL-6 and chemokines, they dysmorph myeloid differentiation and contribute to T-cell exclusion; and by supplying pro-angiogenic and trophic factors, they sustain tumour proliferation and vascular remodelling.[38] Inflammatory signalling within CAFs, especially NF-κB-driven

IL-6 output, sustains STAT3 activity in neighbouring tumour and immune cells, while mechanical feedbacks from the stiffened matrix reinforce these chronic programmes.

IL-1β is one of the key pathways of this inflammatory construction logic as an inflammasome-gated amplifier of chronic tumour-promoting inflammation. NF-κB-dependent priming induces pro-IL-1β expression, and activation of inflammasomes such as NLRP3 leads to caspase-1–mediated maturation and release of IL-1β from myeloid and stromal sources exposed to DAMPs and PAMPs. Engagement of IL-1R then signals through MyD88–IRAK–TRAF6 to activate NF-κB and MAPKs, boosting production of IL-6 and chemokines, recruiting neutrophils and macrophages, and enhancing angiogenesis and matrix remodelling. When sustained, this axis converts acute defensive inflammation into the chronic, myeloid-skewed state that supports tumour growth, invasion, and immune suppression.[39]

NF-κB functions as the master integrator of survival signalling and inflammatory fields across tumour and stromal compartments. Activated by IL-1, TLRs, TNF, and diverse oncogenic stresses, NF-κB coordinates transcription of genes that promote proliferation and survival, induce cytokines and chemokines, and upregulate checkpoint ligands. It primes inflammasome components—closing a loop back to IL-1β—and induces IL-6 family cytokines that propagate JAK–STAT3 activation, shaping myeloid recruitment and differentiation. Persistent NF-κB activity correlates with immune evasion, angiogenesis, EMT, and resistance to therapy across malignancies, reflecting its central position in the construction network.

STAT3 provides the persistence module that stabilizes immune suppression and tumour plasticity. Activated downstream of gp130-coupled IL-6 family cytokines via JAKs, and reinforced by IL-10 signalling and stress contexts such as hypoxia, STAT3 is engaged in tumour cells, CAFs, and myeloid compartments. STAT3 cooperates with NF-κB at promoters and enhancers to sustain survival, angiogenic, and immunosuppressive transcriptional programmes, while also reinforcing stemness and EMT traits that underwrite invasion and metastatic potential. Its chronic activation is therefore both a readout and a driver of the pathological steady state achieved by the tumour–stroma consortium.[40]

As an interlinked system, the IL-1β→NF-κB→IL-6/STAT3 loop links initiation, consolidation, and maintenance of the malignant niche. Early tumour-intrinsic alterations and microbial or sterile danger signals activate PRRs, IL-1 pathways, and related innate sensors, triggering NF-κB and the first cytokine waves that recruit myeloid cells and activate fibroblasts. As

TAMs and CAFs remodel matrix and vasculature, the loop entrenches immunosuppression, angiogenesis, and fibrosis; hypoxia and metabolite signalling (including lactate and adenosine) further bias immunity and tune transcriptional outputs. Mechanical and inflammatory feedbacks then stabilize immune exclusion and tumour cell plasticity, enabling invasion, intravasation, and metastatic niche conditioning—a mature construction which is co-dependent on a well recruited machinery that sustains tumour cells.[41]

8. Therapeutic implications

The establishment of the timeline of this tripartite pathophysiology is essential to plan efficient multi-target combination and sequential therapies to prevent the progression of this biofilm-induced chronic inflammation and tumourigenesis, and/or eradicate it. Identification of these clinically significant chronological nodes is vital to design strategies to combat the progression in multi-pronged targeted therapies.

8.1 Antibiotics and AMR

Antibiotics are microbial secondary metabolites exhibiting bactericidal and/or bacteriostatic properties. Penicillin, for instance, was hailed as "the wonder drug" after it was used to save countless lives in World War 2. However, over years of the sheer abuse of this "wonder drug" and other types of antibiotics discovered, we are facing major challenges due to the rise in antimicrobial resistance (AMR) that is nothing short of a global catastrophe. Ho et al.[42] estimated over 6 million people could lose their lives due to antimicrobial resistant bacterial infections. Today, the most dreaded bacterial pathogens; ESKAPEE (*Enterococcus faecium*, *Staphylococcus aureus*, *Klebsiella pneumoniae*, *Acinetobacter baumannii*, *Pseudomonas aeruginosa*, *Escherichia coli*, and *Enterobacter* spp.), are prevalent multi-resistant nosocomial strains.[43]

AMR is attributed by various mechanisms, like mutated targets, non-target alternatives, efflux pumps, antibiotic degradation etc., particularly elevated in biofilm-participating cells. These mechanisms, albeit extremely intriguing, lie outside the scope for this book; we suggest Reygaert[44] for further reading. Ongoing work in AMR explores various methods to increase the sensitivity of these bacteria. Most promising of which are antibiotic coated metal nanoparticles[45] and antibiotic-bacteriophage[46] combination therapies.

Different antibiotics have shown conflicting results; where some antibiotics also show anti-cancer properties, while some, on the contrary, show cancer-promoting effects.[47]

8.2 Anti-angiogenic and anti-metastatic therapies

Anti-angiogenic therapy in solid tumors is best understood as modulation of a vascular ecosystem rather than simple VEGF shutdown; its clinical value depends on context (tumor type, line of therapy) and on leveraging a short-lived normalization window to enhance drug delivery and immune access.

In metastatic colorectal cancer (mCRC), several FDA approved antibodies against VEGF/VEGFR2—bevacizumab (anti-VEGF-A), aflibercept (VEGF-trap), and ramucirumab (anti-VEGFR2)—are partnered with cytotoxic backbones.[48,49] Bevacizumab with oxaliplatin- or irinotecan-based chemotherapy improves progression-free survival and modestly overall survival versus chemotherapy alone; after oxaliplatin exposure, aflibercept + FOLFIRI or ramucirumab + FOLFIRI yield second-line overall survival gains. Later lines rely on oral TKIs such as regorafenib and fruquintinib for small but reproducible overall survival benefits. While chemotherapy, in combination with immunotherapy and radiotherapy are effective to some extent, there is a risk of severe side effects,[50] including hepatotoxicity, and GI bleeding.

As an alternative therapeutic option, phytoconstituents pose as potential drug molecules, with ongoing studies that suggest their therapeutic properties with reduced side effects.[51,52]

8.3 Drawbacks and inflammatory "obstacles"

The efficacy of these therapeutic agents, antibiotics and anti-cancer both, are evaluated using various parameters like therapeutic dosage, minimal inhibitory concentration (MIC), minimal bactericidal concentration (MBC) or minimal cytotoxic concentration (MCC), and perhaps most importantly, stability in the blood. Stability in the blood refers to the amount of time required to deplete the concentration of the bioactive molecule in the systemic circulation from a therapeutic range to a sub-therapeutic range.[53]

These molecules, being xenobiotic, are degraded by the liver under its normal detoxification function. High concentrations of such xenobiotics exhibit hepatotoxic effects. Therefore hepato-compatibility is paramount in preventing further complications, as hepatotoxicity is inseparably nestled with chronic inflammation in a positive feedback loop of liver damage-induced inflammation and chronic inflammation-linked liver damage. *In silico,* in vitro and in vivo screenings of novel bioactive compounds to ensure hepato-compatibility is an essential part of drug discovery.[51,54] This becomes all the more important for designing orally administrable drugs to pass the hepatic portal unheeded; which is responsible for detoxification of the majority of the ingested volume absorbed in the intestines: called the "first pass effect",[55] (pp. 861–862, 2010, pp. 994–995).

Although, the interplay of liver function in drug detoxification and inflammation is not so simple. Some studies indicate inflammation can downregulate hepatic DMET (drug metabolising enzymes and transportors), which can further influence the stability of the drug.[56] Some prodrugs, which are actually dependent on hepatoenzymatic modifications for activation of the drug, lose their functionality entirely or partially; while the drugs designed to maintain therapeutic concentrations after hepatic detoxification have an increased risk of accumulation and subsequent non-specific toxicity.

8.4 Immunosuppression and efficacy of therapeutic agents

To circumvent these inflammatory obstacles, a sequential approach can be designed. Targeted immune suppression to reduce chronic inflammation as a precursor for effective antimicrobial and anti-cancer treatment is a risky endeavour. Although, certain strategies, discussed further, have shown remarkable success in treating bacterial infections and cancer conditions in immunodeficient patients.

General immunosuppressants like corticosteroids and NSAIDs (non-steroidal anti-inflammatory drugs) have been a popular choice to tackle chronic inflammation and hypersensitivity in the past. NSAIDs reportedly exhibit anti-cancer properties for certain cancers like colorectal cancer,[57,58] perhaps by a dual action by combining its anti-inflammatory and reported anti-angiogenic activities.[59] Whereas it also increased the risk of cancer progression in other cases.[58]

Corticosteroids, although very much in use, are infamous for causing hormonal imbalances, especially in long-term high-dose cases. Hormonal imbalance, especially in women, is again linked with cancer progression, more specifically, breast cancer.[60]

Modern targeted methods of immune suppression may be more reliable. CAR-T cell therapy against proinflammatory immunocytes like M1 macrophages, neutrophils, etc. It is, however, not void of side-effects. CAR-T cell activation is linked with CRS (cytokine release syndrome), a high-toxicity proinflammatory ramification of the therapy.[61]

Immunosuppressive therapies can be combined with medication strategies designed for immunodeficient individuals might base compelling plans for combination therapies in eradicating cases of biofilm-induced chronic inflammation and tumour.[62]

Z. Liu et al.[63] exploited the localised immunosuppressive nature of the tumour microenvironment to employ reduced-risk anti-cancer CAR-T cells, avoiding the immune suppression requirement.

Emerging therapeutic strategies aim to simultaneously dismantle microbial biofilm integrity and restore anti-tumour immunity. Enzymatic dispersal agents such as Dispersin B or DNase I may increase penetration of checkpoint inhibitors by degrading EPS barriers. Coupling biofilm-targeting nanocarriers with PD-1/PD-L1 blockade, STAT3 inhibitors, or anti-VEGF therapies represents a promising interdisciplinary strategy to collapse the protective tumour–biofilm niche.

Precision biofilm–Tumour targeting

9. Future prospects

That being said, we believe these approaches are easier-said-than-done. Countless strategies can be formulated but these lack any significance without the backing of rigorous quantitative *in silico*, in vitro, and in vivo studies, and indisputably, clinical trials. This problem, with all its complexities and intricacies, opens itself up to researchers of various backgrounds and expertise to work cohesively, not competitively, in developing a plan of action; for we have to be truculent in challenging the pathogens adapting and evolving faster than we can understand them, and not each other.

References

1. Madigan MT, Martinko JM, Stahl D, Clark DP. *Brock Biology of Microorganisms (13th Edition)*. 13th ed. Benjamin Cummings; 2010.
2. Flemming H-C, Wuertz S. Bacteria and archaea on Earth and their abundance in biofilms. *Nat Rev Microbiol*. 2019;17(4):247–260. https://doi.org/10.1038/s41579-019-0158-9
3. Smith K, Perez A, Ramage G, Gemmell CG, Lang S. Comparison of biofilm-associated cell survival following in vitro exposure of meticillin-resistant Staphylococcus aureus biofilms to the antibiotics clindamycin, daptomycin, linezolid, tigecycline and vancomycin. *Int J Antimicrob. Agents*. 2009;33(4):374–378. https://doi.org/10.1016/j.ijantimicag.2008.08.029
4. Yin W, Wang Y, Liu L, He J. Biofilms: the microbial "Protective Clothing" in extreme environments. *Int J Mol Sci*. 2019;20(14):3423. https://doi.org/10.3390/ijms20143423

5. Flemming H-C, Wingender J. The biofilm matrix. *Nat Rev Microbiol.* 2010;8(9):623–633. https://doi.org/10.1038/nrmicro2415
6. Ryan KJ, Ahmad N, Alspaugh JA, Drew WL, Lagunoff M, Pottinger P, et al. *Sherris Medical Microbiology, Seventh Edition.* McGraw Hill Professional; 2018.
7. Abudukelimu A, Barberis M, Redegeld FA, Sahin N, Westerhoff HV. Predictable irreversible switching between acute and chronic inflammation. *Front Immunol.* 2018;9:2018. https://www.frontiersin.org/journals/immunology/articles/10.3389/fimmu.2018.01596.
8. Liu T, Zhang L, Joo D, Sun S-C. NF-κB signaling in inflammation. *Signal Transduct Target Ther.* 2017;2(1):17023. https://doi.org/10.1038/sigtrans.2017.23
9. Park BS, Lee J-O. Recognition of lipopolysaccharide pattern by TLR4 complexes. *Exp & Mol Med.* 2013;45(12):e66–e66. https://doi.org/10.1038/emm.2013.97
10. Popko K, Gorska E, Stelmaszczyk-Emmel A, Plywaczewski R, Stoklosa A, Gorecka D, et al. Proinflammatory cytokines IL-6 and TNF-α and the development of inflammation in obese subjects. *Eur J Med Res.* 2010;15(2):120–122. https://doi.org/10.1186/2047-783X-15-S2-120
11. Wong E, Cohen T, Romi E, Levin M, Peleg Y, Arad U, et al. Harnessing the natural inhibitory domain to control TNFα Converting Enzyme (TACE) activity in vivo. *Sci Rep.* 2016;6(1):35598. https://doi.org/10.1038/srep35598
12. Aggarwal BB. Signalling pathways of the TNF superfamily: a double-edged sword. *Nat Rev Immunol.* 2003;3(9):745–756. https://doi.org/10.1038/nri1184
13. Agashe RP, Lippman SM, Kurzrock R. JAK: not just another kinase. *Mol Cancer Ther.* 2022;21(12):1757–1764. https://doi.org/10.1158/1535-7163.MCT-22-0323
14. Ren K, Torres R. Role of interleukin-1β during pain and inflammation. *Brain Res Rev.* 2009;60(1):57–64. https://doi.org/10.1016/j.brainresrev.2008.12.020
15. Broz P, Dixit VM. Inflammasomes: mechanism of assembly, regulation and signalling. *Nat Rev Immunol.* 2016;16(7):407–420. https://doi.org/10.1038/nri.2016.58
16. Vezzani A, French J, Bartfai T, Baram TZ. The role of inflammation in epilepsy. *Nat Rev Neurol.* 2011;7(1):31–40. https://doi.org/10.1038/nrneurol.2010.178
17. Punt J, Stranford S, Jones P, Owen JA. *Kuby Immunology.* 8th ed. W.H. Freeman; 2018.
18. Wang Y, Branicky R, Noë A, Hekimi S. Superoxide dismutases: dual roles in controlling ROS damage and regulating ROS signaling. *J Cell Biol.* 2018;217(6):1915–1928. https://doi.org/10.1083/jcb.201708007
19. Gochman E, Mahajna J, Shenzer P, Dahan A, Blatt A, Elyakim R, et al. The expression of iNOS and nitrotyrosine in colitis and colon cancer in humans. *Acta Histochemica.* 2012;114(8):827–835. https://doi.org/10.1016/j.acthis.2012.02.004
20. Anand P, Hausladen A, Wang Y-J, Zhang G-F, Stomberski C, Brunengraber H, et al. Identification of S-nitroso-CoA reductases that regulate protein S-nitrosylation. *Proc Natl Acad Sci USA.* 2014;111(52):18572–18577. https://doi.org/10.1073/pnas.1417816112
21. Di Meo S, Reed TT, Venditti P, Victor VM. Role of ROS and RNS sources in physiological and pathological conditions. *Oxid Med Cell Longev.* 2016;2016:1245049. https://doi.org/10.1155/2016/1245049
22. Cobley JN. 50 shades of oxidative stress: a state-specific cysteine redox pattern hypothesis. *Redox Biol.* 2023;67:102936. https://doi.org/10.1016/j.redox.2023.102936
23. Zorov DB, Juhaszova M, Sollott SJ. Mitochondrial reactive oxygen species (ROS) and ROS-induced ROS release. *Physiol. Rev.* 2014;94(3):909–950. https://doi.org/10.1152/physrev.00026.2013
24. Daniel M, Peek GW, Tollefsbol TO. Regulation of the human catalytic subunit of telomerase (hTERT). *Gene.* 2012;498(2):135–146. https://doi.org/10.1016/j.gene.2012.01.095
25. Liu Z, Zhou Z, Dang Q, Xu H, Lv J, Li H, et al. Immunosuppression in tumor immune microenvironment and its optimization from CAR-T cell therapy. *Theranostics.* 2022;12(14):6273–6290. https://doi.org/10.7150/thno.76854

26. Xue W, Yang L, Chen C, Ashrafizadeh M, Tian Y, Sun R. Wnt/β-catenin-driven EMT regulation in human cancers. *Cell Mol Life Sci*. 2024;81(1):79. https://doi.org/10.1007/s00018-023-05099-7
27. Ouyang Y, Liu G, Xu W, Yang Z, Li N, Xie C, et al. Helicobacter pylori induces epithelial-mesenchymal transition in gastric carcinogenesis via the AKT/GSK3β signaling pathway. *Oncol Lett*. 2021;21(2):165. https://doi.org/10.3892/ol.2021.12426
28. Sun L, Xing J, Zhou X, Song X, Gao S. Wnt/β-catenin signalling, epithelial-mesenchymal transition and crosslink signalling in colorectal cancer cells. *Biomed. Pharmacother*. 2024;175:116685. https://doi.org/10.1016/j.biopha.2024.116685
29. Hu K, Babapoor-Farrokhran S, Rodrigues M, Deshpande M, Puchner B, Kashiwabuchi F, et al. Hypoxia-inducible factor 1 upregulation of both VEGF and ANGPTL4 is required to promote the angiogenic phenotype in uveal melanoma. *Oncotarget*. 2016;7(7):7816–7828. https://doi.org/10.18632/oncotarget.6868
30. Chriqui L-E, Cavin S, Perentes JY. Dual implication of endothelial adhesion molecules in tumor progression and cancer immunity. *Cell Adhes & Migr*. 2025;19(1):2472308. https://doi.org/10.1080/19336918.2025.2472308
31. Hansen W. Neuropilin 1 guides regulatory T cells into VEGF-producing melanoma. *OncoImmunology*. 2013;2(2):e23039. https://doi.org/10.4161/onci.23039
32. Saeed AF. Tumor-associated macrophages: polarization, immunoregulation, and immunotherapy. *Cells*. 2025;14(10):741. https://doi.org/10.3390/cells14100741
33. Zhang C, Fei Y, Wang H, Hu S, Liu C, Hu R, et al. CAFs orchestrates tumor immune microenvironment—A new target in cancer therapy? *Front Pharmacol*. 2023;14:*2023*. https://www.frontiersin.org/journals/pharmacology/articles/10.3389/fphar.2023.1113378.
34. Lee C-G, Hwang S, Gwon S-Y, Park C, Jo M, Hong J-E, et al. Bacteroides fragilis toxin induces intestinal epithelial cell secretion of interleukin-8 by the E-Cadherin/β-Catenin/NF-κB dependent pathway. *Biomedicines*. 2022;10(4):827. https://doi.org/10.3390/biomedicines10040827
35. Ramakrishnan S, Anand V, Roy S. Vascular endothelial growth factor signaling in hypoxia and inflammation. *J Neuroimmune Pharmacol*. 2014;9(2):142–160. https://doi.org/10.1007/s11481-014-9531-7
36. Zhao H, Wu L, Yan G, Chen Y, Zhou M, Wu Y, et al. Inflammation and tumor progression: signaling pathways and targeted intervention. *Signal Transduct Target Ther*. 2021;6(1):263. https://doi.org/10.1038/s41392-021-00658-5
37. Basak U, Sarkar T, Mukherjee S, Chakraborty S, Dutta A, Dutta S, et al. Tumor-associated macrophages: an effective player of the tumor microenvironment. *Front Immunol*. 2023;14. https://doi.org/10.3389/fimmu.2023.1295257
38. Pawar JS, Salam MA, Dipto MSU, Al-Amin MY, Salam MT, Sengupta S, et al. Cancer-associated fibroblasts: immunosuppressive crosstalk with tumor-infiltrating immune cells and implications for therapeutic resistance. *Cancers*. 2025;17(15):2484. https://doi.org/10.3390/cancers17152484
39. Rébé C, Ghiringhelli F. Interleukin-1β and cancer. *Cancers*. 2020;12(7):1791. https://doi.org/10.3390/cancers12071791
40. Abaurrea A, Araujo AM, Caffarel MM. The role of the IL-6 cytokine family in epithelial–Mesenchymal plasticity in cancer progression. *Int J Mol Sci*. 2021;22(15):8334. https://doi.org/10.3390/ijms22158334
41. Vlahopoulos SA. Aberrant control of NF-κB in cancer permits transcriptional and phenotypic plasticity, to curtail dependence on host tissue: molecular mode. *Cancer Biol Med*. 2017;14(3):254. https://doi.org/10.20892/j.issn.2095-3941.2017.0029
42. Ho CS, Wong CTH, Aung TT, Lakshminarayanan R, Mehta JS, Rauz S, et al. Antimicrobial resistance: a concise update. *Lancet Microbe*. 2025;6(1) https://doi.org/10.1016/j.lanmic.2024.07.010

43. Oyenuga N, Cobo-Díaz JF, Alvarez-Ordóñez A, Alexa E-A. Overview of antimicrobial resistant ESKAPEE pathogens in food sources and their implications from a one health perspective. *Microorganisms*. 2024;12(10):2084. https://doi.org/10.3390/microorganisms12102084
44. Reygaert WC. An overview of the antimicrobial resistance mechanisms of bacteria. *AIMS Microbiol*. 2018;4(3):482–501. https://doi.org/10.3934/microbiol.2018.3.482
45. Wahab S, Salman A, Khan Z, Khan S, Krishnaraj C, Yun S-I. Metallic nanoparticles: a promising arsenal against antimicrobial resistance-unraveling mechanisms and enhancing medication efficacy. *Int J Mol Sci*. 2023;24(19) https://doi.org/10.3390/ijms241914897
46. Santamaría-Corral G, Senhaji-Kacha A, Broncano-Lavado A, Esteban J, García-Quintanilla M. Bacteriophage–antibiotic combination therapy against Pseudomonas aeruginosa. *Antibiotics*. 2023;12(7):1089. https://doi.org/10.3390/antibiotics12071089
47. Pfab C, Schnobrich L, Eldnasoury S, Gessner A, El-Najjar N. Repurposing of antimicrobial agents for cancer therapy: what do we know? *Cancers*. 2021;13(13):3193. https://doi.org/10.3390/cancers13133193
48. Lai E, Cascinu S, Scartozzi M. Are all anti-angiogenic drugs the same in the treatment of second-line metastatic colorectal cancer? Expert opinion on clinical practice. *Front Oncol*. 2021;11. https://doi.org/10.3389/fonc.2021.637823
49. Li J, Li Z, Wang K. Targeting angiogenesis in gastrointestinal tumors: strategies from vascular disruption to vascular normalization and promotion strategies angiogenesis strategies in GI tumor therapy. *Front Immunol*. 2025;16:*2025*. https://www.frontiersin.org/journals/immunology/articles/10.3389/fimmu.2025.1550752.
50. Pavlidis ET, Pavlidis TE. Role of bevacizumab in colorectal cancer growth and its adverse effects: a review. *World J Gastroenterol*. 2013;19(31):5051–5060. https://doi.org/10.3748/wjg.v19.i31.5051
51. Senrung A, Kumar D, Bhardwaj N, Sharma K, Sharma H, Pandey S, et al. Discovery of novel bioactive compounds in Catharanthus roseus exhibiting anti-angiogenic activity against VEGFR-2 TKD for Glioblastoma tumor growth suppression. *Biomed Therapeutics Lett*. 2024;11(2):912–912. https://doi.org/10.62110/sciencein.btl.2024.v11.912
52. Sharma H, Pandey S, Bhardwaj N, Thakur M, Singh J, Kumar N, et al. *In-silico screening of phytochemicals from Hydnocarpus wightianus for potential therapeutic applications against Glioblastoma. Traditional Knowledge Systems and Biodiversity*. Sooryagatha; 2025:41–59.
53. Chorghade MS. *Drug Discovery and Development*. Wiley-interscience; 2006.
54. Andrade RJ, Robles M, Fernández-Castañer A, López-Ortega S, López-Vega MC, Lucena MI. Assessment of drug-induced hepatotoxicity in clinical practice: a challenge for gastroenterologists. *World J Gastroenterol.: WJG*. 2007;13(3):329–340. https://doi.org/10.3748/wjg.v13.i3.329
55. Tortora GJ, Derrickson BH. *Principles of Anatomy & Physiology*. 13th ed. John Wiley & Sons; 2010.
56. Stanke-Labesque F, Gautier-Veyret E, Chhun S, Guilhaumou R. Inflammation is a major regulator of drug metabolizing enzymes and transporters: consequences for the personalization of drug treatment. *Pharmacology & Therapeutics*. 2020;215:107627. https://doi.org/10.1016/j.pharmthera.2020.107627
57. Maniewska J, Jeżewska D. Non-steroidal anti-inflammatory drugs in colorectal cancer chemoprevention. *Cancers*. 2021;13(4):594. https://doi.org/10.3390/cancers13040594
58. Wong RSY. Role of nonsteroidal anti-inflammatory drugs (NSAIDs) in cancer prevention and cancer promotion. *Adv Pharmacol Sci*. 2019;2019:3418975. https://doi.org/10.1155/2019/3418975
59. Jones MK, Wang H, Peskar BM, Levin E, Itani RM, Sarfeh IJ, et al. Inhibition of angiogenesis by nonsteroidal anti-inflammatory drugs: Insight into mechanisms and implications for cancer growth and ulcer healing. *Nat Med*. 1999;5(12):1418–1423. https://doi.org/10.1038/70995

60. Satpathi S, Gaurkar SS, Potdukhe A, Wanjari MB (n.d.). Unveiling the role of hormonal imbalance in breast cancer development: a comprehensive review. Cureus, 15(7), e41737. https://doi.org/10.7759/cureus.41737.
61. Frey N, Porter D. Cytokine release syndrome with chimeric antigen receptor T cell therapy. *Biol Blood Marrow Transplant: J Am Soc Blood Marrow Transp.* 2019;25(4):e123–e127. https://doi.org/10.1016/j.bbmt.2018.12.756
62. Segundo GRS, Condino-Neto A. Treatment of patients with immunodeficiency: medication, gene therapy, and transplantation. *J de Pediatr.* 2020;97(1):S17–S23. https://doi.org/10.1016/j.jped.2020.10.005
63. Liu J, Xiao Q, Xiao J, Niu C, Li Y, Zhang X, et al. Wnt/β-catenin signalling: function, biological mechanisms, and therapeutic opportunities. *Signal Transduct Target Ther.* 2022;7(1):3. https://doi.org/10.1038/s41392-021-00762-6

CHAPTER SIX

Biofilms and beyond: The impact of *Porphyromonas gingivalis* on oral squamous cell carcinoma

Souradip Basu*
Department of Biotechnology, University of Engineering and Management, Kolkata, India
*Corresponding author. e-mail address: souradip.basu@iem.edu.in

Contents

Abstract

Oral squamous cell carcinoma (OSCC) appears as a very common type of head and neck cancer across the world in the recent past. The widespread use of contemporary diagnostic and prognostic methods, a low rate of survival (approximately five-year), a

Progress in Molecular Biology and Translational Science, Volume 224
ISSN 1877-1173, https://doi.org/10.1016/bs.pmbts.2026.08.001

relatively higher chances towards recurrence, and a wide range of malignancy along with metastases are still caused by delayed diagnosis and treatment resistance. There is growing evidence that oral microbiome bacteria, including the Gram-negative anaerobic *Porphyromonas gingivalis*, may be important in the development of OSCC by causing periodontitis. In fact, the activity of a dysbiotic microbiota has been linked to the epithelial-to-mesenchymal transition (EMT) and dysregulated immune response. Conversely, organisms that are typically thought of as oral commensals, including *streptococci* from the *mitis* group, are frequently negatively correlated and have anticancer qualities in vivo. This chapter summarizes the role of *P. gingivalis* in OSCC initiation, progression, invasion, and metastatic dissemination, highlighting its contribution to microbial dysbiosis, immune modulation, epithelial–mesenchymal transition, and tumor microenvironment remodelling. Several reports have showed a positive correlation exist between the bacteria present in the periodontal part along with oral carcinoma. This would in turn pave the way for future researchers to target *P. gingivalis* as a potential therapeutic cornerstone to combat the life-threatening diseases.

1. Introduction

For millennia, philosophers have debated the nature of human being; more recently, biologists have taken an interest in this subject. The continuum advancement of the human microbiome stems to the holobionts or symbiomes theory where an intricated integration has been found in context to the multitaxon facets of the biological entities.[1] In fact, the host and microbiome have a close association that leads to co-dependency in order to preserve a homeostatic condition in terms of the organism's health. On the other hand, an expanding number of pathologic disorders can result from a dysbiotic relationship.[2,3] It is clearly evident that a symbiosis exists between the health of a human along with the microbiome, even though causal linkages are still unclear in many cases.[4] By definition, carcinogenesis entails a significant disturbance of tissue and cell homeostasis. Viral components flag top rank in terms of the one-fifth of human malignancies globally, despite the fact that the etiology of cancer is complicated and multiple.[5] Oral squamous cell carcinoma (OSCC) is the most common histological subtype of head and neck squamous cell carcinoma (HNSCC), accounting for the majority of malignancies arising within the oral cavity.[6,7] The highly invasive clinical phenotype of HNSCC dictates the clinical history, progression, future development, mortality, treatment outcomes, long-term prognosis, local as well as nodal recurrence, and patient survival of the disease. According to studies, HNSCC is two to four times more common in males than in women.[8] It is also the seventh most common cancer globally, making up approximately 4.5 % of all human cancers.[9] Over 377,713 patients have been diagnosed with oral squamous cell carcinoma (OSCC), which has led to 177,757 fatalities.[10,11] Crucially, the five-year/tumor-free survival of OSCC patients has not altered in recent years, despite the advancement of contemporary treatment techniques. Smoking, drinking alcohol, and being exposed to carcinogens and pollutants in the environment are common risk factors for the development of HNSCCs, including OSCC. According to data from the National Comprehensive Cancer Network (NCCN), these factors are known to be responsible for about 90% of instances of this kind of cancer. OSCC has also been linked to poor dental hygiene, a diet low in vegetables, and the usage of areca nut products, such as betel quid and betel leaves, as well as slaked lime.[9,12] Tobacco smoking includes around 7000 hazardous chemicals that have been shown to cause cancer, it is a key contributing factor to the formation of HNSCC, including OSCC. Its carcinogenic components, including nitrosamines (TNA), N′-nitrosonornicotine (NNN), and benzo[*a*]pyrene, a major Polycyclic Aromatic Hydrocarbon (PAH), raise the risk of tumour initiation and progression as well as the development of lymph node metastases. Alcohol is a co-carcinogen because it causes epithelial atrophy, which can degrade the lipids in cell membranes and make it easier for carcinogens formed from tobacco to enter the body. Moreover, acetaldehyde, the primary metabolite of ethanol, is a potent mutagen in and of itself. By causing an epithelial-to-mesenchymal transition (EMT)-like phenotype, the combination of alcohol and smoking raises the risk of HNSCC by up to 35 times.[13–15] Unfortunately, current diagnostic and treatment methods based on surgical therapy and radiochemotherapy are still inadequate because many patients with HNSCC are diagnosed at a very advanced stage, i.e., in WHO stages III and IV. In fact, the HNSCC population's five-year cancer-free survival or worldwide overall survival rates are between 40–60%.[16,17] Numerous mutations in four groups of genes are found in tobacco-related tumours, according to data from the Cancer Genome Atlas Network (TCGA): those that control the cell cycle (*CDKN2A and CCND1*), those that determine cell proliferation and survival (*TP53, HRAS, PIK3CA, and EGFR*), those that control cell differentiation (NOTCH1), and a gene that controls the Wnt signalling pathway.[8,18,19] In the course of cancers in this area, mutations of proto-oncogenes, such as c-myc, RAS family genes (*KRAS, HRAS, NRAS, ERB-B, BRAF, HER-2, c-KIT, BCL-2, STAT3*), tumour suppressor genes known as anti-oncogenes (*RB1, P53, PTEN,*

CDKN2A, INK4), and genes regulating the pro-inflammatory tumour microenvironment are also frequently seen.[8,20] Crucially, infection with carcinogenic strains of the human papillomavirus (HPV), mainly HPV-16 and, to a lesser extent, HPV-18, can also result in SCC of the oral cavity and oropharynx. Up to 38–80% of new diagnoses of oropharynx squamous cell carcinomas (OPSCC) are currently caused by these tumours, which have become more common worldwide. The viral oncoproteins HPV16/18 E5, E6, and E7 are the carcinogenic agents in this kind of HNSCC.[21–23] Incorporation of HPV genomic DNA into the genetic material of the host epithelial cells may lead to molecular alterations that can initiate the process of cancer. Oncoproteins E6 and E7 interfere with intracellular signalling pathways that control the cell cycle by degrading cell cycle regulatory proteins such pRB and the tumour suppressor protein p53.[14,24–26] Crucially, the HPV16/18 E6 protein also stimulates the transcription of the human telomerase catalytic subunit (hTERT) and deregulates the c-*myc* oncogene. Tumour cell immortalisation is inhibited by c-*myc* activity, which also reverses the inhibitory effects of CDK $p27^{KIP1}$ and $p21^{CIP1/WAF1}$ and interferes with the function of CDKs, cyclins, and E2F transcription factors.[20,26]

Viral infections contribute substantially to the global cancer burden and are implicated in approximately 15–20% of human malignancies worldwide. Beyond HPV-associated cancers, several oncogenic viruses, including Epstein–Barr virus (EBV), hepatitis B virus (HBV), hepatitis C virus (HCV), human immunodeficiency virus (HIV), human herpesvirus 8 (HHV-8), Merkel cell polyomavirus, and human T-cell leukemia virus type 1 (HTLV-1), have been associated with malignant transformation through mechanisms involving chronic inflammation, immune dysregulation, genomic instability, and modulation of host signalling pathways.[27,28] Although HPV remains as the predominant viral risk factor for oropharyngeal squamous cell carcinoma, increasing evidence suggests that a subset of oral squamous cell carcinomas develops independently of established viral and lifestyle-associated risk factors. Consequently, attention has shifted toward alternative etiological contributors, particularly microbial dysbiosis and chronic periodontal infections.[29,30] Among the microorganisms implicated, *P. gingivalis* has emerged as a potential non-viral driver of oral carcinogenesis owing to its ability to modulate host immunity, alter epithelial cell signalling, and promote a pro-tumorigenic microenvironment.[31] Indeed, the lesions resemble periodontal disease, and smoking and alcohol use do not seem to be linked to gingival squamous cell carcinoma.[32,33] According to a number of studies, periodontitis is a significant risk factor for oral squamous cell carcinoma, and the lesions have also been linked to periodontal infections.[34–36] Several recent papers have analysed the list of species linked to oral squamous cell cancer in great detail.[31,37] While biopsy sections stained with immunohistochemistry reveal the bacteria's intracellular location, tumour scrapings, homogenates, and swabs suggest surface colonisation and/or penetration of the tumour site. Despite these warnings, some patterns have emerged. Oral squamous cell carcinoma is typically positively correlated with traditional periodontal infections such *Porphyromonas gingivalis*, *Fusobacterium nucleatum*, and *Treponema denticola*. Conversely, organisms that are typically thought of as oral commensals, including *streptococci* from the *mitis* group, are frequently negatively correlated and have anticancer qualities in vivo.[30,37,38] *P. gingivalis* is overabundant in esophageal malignant tissue, while salivary *P. gingivalis* levels are linked to the development of esophageal squamous cell carcinoma.[39,40] Furthermore, a poor prognosis is linked to the presence of *P. gingivalis* in lesions of esophageal or oral squamous cell carcinoma.[41–44] The current pattern is one of a dysbiotic microbial community enriched in potentially carcinogenic organisms like *P. gingivalis* and with an under-representation of homeostatic commensals, contributing to the development of oral squamous cell carcinoma and esophageal squamous cell carcinoma, though more research will improve precision in the field.

2. *Porphyromonas gingivalis*: Key player of oral biofilms

2.1 Keystone pathogen concept

Keystone Pathogen Hypothesis significantly exerts a new avenue towards the traditional paradigm of the microbial pathogenicity where abundance of the microbes solely not remain as a key regulator. Based on this concept, *P. gingivalis* though present in lower abundances but exerts severe impact towards the alterations of both the microbial signatures as well as the other factors like host immune interactions.[45] The organism needs to evade the host immune surveillance mechanism to achieve the keystone status that would automatically reshape the balance of the oral microbiome.[46] Beyond complement manipulation, *P. gingivalis* actively rewires innate immune signalling through molecular crosstalk between gingipains, complement component C5a receptor 1 (C5aR1), and Toll-like receptor 2 (TLR2). Gingipains generated by *P. gingivalis* cleave complement component C5 to produce biologically active C5a, resulting in persistent activation of C5aR1.

Simultaneous stimulation of TLR2 by bacterial ligands triggers a signalling cascade that selectively suppresses MyD88-dependent antimicrobial responses while preserving phosphoinositide 3-kinase (PI3K)-mediated inflammatory signalling. This functional uncoupling disconnects bacterial clearance from inflammation, thereby impairing phagocytic killing despite sustained production of pro-inflammatory mediators. Consequently, a nutrient-rich and inflammophilic microenvironment is established that favors the expansion of pathobionts, promotes microbial dysbiosis, and facilitates chronic inflammatory responses. Such immune subversion is considered a defining characteristic of the keystone pathogen activity of *P. gingivalis.*[46,47] However, the induction of the dysbiosis by *P. gingivalis* can be correlated due to the functional redundancy along with the severe spike of the microbial diversity among the pathogenic taxa[48] (Fig. 1). This also leads towards the alterations of the inflammation responses along with the metabolic activity of the host. Collectively, these alterations promote chronic tissue damage, metabolic dysregulation, and persistent inflammation, thereby creating a microenvironment conducive to disease progression and malignant transformation.[49]

2.2 Biofilm formation and persistence

The dynamic community of oral microbiota are known to exist in form of a biofilm within the oral cavity where, *P. gingivalis* acts as a late colonizer.[50] They typically adhere to the preexisting microbial communities occupying the tooth surfaces and gingival tissues.[51] Specific adhesin molecules along with a typical mechanism of co-aggregation interaction with early colonisers like *Streptococci,* intermediate colonizers like *Fusobacterium nucleatum* are known to make them potent to integrate into the biofilm.[52,53] Moreover, the extracellular polymeric substances of biofilm provide them a selective advantage through acting not only as a physical barrier but also limiting the penetration of antimicrobial agents and host immune factors.[46] Like the other biofilm microenvironment where complementary metabolic capabilities are the basic way of biofilm resilience, *P. gingivalis* relies on proteolytic metabolism and benefits from the metabolic byproducts of neighbouring organisms of oral biofilm as well.[54] From the pathogenic viewpoint, prevalence of *P. gingivalis* within these communities ensures continuous stimulation of host immune responses, leading to chronic inflammation and hence tissue damage, epithelial barrier disruption, and increased susceptibility to neoplastic transformation.[49]

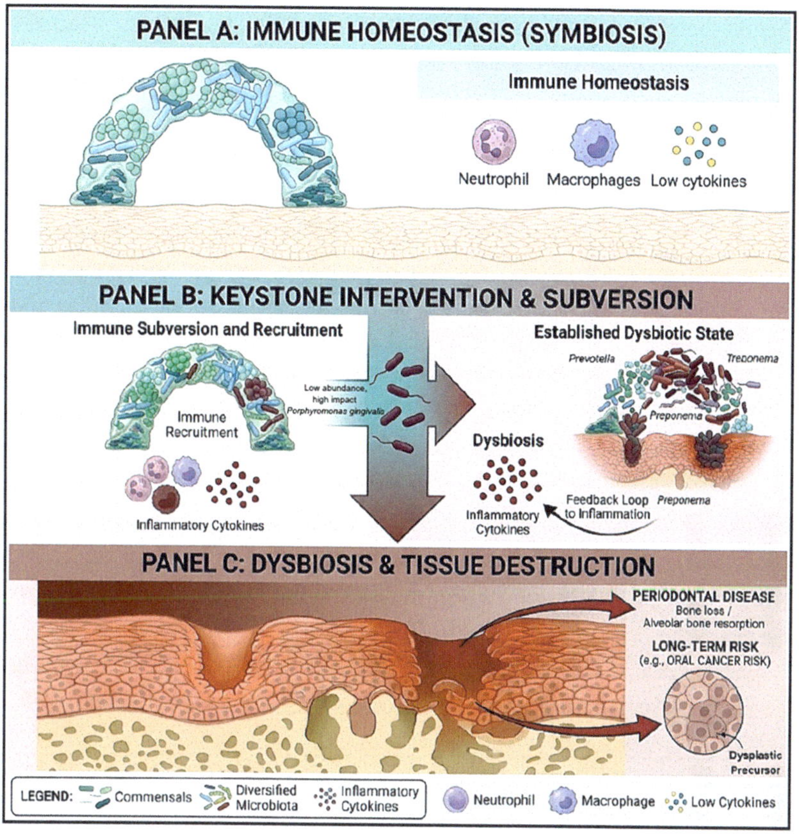

Fig. 1 **Progression from oral symbiosis to dysbiosis and tissue destruction. Panel A:** Immune homeostasis under symbiotic conditions characterized by a diversified microbiota and low cytokine activity. **Panel B:** Keystone pathogen intervention (e.g. *P. gingivalis*) leading to immune subversion, recruitment of inflammatory cells, and the establishment of a dysbiotic state featuring species such as *Treponema* and *Prevotella*. **Panel C:** Clinical consequences of chronic dysbiosis, including periodontal bone loss, alveolar resorption, and the potential for long-term oral cancer risk via dysplastic precursors.

2.3 Virulence factors

The pathogenic potential of *P. gingivalis* is known to rely on a set of virulence factors viz. Gingipains, Lipopolysaccharides (LPS), Fimbriae, Outer Membrane Vesicles (OMVs) enabling their colonization, invasion, and manipulation of host tissues. All these create a microenvironment supporting bacterial persistence and carcinogenic processes altogether.[55] Gingipains, the virulence determinants of class cysteine protease have been reported to modulate host immune system and amplify inflammatory responses through degrading cytokines, immunoglobulins, and extracellular matrix components.[56] Another important mechanism through which *P. gingivalis* promotes immune dysregulation is the expression of structurally heterogeneous lipopolysaccharides (LPS). Unlike canonical enterobacterial LPS, *P. gingivalis* LPS exhibits lipid A heterogeneity, enabling differential activation of TLR2 and TLR4 signaling pathways. This atypical signaling promotes NF-κB-mediated production of pro-inflammatory cytokines, including IL-6, IL-8, and TNF-α, while facilitating immune evasion and persistent inflammation. Consequently, these events contribute to a tumor-promoting microenvironment that supports OSCC progression and therapeutic resistance.[57] Fimbriae as a potent virulence factor not only play a major role in biofilm colonization but also facilitates bacterial internalisation through cytoskeletal rearrangements in host cells. Reports have also identified a cargo of virulence factors containing gingipains, LPS, and DNA released by *P. gingivalis* called Outer Membrane Vesicles (OMVs) via penetrating host cells, these OMVs extends bacterial intracellular communication and sustained immune disruption.[58]

3. Epidemiological and clinical associations with OSCC

As per epidemiological evidences obtained over a period of time, there is a consistent link between the OSCC and presence as well as abundances of *P. gingivalis* in oral tissues.[37,59] Clinical data has also confirmed the higher detection rates of the particular bacterium of interest in tumour tissue compared to the healthy controls.[60] Additionally, its presence has been reported with advanced tumour stage and lymph node metastasis.[61,62] Periodontal disease, characterized by chronic infection with *P. gingivalis* and other anaerobic pathogens, has emerged as a significant risk factor for OSCC. As per reports, patients with chronic periodontitis remain at a significantly higher risk zone for OSCC.[63] Data coming from the salivary diagnostics have further confirmed the association. Higher titre of *P. gingivalis* in saliva and in esophageal carcinoma tissues highlights its potential to act as a non-invasive biomarker. Besides these, few other lifestyle factors like smoking, alcohol consumption, and socioeconomic status can influence the expression as well as the interpretation of epidemiological data.[64,65]

4. Molecular mechanisms linking *P. gingivalis* to OSCC

4.1 Promotion of chronic inflammation

P. gingivalis makes the pivotal pathway of carcinogenesis through generating a sustained inflammatory response within the oral cavity via producing pro-inflammatory cytokines viz. IL-6, IL-8, and TNF-α.[49,66] Uncontrolled cellular proliferation, apoptotic inhibition, induced DNA damages with reactive oxygen and nitrogen species, increased angiogenesis, tissue remodelling are the pivotal steps towards the promotion of chronic inflammation. These mediators sustain a chronic inflammatory microenvironment that promotes cellular proliferation, angiogenesis, genomic instability, and tumor progression.[44]

4.2 Modulation of cell signaling pathways

Molecular alterations in the host cells through manipulation of multiple signaling pathways like NF-κB and MAPK pathways is one of the primary ways through which *P. gingivalis* regulate inflammation and cell survival.[67,68] Modulation of microRNA expression, suppression of tumour suppressor genes such as PDCD4 via upregulation of oncogenic miR-21, resistance to programmed cell death through influencing the PI3K/Akt pathway are another matter of concern in terms of carcinogenic growth.[69–72]

4.3 Epithelial–Mesenchymal transition (EMT)

Epithelial–mesenchymal transition (EMT) is a biological process through which epithelial cells lose their polarity and cell–cell adhesion properties and acquire mesenchymal characteristics, including enhanced migratory capacity, invasiveness, and resistance to apoptosis. EMT is widely recognized as a crucial event in cancer progression, particularly in invasion, metastasis, and therapeutic resistance.[73] *P. gingivalis* promotes EMT through both transcriptional and post-translational mechanisms targeting E-cadherin. Gingipain-mediated proteolytic cleavage directly disrupts epithelial junctions, while initiate the upregulation of certain mesenchymal markers such as vimentin and N-cadherin, along with the activation of EMT-related transcription factors including Snail, Slug, Twist, and ZEB1 is associated with *P. gingivalis* infection.[74–77] All these transcription factors as a whole can result in cytoskeletal reorganization and increased cellular motility through repression of epithelial gene expression while activating mesenchymal gene expression. Modulation of signaling pathways like TGF-β, Wnt/β-catenin, NF-κB, and MAPK has been found to be associated with *P. gingivalis*-induced EMT.[78] Moreover, Chronic exposure to *P. gingivalis* has been shown to enhance stemness markers such as CD44 and ALDH1 in oral epithelial cells, suggesting a role in tumour recurrence and treatment resistance.[79]

4.4 Immune evasion and immunosuppression

Immune evasion and immune modulation are the key way of *P. gingivalis* to establish a chronic pattern of infection unlike other acute infections where strong triggering of immune system is very common. Thus, *P. gingivalis* infection can create a tumour growth favouring chronic inflammatory yet immunosuppressed environment.[68] Manipulation of the complement system and Toll-like receptor signaling are the two major ways of immune evasion used by the bacterium.[57] A unique state called "inflammophilic dysbiosis", is very common is this kind of infection where the ineffective bacterial clearance in sustained inflammatory state is caused by simultaneous activation and inhibition of different arms of the immune response.[80] During carcinogenic condition, *P. gingivalis* has been reported to suppress T-cell mediated immune responses through expansion of myeloid-derived suppressor cells (MDSCs).[81] These MDSCs in turn promote tumour growth by secreting immunosuppressive cytokines like IL-10 and TGF-β.[82] Additionally, *P. gingivalis* infection promotes macrophage polarization through the shifting of pro-inflammatory M1 phenotype to the anti-inflammatory M2 phenotype, supporting angiogenesis, tissue remodelling, suppression of anti-tumour immunity and hence tumour progression as a whole.[83] Downregulation of natural killer (NK) cells and cytotoxic T cells are also found to be associated with *P. gingivalis* infection which aids in tumour surveillance.[84] Chronic periodontal pathogens remain instrumental during the generation of the tumour microenvironment. This would in turn act as a stepping stone for the initiation of the infection along with the carcinogenesis. However, immunity of the host in presence of the microbes would play as a major factor for the progression of the cancer along with the therapeutic response.[75]

4.5 Autophagy and cellular survival

Recent studies have demonstrated that *P. gingivalis* manipulates host autophagic pathways to facilitate intracellular persistence.[68,85] Following invasion of epithelial cells, the bacterium localizes within autophagosome-like compartments and interferes with normal lysosomal fusion. Mechanistically, *P. gingivalis* modulates the PI3K/Akt/mTOR signaling axis, a central regulator of autophagy. Activation of Akt and mTOR signaling suppresses autophagic degradation while preserving intracellular vesicular compartments that can serve as bacterial niches.[86] In parallel, altered autophagic flux reduces apoptosis, enhances cellular stress tolerance, and promotes survival of transformed epithelial cells. Through these mechanisms, *P. gingivalis* establishes a persistent intracellular reservoir while simultaneously creating conditions favorable for tumor cell survival and progression.

5. Tumor microenvironment remodelling

5.1 Interaction with stromal cells

A complex network of stromal cells, immune cells, blood vessels, extracellular matrix components, and signaling molecules are the component of the tumour microenvironment (TME) apart from the cancer cells. This also exerts a significant role in context to the initiation, progression, invasion as well as metastasis of the tumour.[87]

The multiple components of the TME can be exhibited by *P. gingivalis* like the activation of the fibroblasts which in turn can be converted into cancer-associated fibroblasts (CAFs). Afterwards, these CAFs can initiate the secretion of the growth factors, cytokines, and matrix metalloproteinases (MMPs). Finally, tumor cell proliferation, extracellular matrix degradation, and tumor invasion can also be promoted by these molecules.[49,88]

However, the endothelial cells are also being promoted by the bacterium which in turn play a crucial role in context to the development of the vascular endothelial growth factor (VEGF) and other pro-angiogenic factors via angiogenesis. This scenario also provides a constant support of oxygen and nutrients to the developing tumour cells.[89] Apart from that, the filtration of the immune cell has been modulated by *P. gingivalis* towards the tumor microenvironment, which in turn create the support of the tumour growth.[49] Therefore, the stromal environment remains intact that automatically favours the development of the tumour besides the alterations in the epithelial cells as well.

5.2 Extracellular vesicles and migrasomes

Extracellular vesicles (EVs) have emerged as important mediators of intercellular communication within the tumor microenvironment. *P. gingivalis* produces outer membrane vesicles (OMVs) enriched in gingipains, lipopolysaccharides, fimbrial proteins, nucleic acids, and other virulence determinants that can be internalized by epithelial and immune cells.[90] These vesicles activate inflammatory signalling pathways, alter host gene expression, and contribute to immune evasion and tumor-promoting inflammation. Increasing evidence suggests that bacterial OMVs influence epithelial plasticity, cytokine secretion, and angiogenic signalling, thereby facilitating conditions favorable for tumor progression.[56,58]

Migrasomes are recently identified extracellular organelles generated along retraction fibers during cell migration and have been implicated in intercellular signaling and metastatic dissemination. However, direct experimental evidence linking *P. gingivalis* infection to migrasome biogenesis, cargo regulation, or migrasome-mediated metastatic signaling in OSCC remains unavailable.[91] Consequently, the potential involvement of migrasomes in *P. gingivalis*-associated oral carcinogenesis should currently be considered a promising but unexplored area for future investigation.

5.3 Synergistic effects with other microbiota

From the above segments, it is quite evident that *P. gingivalis* acts in a better manner in the polymicrobial environment in context to the carcinogenesis. However, oral biofilms contain a significant amount of complex microbial consortia that can be interacted using the metabolic pathways and quorum sensing.[92] Though there are significant facets which underlie the crosstalk of the oral bacterium. *Fusobacterium nucleatum* exhibits immune evasion along with the promotion of the proliferation of the tumour cells, whereas *Treponema denticola* is associated with the inflammation and destruction of the tissues. Therefore, a synergistic pathogenic network can be created by these two organisms which ultimately lead towards the damage of the DNA along with tumour progression.[93,94] It is also eminent that a collective consortium of pathogenic microbes is more powerful towards the carcinogenesis rather than a single one. This would in turn provide an insight towards the development of the microbiome-based therapeutic strategies.

6. From biofilms to cancer: A conceptual framework

Periodontal biofilm infection can be marked as a continuous process that can lead to oral cancer. This progression is completely based upon the interactions of the microbial, immunological, and genetic factors for a prolonged period.[95,96] The dysbiosis event initiates from the prevalence of the pathogenic bacteria like *P. gingivalis* in context to the oral biofilms.[46,47] Therefore, it leads towards the chronic inflammation that can be attributed with the significant rise of the cytokines as well as the reactive oxygen species (ROS).[97] However, epithelial cells can act as a reservoir of genetic and epigenetic changes that automatically enhance the chance of acquiring mutations in tumour suppressor genes as well as activating the oncogenes.[98] This also leads towards the step of malignant transformation due to the persistent inflammation pressure.[96] Finally, the clinicians can get an overview of the prolonged phase of microbial infection as well as the inflammation in context to the cancer development using this portrayed framework.

7. Diagnostic and therapeutic implications

7.1 Biomarker potential

The presence of *P. gingivalis* in saliva, plaque, or tumor tissues has been proposed as a potential biomarker for oral cancer risk assessment and early diagnosis.[99] The collection of saliva remains non-invasive as well as inexpensive which attract the mode of this particular diagnostic regime. Molecular detection methods such as PCR, quantitative PCR (qPCR), and metagenomic sequencing can be used to detect and quantify *P. gingivalis*.[100] In addition to the bacterium itself, its virulence factors, inflammatory cytokines, and microbial metabolites may serve as potential biomarkers for early cancer detection.[49] Early diagnosis is crucial for improving survival rates in oral cancer, and microbiome-based biomarkers may play an important role in future diagnostic strategies.

7.2 Targeting microbial pathways

Understanding the role of *P. gingivalis* in carcinogenesis opens new avenues for therapeutic intervention. Potential strategies include antimicrobial treatments targeting periodontal pathogens, inhibitors of gingipains, probiotics to restore microbial balance, and anti-inflammatory therapies to reduce chronic inflammation. Biofilm disruption strategies are particularly important because bacteria within biofilms are highly resistant to antibiotics.[101,102] Novel approaches such as antimicrobial peptides, nanoparticles, and quorum sensing inhibitors are currently being investigated. Targeting host–microbe interactions rather than the bacteria alone may also represent a promising therapeutic strategy.[103]

7.3 Integration with precision medicine

The concept of precision medicine is transforming modern healthcare by shifting medical practice from a generalized treatment approach to one that considers individual variability in genetics, environment, lifestyle in context to the microbiome.[104] In the context of oral squamous cell carcinoma (OSCC), integrating microbial factors- particularly periodontal pathogens such as *Porphyromonas gingivalis-* into precision medicine frameworks represents an emerging and promising direction in oncology and oral health research.[45]

Traditionally, OSCC treatment strategies have been based on tumor stage, histopathological grading, and anatomical location. However, clinical outcomes often vary significantly among patients with similar tumor characteristics, indicating that additional biological factors influence disease progression and treatment response.[105] Recent advances in genomics, transcriptomics, proteomics, and metagenomics have revealed that tumor development is not solely driven by host genetic mutations but is also influenced by the tumor microenvironment, immune responses, and microbial communities associated with the tumor site.[106] Therefore, integrating microbiome profiling into precision medicine could provide a more comprehensive understanding of OSCC pathogenesis and patient-specific disease progression.[107]

Another important dimension of precision medicine is personalized therapeutic intervention. Patients with OSCC associated with periodontal pathogens may benefit from combined treatment approaches that include not only surgery, chemotherapy, or radiotherapy but also periodontal therapy, antimicrobial treatment, anti-inflammatory therapy, and microbiome modulation strategies such as probiotics or prebiotics.[102] Such integrative treatment approaches may help reduce inflammation, decrease microbial load, and improve overall treatment outcomes.

Furthermore, advances in artificial intelligence and machine learning are enabling the integration of large datasets, including genomic, transcriptomic, microbiome, and clinical data. Such integrative analyses may help identify predictive biomarkers, therapeutic targets, and disease progression models specific to individual patients.[108,109] In the future, clinicians may use combined molecular and microbial profiles to design personalized treatment plans for OSCC patients.[109]

Another promising area is microbiome-targeted therapy, which may include selective antimicrobial agents targeting periodontal pathogens, inhibitors of

Table 1 Diagnostic, therapeutic and precision medicine purview towards the mapping of oral squamous cell carcinoma (OSCC) associated with *Porphyromonas gingivalis*.

Category	Approach	Target/marker	Method/technology	Clinical significance
Diagnostic Biomarkers	Salivary detection of *P. gingivalis*	Bacterial DNA	PCR, qPCR	Non-invasive early detection and risk assessment
	Microbial community profiling	Dysbiotic oral microbiome	16S rRNA sequencing, Metagenomics	Identification of high-risk microbial signatures
	Virulence factor detection	Gingipains, LPS	ELISA, Proteomics	Indicates pathogenic activity rather than mere presence
	Inflammatory biomarkers	IL-6, IL-8, TNF-α	Cytokine assays	Reflects chronic inflammation linked to carcinogenesis
	Microbial metabolites	Short-chain fatty acids, polyamines	Metabolomics	Potential early biochemical indicators of tumor environment
Therapeutic Strategies	Antimicrobial therapy	Periodontal pathogens	Antibiotics, antimicrobial peptides	Reduces microbial load and chronic infection
	Gingipain inhibition	Bacterial proteases	Small molecule inhibitors	Prevents tissue destruction and immune modulation

(continued)

Table 1 Diagnostic, therapeutic and precision medicine purview towards the mapping of oral squamous cell carcinoma (OSCC) associated with *Porphyromonas gingivalis*. (*cont'd*)

Category	Approach	Target/marker	Method/technology	Clinical significance
	Biofilm disruption	Polymicrobial biofilm matrix	Nanoparticles, quorum sensing inhibitors	Improves antimicrobial penetration and treatment efficacy
	Microbiome modulation	Oral microbial balance	Probiotics, prebiotics	Restores microbial homeostasis
	Anti-inflammatory therapy	Chronic inflammation	NSAIDs, cytokine inhibitors	Reduces inflammation-driven carcinogenesis
Precision Medicine Approaches	Microbiome profiling	Individual microbial composition	Next-generation sequencing	Personalized risk assessment
	Host genetic susceptibility analysis	Immune and inflammatory gene polymorphisms	Genomics	Identifies high-risk individuals
	Treatment response prediction	Microbiome–drug interactions	Multi-omics analysis	Helps select effective therapy
	Personalized combined therapy	Host + microbiome factors	Integrated clinical approach	Improves treatment outcomes
	Microbiome-targeted therapy	Dysbiosis	Targeted antimicrobials, microbiome transplantation	Precision treatment based on microbial profile

Table 2 Mechanism and clinical application of *P. gingivalis* during oral squamous cell carcinoma (OSCC).

Biological function	Molecular target	Mechanism	Clinical relevance
Chronic inflammation	IL-6, IL-8, TNF-α	NF-κ β activation	Risk stratification
Immune evasion	C5aR1/TLR2	Impaired bacterial clearance	Immunotherapy target
EMT induction	E-cadherin (CDH1)	Gingipain cleavage; Snail/ Slug activation	Predict metastasis
Cell survival	PI3K/Akt/ mTOR	Altered autophagic flux	Therapeutic target
Immune suppression	MDSCs, IL-10, TGF-β	Tumor immune escape	Combination immunotherapy
Checkpoint regulation	PD-L1	Immune exhaustion	Immune checkpoint therapy
Angiogenesis	VEGF	Endothelial activation	Anti-angiogenic therapy
Biofilm persistence	Gingipains, fimbriae	Dysbiosis and chronic infection	Biofilm disruption therapy
OMV-mediated signaling	Gingipains, LPS cargo	Cell-cell communication	Biomarker development

bacterial virulence factors such as gingipains, biofilm disruption strategies, and microbiome transplantation approaches aimed at restoring a healthy oral microbial community.[110] These approaches align with precision medicine principles by targeting disease mechanisms specific to individual patients rather than applying generalized treatments. In summary, the integration of microbiome research into precision medicine represents a significant advancement in understanding and managing oral squamous cell carcinoma.[111]

The following tables are depicting the above-mentioned cited claims from different studies pertaining to this field which would in turn provide a more holistic overview for the readers (Tables 1, 2).

8. Future directions

Oral microbiome and its association with cancer remains as a domain of interest where plethora of questions still to be answered. *P. gingivalis* is a key player to detect the tumour tissues, while further investigation has been required to assess whether it has a direct role in initiation of the cancer. However, we also mentioned that a polymicrobial environment is also a signature hallmark during the tumor microenvironment colonization. Periodontal infection may also precede the development of the cancer. Therefore, more longitudinal studies are essential to validate this particular claim. Apart from that, standard microbiological procedures along with quantification of the results would remain essential to maintain the reproducibility of the studies across the wider cohorts.

9. Conclusion

The current understanding stems that *Porphyromonas gingivalis*, a keystone pathogenic bacterium in the human oral microbiota. Along with describing the role of *P. gingivalis* and other cooperating/counteracting pathogens like *F. nucleatum* in periodontitis and the start of carcinogenesis, the work also examines the mechanisms and signalling pathways that contribute to preserving the proper morphology and function of oral tissues. These pathogens have been shown to affect host immune response, cell cycling and proliferation, apoptosis, pro-inflammatory effects, epithelial-to-mesenchymal transition, and tumour spread. The review provides a more comprehensive understanding of the carcinogenic risk of *P. gingivalis* and the main ways how it affects oral cancer, the most prevalent type of head and neck cancer. It offers a thorough evaluation of the examined biological material in terms of cohort kinds, diagnostic techniques, and the function of virulence factors like gingipains and FimA. Thus, it makes clear that the possible involvement of *P. gingivalis* in OSCC cancer cell genesis, growth, and metastasis. Additionally, it shows the possibility of controlling intracellular checkpoints and generating tumour antigens, the impact of periodontal pathogens on host immune cell function, and significant regulatory signalling pathways for carcinogenesis.

Acknowledgement

I would like to acknowledge AICTE Post-Doctoral Fellowship (APDF) committee for providing me the necessary financial support for pursuing my research.

References

1. Rosenberg E, Zilber-Rosenberg I. The hologenome concept of evolution: medical implications. *Rambam Maimonides Med J*. 2019;10(1):e0005. https://doi.org/10.5041/RMMJ.10359
2. Thomas S, Izard J, Walsh E, et al. The host microbiome regulates and maintains human health: a primer and perspective for non-microbiologists. *Cancer Res*. 2017;77(8):1783–1812. https://doi.org/10.1158/0008-5472.CAN-16-2929
3. Sarkar A, McInroy CJA, Harty S, et al. Microbial transmission in the social microbiome and host health and disease. *Cell*. 2024;187(1):17–43. https://doi.org/10.1016/j.cell.2023.12.014
4. Aggarwal N, Kitano S, Puah GRY, Kittelmann S, Hwang IY, Chang MW. Microbiome and human health: current understanding, engineering, and enabling technologies. *Chem Rev*. 2023;123(1):31–72. https://doi.org/10.1021/acs.chemrev.2c00431
5. Torres MKDS, Pereira Neto GDS, Cayres Vallinoto IMV, Reis LO, Vallinoto ACR. The impact of oncogenic viruses on cancer development: a narrative review. *Biology*. 2025;14(7):797. https://doi.org/10.3390/biology14070797
6. Charap AJ, Enokida T, Brody R, et al. Landscape of natural killer cell activity in head and neck squamous cell carcinoma. *J Immunother. Cancer*. 2020;8(2):e001523. https://doi.org/10.1136/jitc-2020-001523
7. Sarradin V, Siegfried A, Uro-Coste E, Delord JP. Classification de l'OMS 2017 des tumeurs de la tête et du cou: principales nouveautés et mise à jour des méthodes diagnostiques [WHO classification of head and neck tumours 2017: Main novelties and update of diagnostic methods]. *Bull du Cancer*. 2018;105(6):596–602. https://doi.org/10.1016/j.bulcan.2018.04.004
8. Cancer Genome Atlas Network. Comprehensive genomic characterization of head and neck squamous cell carcinomas. *Nature*. 2015;517(7536):576–582. https://doi.org/10.1038/nature14129
9. Sung H, Ferlay J, Siegel RL, et al. Global cancer statistics 2020: GLOBOCAN estimates of incidence and mortality worldwide for 36 cancers in 185 countries. *CA: a Cancer J Clinicians*. 2021;71(3):209–249. https://doi.org/10.3322/caac.21660
10. Bray F, Laversanne M, Sung H, et al. Global cancer statistics 2022: GLOBOCAN estimates of incidence and mortality worldwide for 36 cancers in 185 countries. *CA: a Cancer J Clinicians*. 2024;74(3):229–263. https://doi.org/10.3322/caac.21834
11. Gormley M, Creaney G, Schache A, Ingarfield K, Conway DI. Reviewing the epidemiology of head and neck cancer: definitions, trends and risk factors. *Br dental J*. 2022;233(9):780–786. https://doi.org/10.1038/s41415-022-5166-x
12. Rindi G, Klimstra DS, Abedi-Ardekani B, et al. A common classification framework for neuroendocrine neoplasms: an International Agency for Research on Cancer (IARC) and World Health Organization (WHO) expert consensus proposal. *Mod Pathol: an Off J U S Can Acad Pathology, Inc*. 2018;31(12):1770–1786. https://doi.org/10.1038/s41379-018-0110-y
13. Du E, Mazul AL, Farquhar D, et al. Long-term survival in head and neck cancer: impact of site, stage, smoking, and human papillomavirus status. *Laryngoscope*. 2019;129(11):2506–2513. https://doi.org/10.1002/lary.27807
14. Sun Z, Sun X, Chen Z, Du J, Wu Y. Head and neck squamous cell carcinoma: risk factors, molecular alterations, immunology and peptide vaccines. *Int J Peptide Res Ther*. 2022;28(1):19. https://doi.org/10.1007/s10989-021-10334-5
15. Miranda-Galvis M, Loveless R, Kowalski LP, Teng Y. Impacts of environmental factors on head and neck cancer pathogenesis and progression. *Cells*. 2021;10(2):389. https://doi.org/10.3390/cells10020389

16. Goyal N, Day A, Epstein J, et al. Head and neck cancer survivorship consensus statement from the American head and neck society. *Laryngoscope Investig Otolaryngol.* 2021;7(1):70–92. https://doi.org/10.1002/lio2.702
17. Chen Y, Zhong NN, Cao LM, Liu B, Bu LL. Surgical margins in head and neck squamous cell carcinoma: a narrative review. *Int. J. Surg. (London, Engl.).* 2024;110(6):3680–3700. https://doi.org/10.1097/JS9.0000000000001306
18. Saada-Bouzid E, Peyrade F, Guigay J. Molecular genetics of head and neck squamous cell carcinoma. *Curr. OpOncol.* 2019;31(3):131–137. https://doi.org/10.1097/CCO.0000000000000536
19. Zhang Y, Koneva LA, Virani S, et al. Subtypes of HPV-positive head and neck cancers are associated with HPV characteristics, copy number alterations, PIK3CA mutation, and pathway signatures. *Clin. Cancer Res: an Off J. Am. Assoc. Cancer Res.* 2016;22(18):4735–4745. https://doi.org/10.1158/1078-0432.CCR-16-0323
20. Alsahafi E, Begg K, Amelio I, et al. Clinical update on head and neck cancer: molecular biology and ongoing challenges. *Cell Death & Dis.* 2019;10(8):540. https://doi.org/10.1038/s41419-019-1769-9
21. Leemans CR, Snijders PJF, Brakenhoff RH. The molecular landscape of head and neck cancer. *Nat. Rev. Cancer.* 2018;18(5):269–282. https://doi.org/10.1038/nrc.2018.11
22. Dong H, Shu X, Xu Q, et al. Current status of human papillomavirus-related head and neck cancer: from viral genome to patient care. *Virologica Sin.* 2021;36(6):1284–1302. https://doi.org/10.1007/s12250-021-00413-8
23. Wittekindt C, Wagner S, Bushnak A, et al. Increasing Incidence rates of oropharyngeal squamous cell carcinoma in germany and significance of disease burden attributed to human papillomavirus. *Cancer Prev. Res. (Philadelphia, Pa).* 2019;12(6):375–382. https://doi.org/10.1158/1940-6207.CAPR-19-0098
24. de Freitas AC, de Oliveira THA, Barros Jr MR, Venuti A. hrHPV E5 oncoprotein: immune evasion and related immunotherapies. *J. Exp. & Clin. Cancer Res: CR.* 2017;36(1):71. https://doi.org/10.1186/s13046-017-0541-1
25. Pal A, Kundu R. Human papillomavirus E6 and E7: the cervical cancer hallmarks and targets for therapy. *Front Microbiol.* 2020;10:3116. https://doi.org/10.3389/fmicb.2019.03116
26. Canning M, Guo G, Yu M, et al. Heterogeneity of the head and neck squamous cell carcinoma immune landscape and its impact on immunotherapy. *Front Cell Dev. Biol.* 2019;7:52. https://doi.org/10.3389/fcell.2019.00052
27. Lowy DR, Schiller JT. Reducing HPV-associated cancer globally. *Cancer Prev. Res. (Philadelphia, Pa).* 2012;5(1):18–23. https://doi.org/10.1158/1940-6207.CAPR-11-0542
28. Gan Y-X, Jiang X, Wang Z-Y, et al. HERV6196 as an enhancer with oncogenic potential in rectal cancer. *Microbiol. Spectr.* 2026;14(1):e0078825. https://doi.org/10.1128/spectrum.00788-25
29. Börnigen D, Ren B, Pickard R, et al. Alterations in oral bacterial communities are associated with risk factors for oral and oropharyngeal cancer. *Sci. Rep.* 2017;7(1):17686. https://doi.org/10.1038/s41598-017-17795-z
30. Ganly I, Yang L, Giese RA, et al. Periodontal pathogens are a risk factor of oral cavity squamous cell carcinoma, independent of tobacco and alcohol and human papillomavirus. *Int. J. Cancer.* 2019;145(3):775–784. https://doi.org/10.1002/ijc.32152
31. Perera M, Al-Hebshi NN, Speicher DJ, Perera I, Johnson NW. Emerging role of bacteria in oral carcinogenesis: a review with special reference to perio-pathogenic bacteria. *J. Oral Microbiol.* 2016;8:32762. https://doi.org/10.3402/jom.v8.32762

32. Yamanaka W, Takeshita T, Shibata Y, et al. Compositional stability of a salivary bacterial population against supragingival microbiota shift following periodontal therapy. *PLoS One*. 2012;7(8):e42806. https://doi.org/10.1371/journal.pone.0042806
33. Koduganti RR, Sehrawat S, Reddy PV. Gingival squamous cell carcinoma: a diagnostic impediment. *J. Indian Soc. Periodontol*. 2012;16(1):104–107. https://doi.org/10.4103/0972-124X.94615
34. Komlós G, Csurgay K, Horváth F, Pelyhe L, Németh Z. Periodontitis as a risk for oral cancer: a case-control study. *BMC Oral Health*. 2021;21(1):640. https://doi.org/10.1186/s12903-021-01998-y
35. Meyer MS, Joshipura K, Giovannucci E, Michaud DS. A review of the relationship between tooth loss, periodontal disease, and cancer. *Cancer Causes & Control: CCC*. 2008;19(9):895–907. https://doi.org/10.1007/s10552-008-9163-4
36. Naavaal S, Garcia DT, Deng X, Bandyopadhyay D. Association between periodontal disease and oral cancer screening among US adults: NHANES 2011-2014. *Community Dent Oral Epidemiol*. 2022;50(3):216–224. https://doi.org/10.1111/cdoe.12655
37. Lamont RJ, Fitzsimonds ZR, Wang H, Gao S. Role of Porphyromonas gingivalis in oral and orodigestive squamous cell carcinoma. *Periodontology 2000*. 2022;89(1):154–165. https://doi.org/10.1111/prd.12425
38. Shin JM, Luo T, Kamarajan P, Fenno JC, Rickard AH, Kapila YL. Microbial communities associated with primary and metastatic head and neck squamous cell carcinoma - a high fusobacterial and low streptococcal signature. *Sci. Rep*. 2017;7(1):9934. https://doi.org/10.1038/s41598-017-09786-x
39. Peters BA, Wu J, Pei Z, et al. Oral microbiome composition reflects prospective risk for esophageal cancers. *Cancer Res*. 2017;77(23):6777–6787. https://doi.org/10.1158/0008-5472.CAN-17-1296
40. Yu G, Gail MH, Shi J, et al. Association between upper digestive tract microbiota and cancer-predisposing states in the esophagus and stomach. *Cancer Epidemiol. Biomarkers & Prev: a Publ. Am. Assoc. Cancer Research, Cosponsored Am. Soc. Preventive Oncol*. 2014;23(5):735–741. https://doi.org/10.1158/1055-9965.EPI-13-0855
41. Gao S, Li S, Ma Z, et al. Presence of Porphyromonas gingivalis in esophagus and its association with the clinicopathological characteristics and survival in patients with esophageal cancer. *Infect. Agents Cancer*. 2016;11:3. https://doi.org/10.1186/s13027-016-0049-x
42. Guo ZC, Jumatai S, Jing SL, Hu LL, Jia XY, Gong ZC. Bioinformatics and immunohistochemistry analyses of expression levels and clinical significance of CXCL2 and TANs in an oral squamous cell carcinoma tumor microenvironment of *Prophyromonas gingivalis* infection. *Oncol. Lett*. 2021;21(3):189. https://doi.org/10.3892/ol.2021.12450
43. Chen MF, Lu MS, Hsieh CC, Chen WC. Porphyromonas gingivalis promotes tumor progression in esophageal squamous cell carcinoma. *Cell Oncol. (Dordrecht, Neth)*. 2021;44(2):373–384. https://doi.org/10.1007/s13402-020-00573-x
44. Gao S, Liu Y, Duan X, et al. Porphyromonas gingivalis infection exacerbates oesophageal cancer and promotes resistance to neoadjuvant chemotherapy. *Br. J. Cancer*. 2021;125(3):433–444. https://doi.org/10.1038/s41416-021-01419-5
45. Gorr S-U, Chen R, Abrahante JE, Joyce PBM. The oral pathogen Porphyromonas gingivalis gains tolerance to the antimicrobial peptide DGL13K by synonymous mutations in hagA. *PLOS ONE*. 2024;19(10):e0312200. https://doi.org/10.1371/journal.pone.0312200
46. Sharma S, Mohler J, Mahajan SD, Schwartz SA, Bruggemann L, Aalinkeel R. Microbial biofilm: a review on formation, infection, antibiotic resistance, control measures, and innovative treatment. *Microorganisms*. 2023;11(6):1614. https://doi.org/10.3390/microorganisms11061614

47. Śmiga M, Smalley JW, Ślęzak P, et al. Glycation of host proteins increases pathogenic potential of Porphyromonas gingivalis. *Int. J. Mol. Sci.* 2021;22(21):12084. https://doi.org/10.3390/ijms222112084
48. Li M, Cui J, Qu R, et al. Porphyromonas gingivalis induces intestinal inflammation through gingipain-dependent gut microbiome dysbiosis. *Microbiome.* 2026. https://doi.org/10.1186/s40168-026-02389-7
49. Starska-Kowarska K. The role of Porphyromonas gingivalis in oral carcinogenesis and progression by remodelling the tumour microenvironment: a narrative review. *Cancers.* 2025;17(21):3478. https://doi.org/10.3390/cancers17213478
50. Tan J, Lamont GJ, Scott DA. Tobacco-enhanced biofilm formation by Porphyromonas gingivalis and other oral microbes. *Mol. Oral Microbiol.* 2024;39(5):270–290. https://doi.org/10.1111/omi.12450
51. Łasica A, Golec P, Laskus A, Zalewska M, Gędaj M, Popowska M. Periodontitis: etiology, conventional treatments, and emerging bacteriophage and predatory bacteria therapies. *Front Microbiol.* 2024;15:1469414. https://doi.org/10.3389/fmicb.2024.1469414
52. Gerits E, Verstraeten N, Michiels J. New approaches to combat Porphyromonas gingivalis biofilms. *J. Oral Microbiol.* 2017;9(1):1300366https://doi.org/10.1080/20002297.2017.1300366
53. Wang B, Deng J, Donati V, et al. The roles and interactions of *Porphyromonas gingivalis* and *Fusobacterium nucleatum* in oral and gastrointestinal carcinogenesis: a narrative review. *Pathog. (Basel, Switz).* 2024;13(1):93. https://doi.org/10.3390/pathogens13010093
54. Ding J, Tan L, Wu L, et al. Regulation of tryptophan-indole metabolic pathway in *Porphyromonas gingivalis* virulence and microbiota dysbiosis in periodontitis. *npj Biofilms Microbiomes.* 2025;11:37. https://doi.org/10.1038/s41522-025-00669-y
55. How KY, Song KP, Chan KG. Porphyromonas gingivalis: an overview of periodontopathic pathogen below the gum line. *Front Microbiol.* 2016;7:53. https://doi.org/10.3389/fmicb.2016.00053
56. Sharaf S, Hijazi K. Modulatory mechanisms of pathogenicity in Porphyromonas gingivalis and other periodontal pathobionts. *Microorganisms.* 2022;11(1):15. https://doi.org/10.3390/microorganisms11010015
57. Yamashita Y, Shigeishi H, Yokoyama S, et al. Porphyromonas gingivalis-derived lipopolysaccharide confers chemotherapy resistance and migratory ability on oral cancer cells by activating toll-like receptor 4 signaling pathway. *Mol. Biol. Rep.* 2026;53(1):551. https://doi.org/10.1007/s11033-026-11713-1
58. Wu Z, Long W, Yin Y, et al. Outer membrane vesicles of Porphyromonas gingivalis: recent advances in pathogenicity and associated mechanisms. *Front Microbiol.* 2025;16:1555868. https://doi.org/10.3389/fmicb.2025.1555868
59. Lafuente Ibáñez de Mendoza I, Maritxalar Mendia X, Garcia de la Fuente AM, Quindos Andres G, Aguirre Urizar JM. Role of Porphyromonas gingivalis in oral squamous cell carcinoma development: a systematic review. *J. Periodontal Res.* 2020;55(1):13–22. https://doi.org/10.1111/jre.12691
60. Kong J, Yuan X, Wang J, et al. Frequencies of Porphyromonas gingivalis detection in oral-digestive tract tumors. *Pathol Oncol Res.* 2021;27:628942.
61. Olsen I, Yilmaz Ö. Possible role of Porphyromonas gingivalis in orodigestive cancers. *J. Oral Microbiol.* 2019;11(1):1563410. https://doi.org/10.1080/20002297.2018.1563410
62. Liu Y, Yuan X, Chen K, et al. Clinical significance and prognostic value of Porphyromonas gingivalis infection in lung cancer. *Transl. Oncol.* 2021;14(1):100972. https://doi.org/10.1016/j.tranon.2020.100972
63. Baima G, Minoli M, Michaud DS, et al. Periodontitis and risk of cancer: mechanistic evidence. *Periodontology 2000.* 2024;96(1):83–94. https://doi.org/10.1111/prd.12540

64. Vats R, Yadav P, Bano A, Wadhwa S, Bhardwaj R. Salivary biomarkers in non-invasive oral cancer diagnostics: a comprehensive review. *J. Appl. Oral Sci.* 2024;32:e20240151. https://doi.org/10.1590/1678-7757-2024-0151
65. Li M, Shao D, Fan Z, et al. Non-invasive early detection on esophageal squamous cell carcinoma and precancerous lesions by microbial biomarkers combining epidemiological factors in China. *J. Gastroenterol.* 2024;59(7):531–542. https://doi.org/10.1007/s00535-024-02117-8
66. Farhad SZ, Karbalaeihasanesfahani A, Dadgar E, Nasiri K, Esfahaniani M, Nabi Afjadi M. The role of periodontitis in cancer development, with a focus on oral cancers. *Mol. Biol. Rep.* 2024;51(1):814. https://doi.org/10.1007/s11033-024-09737-6
67. Mei F, Xie M, Huang X, et al. Porphyromonas gingivalis and its systemic impact: current status. *Pathogens.* 2020;9(11):944. https://doi.org/10.3390/pathogens9110944
68. Zheng S, Yu S, Fan X, et al. Porphyromonas gingivalis survival skills: immune evasion. *J. Periodontal Res.* 2021;56(6):1007–1018. https://doi.org/10.1111/jre.12915
69. Li J, Li L, Wang X, Xiao L. Porphyromonas gingivalis inhibition of MicroRNA-205-5p expression modulates proinflammatory cytokines in gingival epithelial cells. *Biochemical Genet.* 2020;58(4):566–579. https://doi.org/10.1007/s10528-020-09957-y
70. Aravindraja C, Vekariya KM, Botello-Escalante R, Rahaman SO, Chan EK, Kesavalu L. Specific microRNA signature kinetics in Porphyromonas gingivalis-induced periodontitis. *Int. J. Mol. Sci.* 2023;24(3):2327. https://doi.org/10.3390/ijms24032327
71. Feng Y, Liu M, Liu Y, Li H. Invasion of human dental pulp fibroblasts by Porphyromonas gingivalis leads to autophagy via the phosphoinositide 3-kinase/Akt/mammalian target of rapamycin signaling pathway. *J. Oral Biosci.* 2024;66(4):10–18. https://doi.org/10.1016/j.job.2024.08.004
72. Ren J, Zou H, Sun S, et al. MiR-21 modulates Pg-LPS induced apoptosis and inflammatory response in HUVECs via NF-κB/iNOS/NO pathway by targeting PDCD4. *Non-coding RNA Res.* 2025. https://doi.org/10.1016/j.ncrna.2025.10.001
73. Huang Y, Hong W, Wei X. The molecular mechanisms and therapeutic strategies of EMT in tumor progression and metastasis. *J. Hematol & Oncol.* 2022;15(1):129. https://doi.org/10.1186/s13045-022-01347-8
74. Asl ZK, Mottaghi M, Farshad F, Azmoudeh F. Porphyromonas gingivalis modulates E-cadherin expression: a systematic review of possible involved mechanisms. *Med. Microecol.* 2025:100146. https://doi.org/10.1016/j.medmic.2025.100146
75. Lialios P, Alimperti S. Role of E-cadherin in epithelial barrier dysfunction: implications for bacterial infection, inflammation, and disease pathogenesis. *Front Cell Infect. Microbiol.* 2025;15:1506636. https://doi.org/10.3389/fcimb.2025.1506636
76. Zou Q, Xie H, Yang W, et al. Advances in the study of the relationship between Porphyromonas gingivalis and various diseases. *Front Cell Dev. Biol.* 2025;13:1480233. https://doi.org/10.3389/fcell.2025.1480233
77. Akram HM, Saliem SS. Turning barrier cells into invaders: the epithelial-mesenchymal transition signature of Porphyromonas gingivalis. *Nov Res. Microbiol. J.* 2026;10(2):146–155. https://doi.org/10.17582/journal.nrmj/2026/10.2.146.155
78. Ma Y, Yu Y, Yin Y, et al. Potential role of epithelial–mesenchymal transition induced by periodontal pathogens in oral cancer. *J. Cell Mol. Med.* 2024;28(1):e18064. https://doi.org/10.1111/jcmm.18064
79. Zang W, Geng F, Liu J, et al. Porphyromonas gingivalis potentiates stem-like properties of oral squamous cell carcinoma by modulating SCD1-dependent lipid synthesis via NOD1/KLF5 axis. *Int. J. Oral Sci.* 2025;17(1):15. https://doi.org/10.1038/s41368-024-00342-8
80. Yu X, Mankia K, Do T, Meade J. *Oral microbiome dysbiosis and citrullination in rheumatoid arthritis. Oral Microbiome: Symbiosis, Dysbiosis and Microbiome Interventions for Maintaining Oral and Systemic Health.* Springer; 2025:185–199.

81. Wen L, Mu W, Lu H, et al. Porphyromonas gingivalis promotes oral squamous cell carcinoma progression in an immune microenvironment. *J. Dental Res.* 2020;99(6):666–675. https://doi.org/10.1177/0022034520909312
82. Gneo L, Rizkalla N, Hejmadi R, Mussai F, de Santo C, Middleton G. TGF-β orchestrates the phenotype and function of monocytic myeloid-derived suppressor cells in colorectal cancer. *Cancer Immunol. Immunother.* 2022;71(7):1583–1596. https://doi.org/10.1098/rsob.200111
83. Chen X, Dou J, Fu Z, et al. Macrophage M1 polarization mediated via the IL-6/STAT3 pathway contributes to apical periodontitis induced by Porphyromonas gingivalis. *J. Appl. Oral Sci.* 2022;30:e20220316. https://doi.org/10.1590/1678-7757-2022-0316
84. Ren J, Han X, Lohner H, et al. P. gingivalis infection upregulates PD-L1 expression on dendritic cells, suppresses CD8+ T-cell responses, and aggravates oral cancer. *Cancer immunol. Res.* 2023;11(3):290–305. https://doi.org/10.1158/2326-6066.CIR-22-0541
85. Liu M, Shao J, Zhao Y, Ma B, Ge S. Porphyromonas gingivalis evades immune clearance by regulating lysosome efflux. *J. Dental Res.* 2023;102(5):555–564. https://doi.org/10.1177/00220345221146097
86. Zheng C, Huang J, Xu S, et al. Porphyromonas gingivalis hijacks mitophagy and lysosomal function to persist in endothelial cells. *Front Cell Infect. Microbiol.* 2025;15:1613366. https://doi.org/10.3389/fcimb.2025.1613366
87. Li C, Dong X, Li B. Tumor microenvironment in oral squamous cell carcinoma. *Front Immunol.* 2024;15:1485174. https://doi.org/10.3389/fimmu.2024.1485174
88. Zhang H, Yue X, Chen Z, et al. Define cancer-associated fibroblasts (CAFs) in the tumor microenvironment: new opportunities in cancer immunotherapy and advances in clinical trials. *Mol. Cancer.* 2023;22(1):159. https://doi.org/10.1186/s12943-023-01860-5
89. Moe KT, Tan KSW. Mechanistic insights on microbiota-mediated development and progression of esophageal cancer. *Cancers.* 2024;16(19):3305.
90. Zhang Z, Liu D, Liu S, Zhang S, Pan Y. The role of Porphyromonas gingivalis outer membrane vesicles in periodontal disease and related systemic diseases. *Front Cell Infect. Microbiol.* 2021;10:585917. https://doi.org/10.3389/fcimb.2020.585917
91. Zhang X, Yao L, Meng Y, Li B, Yang Y, Gao F. Migrasome: a new functional extracellular vesicle. *Cell Death Discov.* 2023;9(1):381. https://doi.org/10.1038/s41420-023-01673-x
92. Joshi RV, Gunawan C, Mann R. We are one: multispecies metabolism of a biofilm consortium and their treatment strategies. *Front Microbiol.* 2021;12:635432. https://doi.org/10.3389/fmicb.2021.635432
93. Wu Z, Ma Q, Guo Y, You F. The role of Fusobacterium nucleatum in colorectal cancer cell proliferation and migration. *Cancers.* 2022;14(21):5350. https://doi.org/10.3390/cancers14215350
94. Pisani F, Pisani V, Arcangeli F, Harding A, Singhrao SK. Treponema denticola has the potential to cause neurodegeneration in the midbrain via the periodontal route of infection—narrative review. *Int. J. Environ. Res. Public Health.* 2023;20(11):6049. https://doi.org/10.3390/ijerph20116049
95. Lasserre JF, Brecx MC, Toma S. Oral microbes, biofilms and their role in periodontal and peri-implant diseases. *Mater (Basel, Switz).* 2018;11(10):1802. https://doi.org/10.3390/ma11101802
96. Acharya S, Hegde U, Acharya AB, Nitin P. Dysbiosis linking periodontal disease and oral squamous cell carcinoma-A brief narrative review. *Heliyon.* 2024;10(11):e32259. https://doi.org/10.1016/j.heliyon.2024.e32259

97. Zhao W, Zhang Y, Chen J, Hu D. Revolutionizing oral care: reactive oxygen species (ROS)-Regulating biomaterials for combating infection and inflammation. *Redox Biol.* 2025;79:103451. https://doi.org/10.1016/j.redox.2024.103451
98. Berman H, Zhang J, Crawford YG, et al. Genetic and epigenetic changes in mammary epithelial cells identify a subpopulation of cells involved in early carcinogenesis. *Cold Spring Harb symposia Quant. Biol.* 2005;70:317–327. https://doi.org/10.1101/sqb.2005.70.051
99. Alsaeedi SM, Aggarwal S. The holistic review on occurrence, biology, diagnosis, and treatment of oral squamous cell carcinoma. *Cureus.* 2022;14(10):e30226. https://doi.org/10.7759/cureus.30226
100. Khalaf MM. Molecular identification of Porphyromonas gingivalis in clinical dentistry. *Int. J. Med. Sci Dental Health.* 2025;11(10):115–122. https://doi.org/10.55640/ijmsdh-11-10-13
101. Baddouri L, Hannig M. Probiotics as an adjunctive therapy in periodontitis treatment-reality or illusion-a clinical perspective. *NPJ Biofilms Microbiomes.* 2024;10(1):148. https://doi.org/10.1038/s41522-024-00614-5
102. Pedrosa ME, Martin V, Fernandes MH, Gomes PS. Gingipain inhibitors as an innovative therapy for periodontal and associated-systemic diseases: a systematic review. *Clin. Oral Investig.* 2025;29(9):418. https://doi.org/10.1007/s00784-025-06472-5
103. Niño-Vega GA, Ortiz-Ramírez JA, López-Romero E. Novel antibacterial approaches and therapeutic strategies. *Antibiotics (Basel, Switz).* 2025;14(4):404. https://doi.org/10.3390/antibiotics14040404
104. Naithani N, Sinha S, Misra P, Vasudevan B, Sahu R. Precision medicine: concept and tools. *Med. Journal, Armed Forces India.* 2021;77(3):249–257. https://doi.org/10.1016/j.mjafi.2021.06.021
105. Lin NC, Hsu JT, Tsai KY. Survival and clinicopathological characteristics of different histological grades of oral cavity squamous cell carcinoma: a single-center retrospective study. *PLoS one.* 2020;15(8):e0238103. https://doi.org/10.1371/journal.pone.0238103
106. Liu Y, Gao F, Cheng Y, Qi L, Yu H. Applications and advances of multi-omics technologies in gastrointestinal tumors. *Front Med.* 2025;12:1630788. https://doi.org/10.3389/fmed.2025.1630788
107. Kim SM, Praveen Z, Kim YH, et al. Functional profiling of the oral microbiome reveals microbial and oncogenic signatures in never-smoking female patients with oral squamous cell carcinoma. *J. Oral Microbiol.* 2025;17(1):2594842. https://doi.org/10.1080/20002297.2025.2594842
108. Nam Y, Kim J, Jung SH, et al. Harnessing artificial intelligence in multimodal omics data integration: paving the path for the next frontier in precision medicine. *Annu. Rev. Biomed data Sci.* 2024;7(1):225–250. https://doi.org/10.1146/annurev-biodatasci-102523-103801
109. Shen Y, Zhang P, Luo J, et al. Artificial intelligence drives advances in multi-omics analysis and precision medicine for sepsis. *Biomedicines.* 2026;14(2):261. https://doi.org/10.3390/biomedicines14020261
110. Seo H, Kim S, Beck S, Song HY. Perspectives on microbiome therapeutics in infectious diseases: a comprehensive approach beyond immunology and microbiology. *Cells.* 2024;13(23):2003. https://doi.org/10.3390/cells13232003
111. Salah AN, Doghish YA, Abbass SO, et al. Microbiota-based therapies in oral health and disorders. *Folia microbiol.* 2025;70(6):1217–1240. https://doi.org/10.1007/s12223-025-01324-x

CHAPTER SEVEN

Microbial allies in malignancy: Role of *Porphyromonas gingivalis* and *Candida* biofilms in oral squamous cell carcinoma

Saikat Mazumder[a,b,*], Debasmita Bhattacharya[c], Sudhriti Maity[b], Trisha Dutta[b], and Sayan Chatterjee[d]
[a]Department of Biotechnology, Institute of Engineering and Management, Kolkata, University of Engineering and Management, Kolkata, West Bengal, India
[b]Department of Food Technology, Guru Nanak Institute of Technology, Kolkata, West Bengal, India
[c]Department of Basic Science and Humanities, Institute of Engineering and Management, Kolkata, University of Engineering and Management, Kolkata, West Bengal, India
[d]University School of Biotechnology, Guru Gobind Singh Indraprastha University, Dwarka, New Delhi, India
*Corresponding author. e-mail address: saikat.23.1989@gmail.com; saikat.mazumder@gnit.ac.in

Contents

Progress in Molecular Biology and Translational Science, Volume 224
ISSN 1877-1173, https://doi.org/10.1016/bs.pmbts.2026.07.009

Abstract

Oral squamous cell carcinoma (OSCC) is the most common type of oral cancer, and it is becoming clear that it is a multifactorial disease condition that is affected by microbial dysbiosis. *Porphyromonas gingivalis* and *Candida spp.* are keystone periodontal pathogens that have become key disease initiators and progressors in the formation of polymicrobial biofilms in periodontal disease. The biofilms are hypervirulent, resistant, and cooperate metabolically, leading to chronic inflammation and epithelial remodelling. *P. gingivalis,* in a mechanistic way, regulates host signaling pathways such as NF- kB, PI3K/Akt, and MAPK to induce anti-apoptotic and proliferative responses, and Candida biofilms trigger carcinogenesis by producing acetaldehyde, disrupting the epithelial barrier, and activating oncogenic pathways. This interaction between bacteria and fungal elements is further synergistic and leads to immune evasion, redox dysbalance, and chronic inflammation, which form a tumor-promoting microenvironment. The implication of emerging evidence also includes the effects of microbial metabolites, quorum sensing, and biofilm architecture in determining the OSCC niche. Notably, such polymicrobial interactions offer new therapeutic avenues, such as anti- biofilm therapy, modulation of microbiomes, and targeted molecular therapy. This chapter thoroughly addresses the mechanistic contributions of *P. gingivalis* and *Candida* biofilms to OSCC pathogenesis with a specific focus on their synergistic contribution and potential translational implications in diagnosis, prevention, and treatment.

1. Introduction

Oral squamous cell carcinoma (OSCC) accounts for about 90 % of all orally occurring malignancies and is a significant health issue worldwide because of its high occurrence and mortality rates, especially in developing countries like South and Southeast Asian countries. Risk factors such as tobacco use, alcohol consumption, and betel quid chewing were long-established risk factors in OSCC.[1] Nevertheless, recent developments in cancer biology have demonstrated the important role of the oral microbiome in carcinogenesis and have led to a paradigm shift in the approach to carcinogenesis to a more complex, multifactorial one.[2] The oral cavity has a very diverse microbial ecosystem, which consists of bacteria, fungi, viruses, and archaea. These microorganisms are normally in a state of balance, but when this balance is disturbed, a situation referred to as dysbiosis, then the microorganisms will facilitate the development of diseases, including cancer.[3] The idea of a microbiome-mediated carcinogenesis is gaining more and more evidence, with microbial communities having a role in the initiation and progression of tumors via chronic inflammation, metabolic activity, and immune regulation.[4]

One of the most common features of oral microorganisms is that they are organized into biofilms, which are structured communities that are entrenched in an extracellular polymeric substance (EPS) matrix. Formation of biofilms promotes survival of microorganisms by providing resistance to host immune and antimicrobial defense as well as interspecies communication and metabolic interactions.[5,6] Notably, biofilms facilitate the interactions between polymicrobes, which are now considered important determinants of pathogenicity. This has resulted in the development of the polymicrobial synergy and dysbiosis (PSD) model, which describes the effects of cooperative interactions between microbial species that have effects that are more severe than those of single pathogens.[7] *Porphyromonas gingivalis*, *Candida spp.*, and *Candida albicans* in particular are some of the microbial species that have been implicated in OSCC, which have come under the focus of attention.[8] *P. gingivalis* is a Gram-negative, anaerobic bacterium and the key pathogen of periodontitis that can regulate host immune responses, disrupt epithelial barriers, and activate oncogenic signaling pathways, including NF-kB and PI3K/Akt. The activities enhance chronic inflammation, prevent apoptosis, and facilitate cellular proliferation, thus facilitating tumorigenesis.[9]

Simultaneously, *Candida albicans* is an opportunistic fungal pathogen that may change its status to be a virulent organism in a dysbiotic environment. It develops strong biofilms and considers virulence factors, including hydrolytic enzymes and carcinogenic metabolites, such as acetaldehyde, which can cause DNA damage and genomic instability.[10,11] Moreover, *Candida* biofilms also play a role in disrupting the epithelial barrier, oxidative stress, and inflammatory reactions, which further promote malignant transformation. Recent research focuses on the fact that the interaction of *P. gingivalis* and *Candida* in polymicrobial biofilms has a greater impact on oral carcinogenesis than either of the organisms.[12] These interkingdom interactions include co-adhesion, metabolic collaboration, and coordinated regulation of virulence factors, which lead to increased pathogenicity and persistence in the host. This synergy encourages prolonged inflammatory signaling, immune evasion, and the development of a tumor-promoting microenvironment.[13]

Microbial biofilms are also essential in the formation of the tumor microenvironment (TME) by affecting the production of cytokines, angiogenesis, and immune cells. Perennial microbial stimulation causes augmented reactive oxygen species (ROS) production, which triggers damage to DNA and genomic instability, important occurrences in cancer development.[14] Moreover, microbial metabolites and signaling molecules can regulate the expression of host genes and epigenetics, which confirms carcinogenic pathways. Based on these findings, there is increased interest in addressing microbial communities to prevent and treat cancer.[15] Anti-biofilm agents, microbiome modulation, and natural bioactive compounds are some of the strategies that are being investigated to discontinue the interaction between pathogens and reestablish microbial balance.[16] In this regard, the current chapter explores the functions of polymicrobial biofilms composed of Porphyromonas gingivalis and Candida spp. in OSCC, their synergistic relationships, the molecular mechanisms, and possible therapeutic consequences.

2. Oral microbiome and OSCC

2.1 Microbial role in the pathogenesis of OSCC

The oral microbiome has a complex role in OSCC, which can be inflammatory, metabolic, and genetic. Chronic inflammation is a major mediator of carcinogenesis, and dysbiotic microbial communities can mediate chronic activation of inflammatory pathways, including NF-kB, and result in pro-inflammatory cytokine production, including IL-6, TNF- alpha, and IL-1 beta. This chronic inflammatory condition stimulates cell growth, suppresses apoptosis, and stimulates angiogenesis.[17,18] Besides inflammation, oral microorganisms also generate a host of carcinogenic metabolites. Indicatively, Candida spp. can break down ethanol into acetaldehyde, a known carcinogen that causes DNA damage and mutagenesis. In the same fashion, bacterial species produce reactive oxygen species (ROS) and nitrogen species, which lead to oxidative stress and genomic instability.[19] Microorganisms also make changes in the host cell signaling pathways, which are involved in the development of cancer. *P. gingivalis* has been found to activate PI3K/Akt and MAPK pathways that lead to cell survival and proliferation, and *Fusobacterium nucleatum* can activate Wnt/catenin signaling, promoting tumor growth and invasion. These communicational activities emphasize that the microbiome can directly control the oncogenic processes.[20]

2.2 Polymicrobial synergy and PSD model

Recent growth has resulted in the evolution of the polymicrobial synergy and dysbiosis (PSD) model, which offers a framework with which to explain microbe-mediated diseases, such as OSCC.[21] This model claims that some key-stone pathogens like *P. gingivalis* are able to coordinate the alterations of the microbial community, which results in dysbiosis and increases the pathogenicity. Here, the interaction between bacteria and fungi is of special interest. The *Candida spp.* are capable of interacting with bacterial pathogens to create heterotypic biofilms, which have greater structural complexity and virulence. Such interkingdom interactions are beneficial to the survival of microbes, evasion of immunity, and inflammatory response, which are useful in the development and progression of tumors.[22]

2.3 Tumor microenvironment modulation

The oral microbiome has a key role in determining the tumor microenvironment (TME) of OSCC. The TME is affected by microbial communities that regulate the cytokine profiles, immune cells infiltration, and angiogenesis.[23,24] Long-term exposure to microbial antigens and metabolites causes the buildup of ROS and other reactive species, which causes damage to DNA and epigenetic changes. In addition, immune modulation caused by microbes, such as activation of regulatory T cells and immune checkpoint pathways, also plays a role in tumor immune evasion. These alterations result in a microenvironment that promotes tumor development, invasion, and metastasis.[25]

3. Porphyromonas gingivalis in OSCC

Porphyromonas gingivalis has proven to be a key pathogen bridging the connection between chronic periodontitis and oral squamous cell carcinoma (OSCC), as more than an opportunistic colonizer but a proactive regulator of the tumorigenic process. Its exclusive ability to reorganize host cellular pathways, to overcome immune surveillance, and to maintain chronic inflammation makes it an important key to microbe-associated carcinogenesis.[26,27]

3.1 The presence of the virulence arsenal and keystone pathogenicity in the isolate

Contrary to classical pathogens, *P. gingivalis* has disproportionate impacts on the microbial community and host responses, although it is relatively low-abundance, which is in line with the polymicrobial synergy and dysbiosis (PSD) model. It has a complex arsenal that mediates its virulence, including gingipains (RgpA, RgpB, Kgp), atypical lipopolysaccharide (LPS), fimbriae, and outer membrane vesicles (OMVs). The pathogenicity it exhibits is centered on the gingipains, which are multi-purpose proteins that degrade and inactivate the extracellular matrix proteins, receptor signaling, and immune mediators.[28,29] *P. gingivalis* LPS adapts to the host immune response by heterogeneity of its structure that facilitates the selective activation or inhibition of Toll-like receptor pathways. OMVs also increase their impact by the direct delivery of virulence factors into host cells, increasing the intensity of pathogenic signaling, even at the site of infection.[30–32]

3.2 The reprogramming of oncogenic signaling networks

One characteristic of *P. gingivalis* in OSCC is that it is able to reprogram host signaling pathways to a pro-tumorigenic condition. Constant infection results in constitutive stimulation of NF-KB, which results in chronic synthesis of pro-inflammatory cytokines (IL-6, IL-8, TNF- alpha) and the creation of an inflammatory environment that is supportive of tumors.[33] At the same time, *P. gingivalis* invigorates the PI3K/Akt pathway, which is a key controller of cell survival, metabolism, and growth. This pathway is inhibited by activating the anti-apoptotic protein Bcl-2 and inhibiting the pro-apoptotic factors (e.g., caspase-9) by activation of this pathway. It is also the bacterium that activates the MAPK/ERK cascade, which promotes the proliferation of the cell and its adaptation to stress.[34] Notably, prolonged establishment of activation of the STAT3 signaling enhances the incorporation of inflammatory and oncogenic signals, facilitating angiogenesis, immune suppression, and tumor development. Such concerted signaling responses are useful in modulating the epithelial cells to a malignant phenotype.[35,36]

3.3 Immune subversion and low-grade persistent inflammation

P. gingivalis is the only pathogen that is adapted to avoid being cleared by the immune system and maintains a chronic non-resolving inflammation, which is typical of cancer. It interferes with Toll-like receptor signaling, especially the TLR2 TLR4 crosstalk, which results in the unregulated innate immune responses.[34,37] Simultaneously, gingipains destroy important elements of the complement (C3, C5), which influences opsonization and phagocytosis. This immunological evasion enables *P. gingivalis* to remain in the tissues of the host, in low-grade inflammation, which constantly brings the carcinogenic processes to life. Persistent discharge of cytokines is not only known to stimulate proliferation of the epithelial cells but also angiogenesis and remodelling of the stroma, stimulating the growth of the tumour.[38,39]

3.4 Apoptotic resistance and uncontrolled proliferation

Inhibition of apoptosis and induction of uncontrolled growth of cells is one of the most vital oncogenic impacts of *P. gingivalis*. The bacteria disrupt intrinsic apoptotic pathways by regulating mitochondrial signaling and inhibiting caspase activation. This allows the survival of cells with genetic damage.[27] Simultaneously, *P. gingivalis* activates major cell cycle control proteins, such as cyclin D1 and c-Myc, which cause an advancement across the G1/S transition and hyperproliferation. These changes interfere with homeostasis of the normal cells and play a role in tumor formation and development.[38]

3.5 Induction of epithelial–mesenchymal transition (EMT)

There is growing evidence that *P. gingivalis* does induce epithelial–mesenchymal transition (EMT), which is essential in tumor invasion and metastasis. Infection triggers the expression of EMT-related transcription factors (Snail, Slug, and Twist), resulting in a down-regulation of E-cadherin and up-regulation of mesenchymal markers (vimentin).[27] These alterations in the molecules lead to increased motility, invasiveness, and resistance to apoptosis of cells, thus promoting the dissemination of metastasis. The fact that *P. gingivalis* can cause the

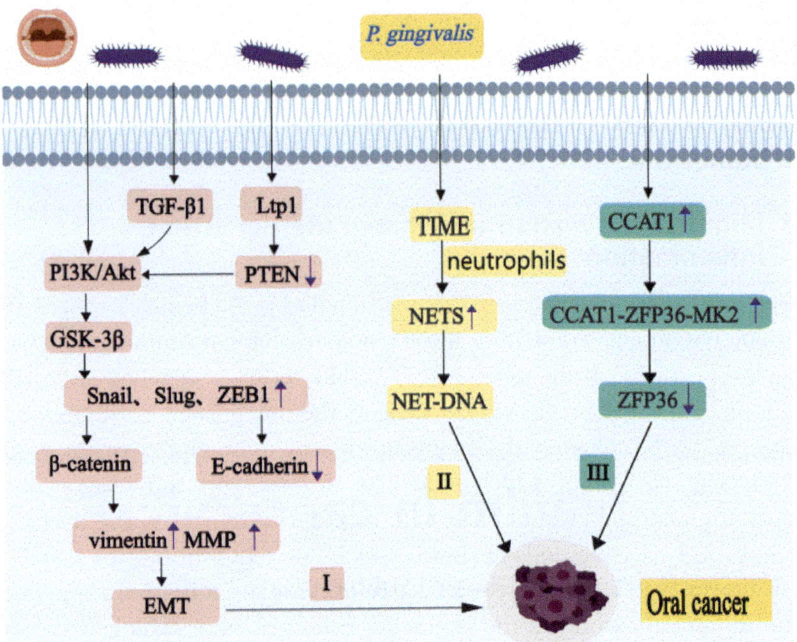

Fig. 1 *P. gingivalis* facilitates the formation and progression of oral cancer by controlling EMT, ZFP36, and NETs. The activation of the PI3K/Akt pathway is enhanced by *P. gingivalis*. It stimulates phosphorylation of GSK-3 2, which causes an increase in Snail, Slug, and ZEB1. *P. gingivalis* can also mediate TGF-B1 and LMWTP to control protein kinase Akt. These end up causing EMT to be witnessed. II, Neutrophils use NETs in response to stimulation by *P. gingivalis*, and they can induce cancer cell migration via the chemotactic action of NET-DNA. III, *P. gingivali* increases the expression of CCAT1. It is capable of developing a complex with MK2 and ZFP36. This intricate mechanism eventually compensates for the repressive actions of ZFP36 on the main downstream oncogenes. Abbreviations: PI3K, phosphatidylinositol 3-kinase; GSK-32, glycogen synthase kinase-32; TGF-β1, transforming growth factor-β1; LMWTP, low molecular weight tyrosine phosphatase; EMT, epithelial-mesenchymal transition; NETs, neutrophil.[41]

EMT indicates its contribution to the initiation of tumors and to the progression of cancer.[40] *P. gingivalis* induces the production of TGF-β1 and the activity of its low molecular weight tyrosine phosphatase (Ltp1) that, in turn, inhibits the activity of the tumor suppressor phosphatase and tensin homolog (PTEN). When PTEN activity is lost, the PI3K/Akt pathway is then turned on. Activated Akt phosphorylates glycogen synthase kinase-3β (GSK-3β), which stabilizes and increases the nuclear localization of β-catenin and activation of the EMT transcription factors Snail, Slug, and ZEB1. These transcription factors are essential for silencing E-cadherin expression and upregulating other mesenchymal markers like vimentin and matrix metalloproteinases (MMPs), which facilitate cellular motility, cell invasion, and dissemination to distant sites (Fig. 1).

3.6 Tumor microenvironment remodelling

In addition to its direct effects on cells, P. gingivalis is involved in the tumor microenvironment (TME) remodeling. It facilitates the release of pro-angiogenic factors, including vascular endothelial growth factor (VEGF), which facilitates neovascularization. It also affects the dynamics of immune cells by favouring the build-up of immunosuppressive cell groups, such as regulatory T cells and myeloid-derived suppressor cells.[27] This re-engineered microenvironment is conducive to tumor survival, evasion of the immune system, and resistance to treatment. It is important to note that *P. gingivalis* also promotes biofilm formation and interacts with other microorganisms, such as *Candida spp.*, and adds to its capacity to increase its oncogenic potential in polymicrobial communities.[39]

4. Candida biofilms in OSCC

Candida spp. and especially *Candida albicans* are now increasingly being recognized as active rather than passive contributors to oral carcinogenesis. Their capacity to develop strong biofilms, generate carcinogenic metabolites, and regulate the immune and signaling pathways of the hosts makes them central to the pathogenesis and progression of oral squamous cell carcinoma (OSCC).[42,43]

4.1 Biofilm formation and structural complexity

One characteristic of *C. albicans* is the ability to develop highly structured multi-layered biofilms of yeast cells, pseudohyphae, and hyphae surrounded by an extracellular matrix. It is a polysaccharide and protein-filled lipid-rich and extracellular DNA-rich matrix that gives the bacteria mechanical stability, defense against antifungal molecules, and the host immune system.[10] The formation of biofilm is a stepwise process, with adhesion, proliferation, maturation, and dispersion being the stages of biofilm development, which are governed by sophisticated genetic and environmental signals. Formation of hyphae is a major component of biofilm structure and serves as a basis of tissue invasion and scaffolding of microbial interactions. Notably, the entire *Candida spp.* biofilms have a much greater resistance to antifungal therapies than planktonic cells, which can lead to long colonization and chronic infection.[44,45]

4.2 Virulence factors and damage to epithelial damage

C. albicans has a plethora of virulence factors directly linked to epithelial damage and carcinogenesis. The most notable of these are adhesins (Als family proteins), secreted hydrolytic enzymes (aspartyl proteases and phospholipases), and the cytolytic peptide toxin candidalysin.[11] The peptide toxin candidalysin, which is produced as a result of the ECE1 gene, interferes with epithelial cell membranes, causes calcium influx, and activates pro-inflammatory signaling pathways. This results in epithelial cell injury, permeability, and sensitivity to carcinogenic insults. Moreover, enzymes of hydrolysis break down the components of the host tissues, which increases invasion and spread.[46,47]

4.3 Metabolites and genotoxicity in carcinogenic progression

Production of carcinogenic metabolites, especially acetaldehyde, is one of the most important contributions of Candida to OSCC. In ethanol metabolism, *C. albicans* breaks down ethanol to acetaldehyde through the alcohol dehydrogenase enzymes.[48] Acetaldehyde is a proven carcinogen that creates DNA adducts, mutations, and disrupts DNA repair mechanisms. Besides acetaldehyde, Candida biofilms also promote the formation of reactive oxygen species (ROS) and reactive nitrogen species, resulting in oxidative stress and genomic instability. The continuous exposure to these genotoxic agents facilitates the buildup of mutations and aids in malignant transformation.[49]

4.4 Host signaling pathway modulation

C. albicans biofilms are active in regulating inflammatory and cancer progression host signaling. Activation of MAPK and NF-kB pathways by infection results in the production of pro-inflammatory cytokines (IL-6 and TNF-alpha).[50,51] Persistent stimulation of these pathways forms a pro-tumorigenic microenvironment with persistent inflammation. Moreover, *C. albicans* was demonstrated to affect epithelial proliferation and survival pathways, which added to hyperplasia and dysplasia. EGFR and down-stream signaling cascades activation stimulates cell growth and anti-apoptotic effect, which further enhances tumor development.[52]

4.5 Immune modulation and chronic inflammation

The host immune system and *Candida spp.* biofilms are very complicated in the interaction between the host and the biofilm. Chronic colonization causes immune dysregulation and tolerance, whereby the acute infection triggers strong immune responses. Activation of Th17-mediated immunity by *Candida spp.* plays a role in antifungal defense but can also be involved in chronic inflammation in case of its dysregulation.[53] The long-lasting infection is linked with enhanced synthesis of inflammatory cytokines and immune cells recruitment, leading to tissue damage and remodeling. Meanwhile, Candida may avoid immune clearance by forming biofilms and by regulating host immune responses, enabling it to persist in colonization.[52]

4.6 Development of epithelial dysplasia and dysplasia progression

Clinical and experimental research has proven that there is a close relationship between chronic *Candida spp.* infection and oral epithelial dysplasia, which is a precursor of OSCC. Candida biofilms stimulate epithelial hyperplasia, destabilize cell junctions, and cause morphological changes that are in line with the initial neoplastic transformation. Also, Candida-related inflammation and metabolic activity establish a favorable microenvironment to promote tumor progression. A combination of epithelial damage, genotoxic stress, and immune modulation leads to the progression of dysplasia to invasive carcinoma.

4.7 Role of *Candida* in polymicrobial interactions

Significantly, *Candida spp.* is not a single-actor, but it is a member of polymicrobial biofilms, which interact with bacterial pathogens like *Porphyromonas gingivalis*. Its hyphae form a physical basis to allow bacterial adhesion, and metabolic by-products and signaling molecules can be used to obtain interkingdom communication. These interactions make biofilms more stable, resistant to antimicrobial agents, and make them more virulent. The promotion of the interaction between Candida and bacterial species is an important determinant of the pathogenesis of OSCC.[54,55]

5. Polymicrobial synergy: *P. gingivalis–Candida* biofilms

Etiology of oral squamous cell carcinoma (OSCC) has recently been redefined in the context of interkingdom polymicrobial synergy, whereby consortia of bacteria and fungi coordinate a pro-tumor microenvironment by interacting in a coordinated manner at the structural, metabolic, and immunological levels.[56] The interaction of Porphyromonas gingivalis and Candida albicans is one of these, and the interaction exhibits a paradigm of pathogenic synergy, including increased resistance to biofilms, greater virulence, and synergistic expression of oncogenic pathways. This synergistic axis goes beyond additive effects and is a systems-level mediator of carcinogenesis.[57]

5.1 The architecture of interkingdom biofilms: Structural and molecular determinants

Heterotypic biofilms between *P. gingivalis* and *C. albicans* are regulated by certain molecular adhesion-receptor interactions and dynamic architectural organization. *C. albicans* hyphae offer a three-dimensional scaffold, which facilitates colonization by the bacteria, whereas *P. gingivalis* uses fimbrial adhesins (FimA, Mfa1) to attach to fungal surface proteins, Als3, and Hwp1.[54] This communication is not only physical but has a signal-dependent biofilm maturation whereby both organisms regulate gene expression according to co-culture conditions. Transcriptomic studies have shown that in co- biofilm formation, genes involved in adhesion, stress tolerance, and virulence are up-regulated in both species. These polymicrobial biofilms have a compositionally distinct extracellular polymeric substance (EPS) matrix that is rich in β-glucans, extracellular DNA, and proteolytic enzymes that together contribute to mechanical stability and diffusion resistance.[58] Also, the spatial compartmentalization of the biofilm forms micro-niches that have oxygen, pH, and nutrient gradients that allow metabolic specialization and survival in adverse environments. These attributes provide a selective advantage, enabling the consortium to be resistant to host defenses and therapeutic interventions.[59,60]

5.2 Metabolic coupling and chemical communication

One of the characteristics of *P. gingivalis*, *Candida* synergy is the metabolic coupling, which involves the meeting of the metabolic requirements of both organisms. C. albicans also generates metabolites like ethanol, lactate, and pyruvate, which are substrates of *P. gingivalis*, a proteolytic and *saccharolytic bacterium. P. gingivalis*, in turn, produces peptides and amino acids by means of proteolysis with the help of gingipain, which contributes to the growth of fungi and the expansion of biofilms.[54,61] This interdependent metabolism is tightly controlled by quorum-sensing networks and chemical-signaling networks. Farnesol is a quorum-sensing molecule secreted by *C. albicans,* and it regulates the expression of bacterial genes and can affect biofilm structure. On the other hand, fungal morphogenesis, especially the transition of the yeast-to-hypha transition, which is a determinant of virulence, is influenced by bacterial autoinducer molecules and short-chain fatty acids (SCFA).[62,63] Moreover, the metabolic products like acetaldehyde and butyrate build up in the biofilm matrix that forms a chemically reactive microenvironment that directly influences the host epithelial cells. This synergy in metabolism increases the persistence of the microbes and the carcinogenicity.[64]

This coupled metabolism is well-regulated, with quorum-sensing and chemical-signaling networks playing a key role. In addition to the capacity to inhibit fungal filamentation at high concentrations, the secreted quorum-sensing molecule farnesol is also a modulator of the expression of the bacterial genes in polymicrobial biofilms. In oral bacteria, farnesol was found to have a regulatory effect on the expression of genes responsible for quorum sensing (luxS, AI-2 signaling pathways), biofilm formation and adhesion (gtfB, gtfC, and ftf), oxidative stress response (sodA and nox), and virulence-associated genes related to the production of extracellular polysaccharide and carbohydrate metabolism. These transcription changes affect the architecture of the biofilm, persistence, and interaction between the different species.[7,65]

On the other hand, signaling molecules of bacteria can control *C. albicans* morphogenesis. In *C. albicans*, the switch from yeast to hyphae is the main process controlled by the intracellular Ras1–cAMP–protein kinase A (PKA) signaling pathway, and several bacterial autoinducers, such as autoinducer-2 (AI-2) and N-acyl homoserine lactones (AHLs), regulate this pathway. Furthermore, bacterial metabolites, short-chain fatty acids (SCFAs) such as butyrate, acetate, and propionate, regulate the levels of intracellular cAMP and regulate the activity of hyphal transcriptional regulators (e.g., Efg1,

Cph1, and Ume6) that control filamentation and biofilm maturation. Interkingdom quorum sensing directly contributes to heterotypic biofilm development and to the cancerogenic potential, as hyphal formation is the structural scaffold of heterotypic biofilm and tissue invasion.[7,10]

5.3 Virulence and invasion of host tissues synergy

Cooperative expression and functional complementation of pathogenic factors greatly enhance the virulence of polymicrobial biofilms. *P. gingivalis* gingipains break down extracellular matrix proteins, interfere with tight junctions, and inactivate immune effectors, which promotes microbial invasion. At the same time, *C. albicans* releases candidalysin that causes cellular damage to epithelial cells, calcium influx, and danger-associated signaling pathway activation. The interplay of these elements results in the loss of epithelial barrier integrity, permeability, and further penetration of microbial components in subepithelial tissues.[66] This encourages long-term exposure of host cells to microbial antigens and toxins, speeding up the epithelial dysplasia and neoplastic change. Notably, co-infection models have shown polymicrobial biofilms to have increased cytotoxicity, production of inflammatory cytokines, and epithelial proliferation relative to mono-species infections, highlighting the pathogenic power of interkingdom synergy.[67]

5.4 Oncogenic cellular reprogramming and convergent oncogenic signaling

One of the key pathways to *P. gingivalis* and *Candida* synergy is convergent activation of oncogenic signaling networks reprogramming host epithelial cells. Both organisms activate NF-kB, which causes the continued expression of pro-inflammatory cytokines and anti-apoptotic genes. At the same time, *P. gingivalis* stimulates the PI3K/Akt/mTOR pathway to facilitate cell survival, cell metabolic restructuring, and apoptotic resistance.[37] *C. albicans*, however, is highly activated by MAPK and EGFR signaling that promotes epithelial proliferation and epithelial stress response. The intersection of these pathways causes amplification of signals and network stability, which are characteristics of cancer-related signaling. Constant stimulation of STAT3 continues to combine inflammatory and proliferative cues, which increases angiogenesis and immune suppression. There is also some new evidence that microbial factors are capable of modulating epigenetic regulators, such as histone acetyltransferases and DNA methyltransferases, to cause stable changes in gene expression patterns related to malignancy.[68] Fig. 1 shows the molecular mechanisms of how *Porphyromonas gingivalis* is involved in oral squamous cell carcinoma. The three primary pathways are (1) *P. gingivalis* triggers EMT via TGF-β1/Ltp1/PTEN/PI3K/Akt/GSK-3β signaling axis, thereby inducing the expression of Snail, Slug, and ZEB1, downregulating E-cadherin expression, and upregulating expression of Vimentin and MMPs. (2) *P. gingivalis* induces the formation of neutrophil extracellular trap (NET) in the tumor immune environment, and released NET-DNA induces cancer cell migration. (3) *P. gingivalis* upregulates CCAT1, which binds to MK2 and ZFP36, preventing ZFP36 from catalyzing the degradation of oncogenic transcripts and thus favoring the initiation of oral carcinogenesis. Moreover, activation of neutrophils and subsequent NET formation facilitates tumor cell migration and maintains an inflammatory tumor microenvironment. The mechanisms taken together provide examples of how microbial signalling pathways make direct contributions to the progression of oral cancer.

5.5 Immune reprogramming and tumor-promoting inflammation

The complex interplay between immune activation and suppression mediated by polymicrobial biofilms leads to immune reprogramming, which is more advantageous to tumor progression. P. gingivalis inhibits the complement pathways and modulates Toll-like receptor signaling, whereas *C. albicans* triggers Th17-mediated inflammation.[69] This bipolar modulation establishes a chronic non-resolving inflammation condition, where there is increased concentration of IL-6, IL-17, and TNF-alpha. These cytokines stimulate epithelial cell growth, angiogenesis, and immunosuppressive cell recruitment of regulatory T cells and myeloid-derived suppressor cells. In addition, both organisms have been found to increase immune checkpoint molecules (e.g., PD-L1) on both epithelial and immune cells, promoting immune evasion and allowing tumor cells to evade cytotoxic T-cell immunity.[70]

5.6 Oxidative stress, genotoxicity, and epigenetic instability

The polymicrobial consortium contributes greatly to the increase of oxidative and genotoxic load in the oral epithelium. The two organisms cause the generation of reactive oxygen and nitrogen species, directly or indirectly, by using the activation of host immune cells. Damages induced by ROS include base alterations, chromosome instability, and DNA strand breakage.[71] At the same time, acetaldehyde generated by *C. albicans* causes the formation of DNA adducts and prevents the repair processes in the DNA, which further contributes to mutagenesis. In addition to direct genotoxicity, such stressors affect epigenetic landscapes, modifying patterns of DNA methylation and histone modification. This may result in the silencing of tumor suppressor genes and the activation of oncogenes, which strengthens malignant transformation.[72]

5.7 Systems-level effect of tumor microenvironment

P. gingivalis and *Candida* biofilms are ecosystem engineers of the tumor microenvironment (TME) at a systems level. They regulate angiogenesis by activating vascular endothelial growth factor (VEGF), they can change the behavior of stromal cells, and enhance extracellular matrix remodeling.[27,73] The biofilm also provides a feed-forward loop of continuous release of microbial products, perpetuating tissue damage, inflammation, and cellular transformation. Such a dynamic interplay between microbes and host tissues highlights the contribution of polymicrobial biofilms to cancer progression as active agents and not passive observers.[74]

6. Molecular mechanisms of biofilm-mediated carcinogenesis and tumor microenvironment modulation

6.1 Chronic inflammation and signaling via cytokines

Chronic colonization with biofilm results in chronic inflammatory pathway activation, which is characteristic of cancer. Pattern recognition receptors (PRRs) (Toll-like receptors (TLRs) and C-type lectin receptors) on epithelial and immune cells are activated by microbial components, including lipopolysaccharides (LPS), fungal cell wall 8 -glucans, and secreted toxins.[14] This leads to activation of downstream signaling cascades, the most notable of which is NF-kB, which in turn causes the continued production of pro-inflammatory cytokines, including IL-6, IL-1β, TNF-alpha, and IL-17. These cytokines lead to the growth of epithelia, suppress apoptosis, and induce angiogenesis. Immune cells that recruit and release more mediators are also recruited in chronic inflammation, as part of a feed-forward inflammatory loop which promotes tumor initiation and progression.[75]

6.2 Oxidative stress and DNA damage

An adverse effect of chronic inflammation is the production of reactive oxygen species (ROS) and reactive nitrogen species (RNS). These molecules are synthesized as a result of microbial metabolism as well as the immune response of the host. High ROS causes damage to the DNA, such as strand breaks, base alterations, and cross-linking, which causes genomic instability.[76] ROS also triggers redox-sensitive signaling that includes NF-kB and MAPK, and increases further on the inflammatory and proliferative signals. The net result is a greater mutational load, enabling malignant change. Moreover, fungal metabolites like acetaldehyde enhance genotoxic stress by the formation of DNA adducts and interruptions of DNA repair pathways, which also increase the risk of carcinogenesis.[77,78]

6.3 Dysregulation of cell proliferation and apoptosis

Pathogens that are associated with biofilm can adjust the most important signaling pathways that control cell proliferation and survival. Stimulation of the PI3K/Akt/mTOR signaling facilitates cellular phenotypic development, metabolism, and apoptotic resistance, and inhibition of tumor suppressor signaling (e.g., p53 signaling) allows damaged cells to survive.[79] At the same time, the MAPK/ERK pathway is turned on, which triggers the cell cycle development and growth. An increase in the oncogenic transcription factors like c-Myc and cyclin D1 further enhances the G1/S transition, resulting in unregulated division of cells. The inhibition of apoptosis occurs in several ways, which include suppression of caspases and activation of anti-apoptotic proteins like Bcl-2. This is a major step in the development of tumors through this imbalance between cell death and survival.[80,81]

6.4 Epigenetic reprogramming

The new evidence points to the fact that microorganisms attached to biofilms are involved in the regulation of the epigenome, i.e., hereditary modification of gene expression without any change in DNA sequence. Short-chain fatty acids (SCFA) and acetaldehyde are microbial metabolites that affect DNA methylation, histone modification, and chromatin remodeling.[82,83] These modifications may cause silencing of tumor suppressor genes and activation of the oncogenes, and strengthen the pathways that cause cancer. As an illustration, changes in histone acetylation levels can amplify the expression of inflammatory and proliferative genes, but hypermethylation of DNA can silence gene expression of cell cycle regulation and cell apoptosis. Notably, the epigenetic alterations can be reversed, and therefore, they can be considered as therapeutic targets.[84,85]

Recently, there have been reports of stable epigenetic changes that are induced by biofilm-associated microorganisms, which play a role in oral carcinogenesis. It has been suggested that chronic infection with Porphyromonas gingivalis causes an increase in the activity of DNA methyltransferases (DNMT1 and DNMT3B), resulting in the hypermethylation of several tumor suppressor genes, such as CDKN2A (p16INK4A), CDH1 (E-cadherin), and SOCS1 and PTEN silencing. These genes are silenced, which leads to cell-cycle progression, EMT, continuing STAT3 activation, and increased PI3K/Akt signaling.[86,87]

The production of microbial metabolites in polymicrobial biofilms also affects histone modifications. Short-chain fatty acids (such as butyrate) act as inhibitors of histone deacetylase (HDAC), which leads to increased acetylation of histone H3 and H4 and thus changes in the expression of genes involved in inflammation and proliferation, like IL6, CXCL8 (IL8), CCND1 (cyclin D1), and MYC. While transient HDAC inhibition has anti-tumor effects in certain situations, chronic exposure in dysbiotic biofilms can lead to inflammatory signaling and metabolic reprogramming that could be conducive to malignant transformation.[86,88] Furthermore, acetaldehyde produced by *Candida albicans* leads to DNA damage and epigenetic instability, such as abnormal DNA methylation and acetylation of histones. Epigenetic silencing of DNA repair genes (including MGMT) and the pathway of tumour suppressor genes (TP53 and CDKN2A) has been associated with acetaldehyde effects, which promote genomic instability and neoplastic progression. Together, these are epigenetic changes acting at a transcriptional level that induce a stable process of growth, immune evasion, epithelial plasticity, and progression of OSCC.[88]

6.5 Immune modulation and tumor immune evasion

Microorganisms that are part of biofilms are of great importance in immune regulation, which manipulates the tumor microenvironment to be conducive to cancer. Recruitment of immunosuppressive cell populations, such as regulatory T cells (Tregs) and myeloid-derived suppressor cells (MDSCs), is recruited as a result of chronic microbial stimulation.[14] Moreover, the immune checkpoints like programmed death-ligand 1 (PD-L1) expressed by epithelial and immune cells are upregulated, which is inhibitory of the cytotoxic T-cell activity, and tumor cells can evade immune surveillance. Anti-tumor immune responses are also suppressed by cytokines like IL-10 and TGF-B, which strengthen this immune suppression. The outcome is a microenvironment that favorably promotes tumor growth and inhibits effective immune-mediated clearance.[89,90]

6.6 Angiogenesis and remodeling of tumor microenvironment

Angiogenesis is another important process in which tumor growth and metastasis occur in the presence of biofilm-mediated signaling. New blood vessels are formed as a result of the expression of vascular endothelial growth factor (VEGF), which is caused by pro-inflammatory cytokines and factors produced by microbes.[74] At the same time, biofilms affect stromal cells and extracellular elements, which facilitates tissue remodeling and tumor invasion. Microbial and inflammatory cues also stimulate the activation of matrix metalloproteinases (MMPs) that break down the barriers in the extracellular matrix, allowing cancer cells to move.[74]

6.7 Combining molecular pathways

The above molecular processes are very interactive and are not single pathways but a network. Indicatively, ROS are not only damaging DNA but also triggering the NF-kB signaling, with cytokine products of NF-kB able to further promote ROS generation.[91] On the same note, expression of inflammatory, proliferation, and immune response genes can be controlled by the epigenetic changes. Such a systems-level integration leads to a self-perpetuating cycle of carcinogenesis, in which biofilm-related microorganisms feed off tumor-enhancing activities. The interaction between the factors of microbes and host responses ultimately determines the course of OSCC development.[92]

6.8 Tumor microenvironment modulation by polymicrobial biofilms

6.8.1 Cytokine networks and gradients of inflammation

The biofilm-related microorganisms create a pro-tumorigenic cytokine environment due to chronic stimulation of the innate and adaptive immunity. The repeated stimulation of pattern recognition receptors results in increased production of cytokines of IL-6, IL-1b, TNF-alpha, and IL-17 that cumulatively stimulate epithelial growth, angiogenesis, and survival signaling (Bai, Bingbing 2023). These cytokines result in the production of local inflammatory gradients in the TME that can shape the recruitment and polarization of immune cells. It is interesting to note that IL-6 and

STAT3 signaling are central to the connection between inflammation and tumor growth by stimulating growth, suppressing apoptosis, and stimulating angiogenic factors.[93,94]

6.8.2 *Cell reprogramming and suppression by immune cells*

The polymicrobial biofilms promote immune reprogramming, which causes a change in immunosuppression instead of effective anti-tumor immunity. Repeated exposure to microbes stimulates the growth of regulatory T cells (Tregs) and myeloid-derived suppressor cells (MDSCs), which prevent the action of cytotoxic T lymphocytes (CTL).[7] Also, there is a polarization of macrophages in the TME towards an M2-like phenotype, which is typified by tissue remodeling and immunosuppressive capabilities. It is a microbial-mediated polarization that is suppressed by microbial factors and cytokines, including IL-10 and TGF-β, that further suppress anti-tumor responses. The additional immune evasion of upregulation of immune checkpoint molecules, especially PD-L1, on epithelial and immune cells, enables the tumor cells to avoid the immune surveillance and enables their survival.[95,96]

6.8.3 *Angiogenesis/vascular remodeling*

The inflammatory responses and signaling induced by biofilms greatly enhance angiogenesis, an essential need for tumor growth. It is stimulated by microbes and results in the up-regulation of vascular endothelial growth factor (VEGF) and other pro-angiogenic mediators.[97] The newly generated vasculature is usually unstructured and porous, which adds to the hypoxic areas in a tumor. Hypoxia also stimulates the activity of hypoxia-inducible factor-1alpha (HIF-1alpha) that boosts the glycolytic metabolism, angiogenesis, and survival signaling in tumor cells. This forms a self-perpetuating cycle of hypoxia and microbial survival with the anaerobic pathogen P. gingivalis preferring low-oxygen conditions.[98]

6.8.4 *Stromal activation and remodeling of the extracellular matrix*

Not only cancer cells make up the TME, but also stromal components, such as fibroblasts, extracellular matrix (ECM), and immune cells. Polymicrobial biofilms mediate these elements by triggering the activation of cancer-associated fibroblasts (CAFs), which produce the growth factors, cytokines, and matrix-degrading enzymes.[99] Proproduced ECM components are broken down by matrix metalloproteinases (MMPs) that are upregulated in response to microbial and inflammatory cues and allow tumor invasion and metastasis. At the same time, ECM remodeling changes the tissue stiffness and structure, as well as enhancing malignant transformation.[100]

6.8.5 Reprogramming of the TME metabolism

Biofilms also hamper metabolic reprogramming of the TME through the alteration of nutrient supply and production of bioactive metabolites. The short-chain fatty acids (SCFAs), polyamines, and the acetaldehyde produced by fungi affect the metabolism and gene expression of cells. These metabolites have the potential to increase glycolysis (Warburg effect), foster epigenetic changes, and promote anabolic processes needed to support rapid tumor growth. Also, microbes and the host cells may engage in metabolic competition, which may establish nutrient-deprived environments that favor more aggressive tumor phenotypes.[101]

7. Diagnostic and therapeutic implications of polymicrobial biofilm in OSCC

7.1 Bio-markers of microbes in the early diagnosis

Non-invasive diagnostic sampling has a distinct advantage in the oral cavity, hence it is a good location to discover biomarkers. Recent developments in high-throughput sequencing methods have made it possible to identify microbial signatures correlating with OSCC, such as increased levels of *P. gingivalis*, Candida spp., and other taxa with dysbiosis.[102] The salivary and mucosal biofilm studies have indicated that the microorganisms may act as early warning factors of malignant transformation, and in most cases, this can be detected before the onset of clinical signs. Specifically, the existence of polymicrobial biofilms that have enhanced virulence and metabolic activity is associated with the progression of diseases.[14] Functional biomarkers (e.g., microbial metabolites (e.g., acetaldehyde, short-chain fatty acids)) and virulence factors (gingipains, candidalysin), in addition to microbial abundance, offer useful information on the disease activity. A combination of such biomarkers with host-induced biomarkers, such as inflammatory cytokines and genetic changes, makes the diagnosis more accurate.[103]

7.2 Microbial signatures prognostics

In addition to early detection, polymicrobial biofilms have great potential to be used as prognostic factors in OSCC. Research has revealed that an advanced tumor stage, high invasiveness, and poor clinical outcomes are linked to increased colonization of *P. gingivalis* and *Candida spp.*[27] Microbial profiling may also give some information about the tumor behavior, such as the chances of metastasis, as well as response to therapy.

Indicatively, biofilm-related inflammation and immunosuppression are associated with aggressive tumor phenotypes and resistance to conventional therapies.[74] Moreover, microbial communities interacting with host signaling pathways, i.e., NF-kB and STAT3, can be used as an indicator of disease progression. The integration of microbial parameters into the current prognostic models is an attractive approach to enhance the stratification of patients and individualized care.[104]

7.3 Therapeutic biofilm targets: Opportunities and challenges

The innate resistance of biofilms to traditional antimicrobial treatment is one of the biggest problems in treating OSCC associated with biofilms. This extracellular polymeric substance (EPS) matrix is a physical and chemical barrier that restricts the penetration of drugs and shields the microbial cells against host immune attacks.[105] Also, polymicrobial interactions provide superior adaptive resistance since microorganisms in the biofilm are able to share resistance mechanisms and metabolic resources. This puts into focus the importance of therapeutic interventions that extend beyond conventional antimicrobial interventions.[7]

7.4 Anti- biofilm and anti-polymicrobial strategies

New treatment methods are also targeting to interfere with biofilm integrity and cross-kingdom interactions. These are quorum-sensing inhibitors: they block microbial communication pathways to inhibit biofilm formation and expression of virulence.[65] Enzymatic destruction of EPS: DNases, proteases, and polysaccharide-degrading enzymes can be used to break down the biofilm structure. Delivery systems based on nanoparticles: Improve the delivery of antimicrobial agents and allow delivery to biofilms in a targeted manner. This type of strategy aims at destabilizing the structure and functional stability of biofilms, which makes microorganisms more vulnerable to treatment.[106]

7.5 Microbiome modulation and probiotics

Another potential avenue of therapy is to restore microbial balance. Probiotics and synbiotics can competitively inhibit the pathogenic microorganisms, reduce inflammation, and maintain a healthy oral microbiome. Moreover, prebiotic measures that selectively enhance the growth of desirable microbes can be used to prevent dysbiosis. The methods are rather appealing with their safety profile and the possibility of preventing diseases in the long run.[107]

7.6 Natural bioactives and phytochemical interventions

Natural compounds, such as polyphenols, flavonoids, and terpenoids, have shown anti-biofilm, anti-inflammatory, and anticancer effects. NF-κB and PI3K/Akt are some other important signaling pathways that can be inhibited by compounds like curcumin, resveratrol, and epigallocatechin gallate (EGCG), as well as microbial biofilms. These phytochemicals have a dual benefit of acting on microbial communities as well as host signaling pathways, and hence they are good prospects as adjunctive therapy in OSCC.[108,109]

7.7 Molecular and immunotherapeutic targeted therapy

Since microbial biofilms activate oncogenic pathways, targeted therapies to signaling pathways, including PI3K, STAT3, and NF-kB, are of interest. These pathways are inhibitable and could suppress tumor growth and make tumors more sensitive to conventional treatment. Immunotherapeutic interventions, especially the use of immune checkpoint inhibitors against PD-1/PD-L1, can be complemented with microbiome-based interventions, as well. The management of microbial communities could be used to boost immune responses and enhance treatment responses.[110]

8. Conclusion

The accumulating evidence of the association of polymicrobial biofilms with oral squamous cell carcinoma (OSCC) represents a paradigm shift in cancer etiology, from a solely host-centered paradigm to a microbe-host interaction model. The interaction between *Porphyromonas gingivalis* and Candida albicans has become an important axis of tumor initiation, progression, and resistance to therapeutic agents in the synergistic interaction of the diverse oral microbiota. By their ability to establish strong polymicrobial biofilms, these microorganisms coordinate a cascade of molecular and cellular processes that jointly reprogram the oral epithelium to malignancy. Mechanistically, biofilm-related pathogens play a role in the development of carcinogenesis based on chronic inflammation, oxidative stress, epigenetic modifications, and dysregulation of essential oncogenic signaling pathways, such as NF-kB, PI3K/Akt, MAPK, and STAT3. Coupling of these pathways leads to a powerful and self-sustaining carcinogenic network. At the same time, microbial metabolites like acetaldehyde and short-chain fatty acids intensify genotoxic stress as well as regulate the expression of host genes, further stipulating tumorigenic mechanisms.

One of the characteristics of such a polymicrobial paradigm is a significant effect on the tumor microenvironment (TME). Candida biofilms actively create a niche favoring tumor growth and spread by promoting immune evasion, angiogenesis, stromal activation, and metabolic reprogramming. Notably, these biofilms contribute to unremitting sources of pro-tumorigenic stimuli, which keep a dynamic feedback loop that perpetuates disease progression. Clinically, identification of microbial roles in OSCC presents groundbreaking chances in the diagnosis of cancers early, prognostic evaluation, and specific treatment. The most promising biomarkers are microbial signatures and metabolites as non-invasive biomarkers, and anti-biofilm and microbiome-modulating therapies are the most promising approaches. Nonetheless, in spite of these improvements, there are still a number of critical issues.

References

1. Ranganathan K, Kavitha L. Clinical aspects of oral cancer and potentially malignant disorders in South and Southeast Asia. *Oral Dis.* 2025;31:1406–1419.
2. Radaic A, Ganther S, Kamarajan P, Grandis J, Yom SS, Kapila YL. Paradigm shift in the pathogenesis and treatment of oral cancer and other cancers focused on the oralome and antimicrobial-based therapeutics. *Periodontol.* 2021;2000(87):76–93.
3. Bostanghadiri N, Kouhzad M, Taki E, Elahi Z, Khoshbayan A, Navidifar T, et al. Oral microbiota and metabolites: key players in oral health and disorder, and microbiota-based therapies. *Front. Microbiol.* 2024;15.
4. Lopez LR, Bleich RM, Arthur JC. Microbiota effects on carcinogenesis: initiation, promotion, and progression. *Annu. Rev. Med.* 2021;72:243–261.
5. Mazumder S, Bhattacharya D, Nag M, Lahiri D. Bioactive compounds from marine algae and fungi in down-regulating quorum sensing. *Blue Biotechnol.* 2024;1:16.
6. Mazumder S, Lahiri D, Nag M. *Biofilm dynamics in food environment. Biofilm Applications to Revolutionize Food Technology*. Cham: Springer Nature Switzerland; 2025:73–99.
7. Nithyanand P, Boya BR, Lee J, Lee J. Polymicrobial biofilms: interkingdom interactions, resistance and therapeutic strategies. *Microb. Biotechnol.* 2025;18.
8. Sukmana BI, Saleh RO, Najim MA, AL-Ghamdi HS, Achmad H, Al-Hamdani MM, et al. Oral microbiota and oral squamous cell carcinoma: a review of their relation and carcinogenic mechanisms. *Front. Oncol.* 2024;14.
9. How KY, Song KP, Chan KG. Porphyromonas gingivalis: an overview of periodontopathic pathogen below the gum line. *Front. Microbiol.* 2016;7.
10. Ponde NO, Lortal L, Ramage G, Naglik JR, Richardson JP. *Candida albicans* biofilms and polymicrobial interactions. *Crit. Rev. Microbiol.* 2021;47:91–111.
11. Talapko J, Juzbašić M, Matijević T, Pustijanac E, Bekić S, Kotris I, et al. Candida albicans—The virulence factors and clinical manifestations of infection. *J. Fungi.* 2021;7:79.
12. Haverman TM, Laheij AMGA, de Soet JJ, de Lange J, Rozema FR. *Candida* and *Porphyromonas gingivalis*: the effect on wound closure in vitro. *J. Oral Microbiol.* 2017;9:1328266.
13. Weng L, Cui Y, Jian W, Zhang Y, Pang L, Cao Y, et al. Inter-kingdom interactions and environmental influences on the oral microbiome in severe early childhood caries. *Microbiol. Spectr.* 2025;13.

14. Choi E, Murray B, Choi S. Biofilm and cancer: interactions and future directions for cancer therapy. *Int. J. Mol. Sci.* 2023;24:12836.
15. Pérez Escriva P, Tavares Bernardino CCorreia, Letellier E. De-coding the complex role of microbial metabolites in cancer. *Cell Rep.* 2025;44:115358.
16. Mishra R, Panda AK, De Mandal S, Shakeel M, Bisht SS, Khan J. Natural anti-biofilm agents: strategies to control biofilm-forming pathogens. *Front. Microbiol.* 2020;11.
17. Acharya S, Hegde U, Acharya AB, Nitin P. Dysbiosis linking periodontal disease and oral squamous cell carcinoma-A brief narrative review. *Heliyon.* 2024;10:e32259.
18. Nikolić N, Pucar A, Tomić U, Petrović S, Mihailović Đ, Jovanović A, et al. Oral microbiota and carcinogenesis: exploring the systemic impact of oral pathogens. *Pathogens.* 2025;14:1233.
19. Karpiński TM. Role of oral microbiota in cancer development. *Microorganisms.* 2019;7:20.
20. Wu Z, Ma Q, Guo Y, You F. The role of Fusobacterium nucleatum in colorectal cancer cell proliferation and migration. *Cancers (Basel).* 2022;14:5350.
21. Avula H, Chakravarthy Y. Models of periodontal disease pathogenesis. *J. Indian Soc. Periodontol.* 2022;26:204–212.
22. Hajishengallis G, Lamont RJ. Beyond the red complex and into more complexity: the polymicrobial synergy and dysbiosis (PSD) model of periodontal disease etiology. *Mol. Oral Microbiol.* 2012;27:409–419.
23. Kashima K, Saito T, Kajikawa H, Kawashima A, Ueyama A, Uzawa N, et al. Impact of oral anaerobic bacteria on the tumor immune microenvironment and prognosis of oral cancer. *J. Transl. Med.* 2025;23:1267.
24. Urzică R-N, Crețu B, Căruntu A, Bucurica S, Farcasiu A-T, Ciupescu LM, et al. The molecular interplay between oral microbiome and oral cancer pathogenesis. *Int. J. Mol. Sci.* 2025;26:10212.
25. Yu Y, Zhong W. Interaction of microbiome and immunity in tumorigenesis and clinical treatment. *Biomed. Pharmacother.* 2022;156:113894.
26. Guo Z, Jing S, Jumatai S, Gong Z. Porphyromonas gingivalis promotes the progression of oral squamous cell carcinoma by activating the neutrophil chemotaxis in the tumour microenvironment. *Cancer Immunol. Immunother.* 2023;72:1523–1539.
27. Starska-Kowarska K. The role of Porphyromonas gingivalis in oral carcinogenesis and progression by remodelling the tumour microenvironment: a narrative review. *Cancers (Basel).* 2025;17:3478.
28. Hajishengallis G, Lamont RJ. Breaking bad: manipulation of the host response by orphyromonas gingivalis. *Eur. J. Immunol.* 2014;44:328–338.
29. Zenobia C, Hajishengallis G. *Porphyromonas gingivalis* virulence factors involved in subversion of leukocytes and microbial dysbiosis. *Virulence.* 2015;6:236–243.
30. Curtis MA, Garnett JA, Darveau RP. The keystone-pathogen hypothesis updated: the role of *Porphyromonas gingivalis* in periodontitis. *J. Periodontal Res.* 2025.
31. Olsen I, Singhrao SK. Importance of heterogeneity in *Porhyromonas gingivalis* lipopolysaccharide lipid a in tissue specific inflammatory signalling. *J. Oral Microbiol.* 2018;10:1440128.
32. Wu Z, Long W, Yin Y, Tan B, Liu C, Li H, et al. Outer membrane vesicles of Porphyromonas gingivalis: recent advances in pathogenicity and associated mechanisms. *Front. Microbiol.* 2025;16.
33. Lu Z, Cao R, Geng F, Pan Y. Persistent infection with Porphyromonas gingivalis increases the tumorigenic potential of human immortalised oral epithelial cells through inhibition. *Cell Prolif.* 2024;57.
34. Aleksijević LH, Aleksijević M, Škrlec I, Šram M, Šram M, Talapko J. Porphyromonas gingivalis virulence factors and clinical significance in periodontal disease and coronary artery diseases. *Pathogens.* 2022;11:1173.

35. Lee H, Jeong AJ, Ye S-K. Highlighted STAT3 as a potential drug target for cancer therapy. *BMB Rep.* 2019;52:415–423.
36. Tolomeo M, Cascio A. The multifaced role of STAT3 in cancer and its implication for anticancer therapy. *Int. J. Mol. Sci.* 2021;22:603.
37. Hajishengallis G, Diaz PI. Porphyromonas gingivalis: immune subversion activities and role in periodontal dysbiosis. *Curr. Oral Health Rep.* 2020;7:12–21.
38. Lamont RJ, Fitzsimonds ZR, Wang H, Gao S. Role of *Porphyromonas gingivalis* in oral and orodigestive squamous cell carcinoma. *Periodontol.* 2022;2000(89):154–165.
39. Muñoz-Medel M, Pinto MP, Goralsky L, Cáceres M, Villarroel-Espíndola F, Manque P, et al. Porphyromonas gingivalis, a bridge between oral health and immune evasion in gastric cancer. *Front. Oncol.* 2024;14.
40. Kazelnik M, Masri R, Caspi M, Wittenstein A, Tsarfaty I, Borenstein E, et al. *Porphyromonas gingivalis* secreted factors drive epithelial–mesenchymal transition (EMT) through gingipains and an H_2 S-mediated bacterial defense system. *Gut Microbes.* 2026;18.
41. Zou Q, Xie H, Yang W, Xu J, Ying S, Liao X, et al. Advances in the study of the relationship between Porphyromonas gingivalis and various diseases. *Front. Cell Dev. Biol.* 2025;13.
42. Ayuningtyas NF, Mahdani FY, Pasaribu TAS, Chalim M, Ayna VKP, Santosh ABR, et al. Role of Candida albicans in oral carcinogenesis. *Pathophysiology.* 2022;29:650–662.
43. Yu D, Liu Z. The research progress in the interaction between Candida albicans and cancers. *Front. Microbiol.* 2022;13.
44. Malinovská Z, Čonková E, Váczi P. Biofilm formation in medically important candida species. *J. Fungi.* 2023;9:955.
45. Wall G, Montelongo-Jauregui D, Vidal Bonifacio B, Lopez-Ribot JL, Uppuluri P. Candida albicans biofilm growth and dispersal: contributions to pathogenesis. *Curr. Opin. Microbiol.* 2019;52:1–6.
46. Lortal L, Lyon CM, Sprague JL, Sonnberger J, Paulin OKA, Wickramasinghe DN, et al. Candidalysin biology and activation of host cells. *MBio.* 2025;16.
47. Moyes DL, Wilson D, Richardson JP, Mogavero S, Tang SX, Wernecke J, et al. Candidalysin is a fungal peptide toxin critical for mucosal infection. *Nature.* 2016;532:64–68.
48. Fiore M, Minni A, Cavalcanti L, Raponi G, Puggioni G, Mattia A, et al. The impact of alcohol consumption and oral microbiota on upper aerodigestive tract carcinomas: a pilot study. *Antioxidants.* 2023;12:1233.
49. Reddy MGS, Kakodkar P, Nayanar G. Capacity of Candida species to produce acetaldehyde at various concentrations of alcohol. *J. Oral Maxillofac. Pathol.* 2022;26:161–165.
50. Wang X, Zhang W, Wu W, Wu S, Young A, Yan Z. Is Candida albicans a contributor to cancer? A critical review based on the current evidence. *Microbiol. Res.* 2023;272:127370.
51. Xiao K, Liu C, Tu Z, Xu Q, Chen S, Zhang Y, et al. Activation of the NF- κ B and MAPK signaling pathways contributes to the inflammatory responses, but not cell injury, in IPEC-1 cells challenged with hydrogen peroxide. *Oxid. Med. Cell. Longev.* 2020;2020:1–14.
52. Malavika G, Sujith Sri Surya R, Datchanamoorthy M, Kakithakara Vajravelu L, Rahul HL, Priya KSL. Role of Candida albicans in chronic inflammation and the development of oral squamous cell carcinoma. *Cancer Pathog. Ther.* 2025;3:402–410.
53. Mannan M, Nabeela S, Mishra R, Uppuluri P. Host immune response against fungal biofilms. *Curr. Opin. Microbiol.* 2024;81:102520.

54. de Jongh CA, Bikker FJ, de Vries TJ, Werner A, Gibbs S, Krom BP. Porphyromonas gingivalis interaction with Candida albicans allows for aerobic escape, virulence and adherence. *Biofilm*. 2024;7:100172.
55. Taff HT, Mitchell KF, Edward JA, Andes DR. Mechanisms of *Candida* biofilm drug resistance. *Future Microbiol*. 2013;8:1325–1337.
56. Monteiro JS, Kaushik K, de Arruda JAA, Georgakopoulou E, Vieira AT, Silva TA, et al. Fungal footprints in oral cancer: unveiling the oral mycobiome. *Front. Oral Health*. 2024;5.
57. Wronowska E, Guevara-Lora I, Brankiewicz A, Bras G, Zawrotniak M, Satala D, et al. Synergistic effects of Candida albicans and Porphyromonas gingivalis biofilms on epithelial barrier function in a 3D aspiration pneumonia model. *Front. Cell. Infect. Microbiol*. 2025;15.
58. Preda VG, Săndulescu O. Communication is the key: biofilms, quorum sensing, formation and prevention. *Discoveries*. 2019;7:e10.
59. Jo J, Price-Whelan A, Dietrich LEP. Gradients and consequences of heterogeneity in biofilms. *Nat. Rev. Microbiol*. 2022;20:593–607.
60. Rather MA, Gupta K, Mandal M. Microbial biofilm: formation, architecture, antibiotic resistance, and control strategies. *Braz. J. Microbiol*. 2021;52:1701–1718.
61. Oka I, Shigeishi H, Ohta K. Co-Infection of oral Candida albicans and porphyromonas gingivalis is associated with active periodontitis in middle-aged and older Japanese people. *Medicina (B. Aires)*. 2022;58:723.
62. Hall RA, Turner KJ, Chaloupka J, Cottier F, De Sordi L, Sanglard D, et al. The quorum-sensing molecules farnesol/homoserine lactone and dodecanol operate via distinct modes of action in Candida albicans. *Eukaryot. Cell*. 2011;10:1034–1042.
63. Ramage G, Saville SP, Wickes BL, López-Ribot JL. Inhibition of *Candida albicans* biofilm formation by farnesol, a quorum-sensing molecule. *Appl. Environ. Microbiol*. 2002;68:5459–5463.
64. Wu S, Ye X, Wang J, Li M, Fu Y, Wang S. Microbial crosstalk along the oral–gut axis: organ-specific oncogenic adaptations of porphyromonas gingivalis and Fusobacterium nucleatum. *Neoplasia*. 2026;74:101290.
65. Markowska K, Szymanek-Majchrzak K, Pituch H, Majewska A. Understanding quorum-sensing and biofilm forming in anaerobic bacterial communities. *Int. J. Mol. Sci*. 2024;25:12808.
66. Sztukowska MN, Dutton LC, Delaney C, Ramsdale M, Ramage G, Jenkinson HF, et al. Community development between *Porphyromonas gingivalis* and *Candida albicans* mediated by InlJ and Als3. *MBio*. 2018;9.
67. Van Dyck K, Pinto RM, Pully D, Van Dijck P. Microbial interkingdom biofilms and the quest for novel therapeutic strategies. *Microorganisms*. 2021;9:412.
68. Ponde NO, Lortal L, Tsavou A, Hepworth OW, Wickramasinghe DN, Ho J, et al. Receptor-kinase EGFR-MAPK adaptor proteins mediate the epithelial response to Candida albicans via the cytolytic peptide toxin, candidalysin. *J. Biol. Chem*. 2022;298:102419.
69. Anju VT, Busi S, Imchen M, Kumavath R, Mohan MS, Salim SA, et al. Polymicrobial infections and biofilms: clinical significance and eradication strategies. *Antibiotics*. 2022;11:1731.
70. Son YM, Kim J. The microbiome-immune axis therapeutic effects in cancer treatments. *J. Microbiol. Biotechnol*. 2022;32:1086–1097.
71. Ionescu C, Kamal FZ, Ciobica A, Halitchi G, Burlui V, Petroaie AD. Oxidative stress in the pathogenesis of oral cancer. *Biomedicines*. 2024;12:1150.
72. Gainza-Cirauqui ML, Nieminen MT, Novak Frazer L, Aguirre-Urizar JM, Moragues MD, Rautemaa R. Production of carcinogenic acetaldehyde by *Candida albicans* from patients with potentially malignant oral mucosal disorders. *J. Oral Pathol. Med*. 2013;42:243–249.

73. Zhou Y, Luo G-H. *Porphyromonas gingivalis* and digestive system cancers. *World J. Clin. Cases*. 2019;7:819–829.
74. Montanari E, Bernardo G, Le Noci V, Anselmi M, Pupa SM, Tagliabue E, et al. Biofilm formation by the host microbiota: a protective shield against immunity and its implication in cancer. *Mol. Cancer*. 2025;24:148.
75. Mao H, Zhao X, Sun S. NF-κB in inflammation and cancer. *Cell. Mol. Immunol.* 2025;22:811–839.
76. Yu W, Tu Y, Long Z, Liu J, Kong D, Peng J, et al. Reactive oxygen species bridge the gap between chronic inflammation and tumor development. *Oxid. Med. Cell. Longev*. 2022;2022.
77. Mizumoto A, Ohashi S, Hirohashi K, Amanuma Y, Matsuda T, Muto M. Molecular mechanisms of acetaldehyde-mediated carcinogenesis in squamous epithelium. *Int. J. Mol. Sci*. 2017;18:1943.
78. NavaneethaKrishnan S, Rosales JL, Lee K-Y. ROS-mediated cancer cell killing through dietary phytochemicals. *Oxid. Med. Cell. Longev*. 2019;2019:1–16.
79. Sheng G, Li C, Liu D, Wang W, Pang N, Zhang Z. The role and significance of the PI3K/Akt-mTOR signaling pathway in infectious diseases. *Front. Cell. Infect. Microbiol*. 2026;16.
80. Guo Y, Pan W, Liu S, Shen Z, Xu Y, Hu L. ERK/MAPK signalling pathway and tumorigenesis (review). *Exp. Ther. Med*. 2020.
81. Sugiura R, Satoh R, Takasaki T. ERK: a double-edged sword in cancer. ERK-dependent apoptosis as a potential therapeutic strategy for cancer. *Cells*. 2021;10:2509.
82. Stein RA, Riber L. Epigenetic effects of short-chain fatty acids from the large intestine on host cells. *MicroLife*. 2023;4.
83. Woo V, Alenghat T. Epigenetic regulation by gut microbiota. *Gut Microbes*. 2022;14.
84. Audia JE, Campbell RM. Histone modifications and Cancer. *Cold Spring Harb. Perspect. Biol*. 2016;8:a019521.
85. Kanwal R, Gupta S. Epigenetic modifications in cancer. *Clin. Genet*. 2012;81:303–311.
86. Li X, Ren Y, Pei S, Zhao K, Chen G, He Z. Dissecting the role of epigenetic regulation in oral squamous cell carcinoma microenvironment: mechanisms and therapeutics. *Front. Immunol*. 2026;17.
87. Tan X, Tan Z, Xu W, Zhou Y, Wei W, Li C, et al. Association of Porphyromonas gingivalis-infected oral squamous cell carcinoma cell-secreted exosomal miR-3648-1-p5 with tumor progression. *Cancer Cell Int*. 2026;26:163.
88. Malavika G, Sujith Sri Surya R, Datchanamoorthy M, Vajravelu LK, Lathakumari RH, Lekshmi Priya KS. Role of Candida albicans in chronic inflammation and the development of oral squamous cell carcinoma. *Cancer Pathog Ther*. 2025;3:402–410.
89. Lin X, Kang K, Chen P, Zeng Z, Li G, Xiong W, et al. Regulatory mechanisms of PD-1/PD-L1 in cancers. *Mol. Cancer*. 2024;23:108.
90. Singh S, Singh N, Baranwal M, Sharma S, Devi SSK, Kumar S. Understanding immune checkpoints and PD-1/PD-L1-mediated immune resistance towards tumour immunotherapy. *3 Biotech*. 2023;13:411.
91. Morgan MJ, Liu Z. Crosstalk of reactive oxygen species and NF-κB signaling. *Cell Res*. 2011;21:103–115.
92. Cabral LGde.S, Martins IM, Paulo EP de A, Pomini KT, Poyet J-L, et al. Molecular mechanisms in the carcinogenesis of oral squamous cell carcinoma: a literature review. *Biomolecules*. 2025;15:621.
93. Aghaei M, Bahreiny SS, Mahdizade AH, Abbasi A, Magharati A, Yousefi-Avarvand A. Cytokine-mediated tumor sensitization: mechanistic frameworks and therapeutic opportunities in cancer Immunotherapy. *Clin. Exp. Med*. 2026;26:135.
94. Johnson DE, O'Keefe RA, Grandis JR. Targeting the IL-6/JAK/STAT3 signalling axis in cancer. *Nat. Rev. Clin. Oncol*. 2018;15:234–248.

95. Gao J, Liang Y, Wang L. Shaping polarization of tumor-associated macrophages in cancer immunotherapy. *Front. Immunol.* 2022;13.
96. Xu R, Zhao X, Chen X, Zhou H, Cui L. Targeting microbe-mediated macrophage education: a novel paradigm in cancer immunotherapy. *Biomater. Res.* 2025;29.
97. Wang Y, Bai M, Peng Q, Li L, Tian F, Guo Y, et al. Angiogenesis, a key point in the association of gut microbiota and its metabolites with disease. *Eur. J. Med. Res.* 2024;29:614.
98. Muz B, de la Puente P, Azab F, Azab AK. The role of hypoxia in cancer progression, angiogenesis, metastasis, and resistance to therapy. *Hypoxia.* 2015:83.
99. Zhang M, Zhang B. Extracellular matrix stiffness: mechanisms in tumor progression and therapeutic potential in cancer. *Exp. Hematol. Oncol.* 2025;14:54.
100. Niland S, Riscanevo AX, Eble JA. Matrix metalloproteinases shape the tumor microenvironment in cancer progression. *Int. J. Mol. Sci.* 2021;23:146.
101. Mahapatra S, Ankri S. The role of biofilm-derived compounds in microbial and protozoan interactions. *Microorganisms.* 2025;14:64.
102. Shinde DB, Mahore JG, Giram PS, Singh SL, Sharda A, Choyan D, et al. Microbiota of saliva: a non-invasive diagnostic tool. *Indian J. Microbiol.* 2024;64:328–342.
103. D'Enfert C, Kaune A-K, Alaban L-R, Chakraborty S, Cole N, Delavy M, et al. The impact of the fungus-host-microbiota interplay upon *Candida albicans* infections: current knowledge and new perspectives. *FEMS Microbiol. Rev.* 2021;45.
104. Zhang S, Paul S, Kundu P. NF-κB regulation by gut microbiota decides homeostasis or disease outcome during ageing. *Front. Cell Dev. Biol.* 2022;10.
105. Mirghani R, Saba T, Khaliq H, Mitchell J, Do L, Chambi L, et al. Biofilms: formation, drug resistance and alternatives to conventional approaches. *AIMS Microbiol.* 2022;8:239–277.
106. Kim J-S, Lim M-C, Kim S-M, Lee J-Y. Extracellular matrix-degrading enzymes as a biofilm control strategy for food-related microorganisms. *Food Sci. Biotechnol.* 2023;32:1745–1761.
107. Maier T. Oral microbiome in health and disease: maintaining a healthy, balanced ecosystem and reversing dysbiosis. *Microorganisms.* 2023;11:1453.
108. Paul JK, Azmal M, Haque ASNB, Talukder OF, Meem M, Ghosh A. Phytochemical-mediated modulation of signaling pathways: A promising avenue for drug discovery. *Advances in Redox Research.* 2024;13:100113.
109. da Silva AP, da Silva Cordeiro ML, de Queiroz Aquino-Martins VG, Scortecci KC. *Role of flavonoids and their potential application in health. Natural Products.* Berlin, Heidelberg: Springer Berlin Heidelberg; 2026:1–73.
110. Leonov G, Starodubova A, Makhnach O, Goldshtein D, Salikhova D. Intratumoral microbiome: impact on cancer progression and cellular immunotherapy. *Cancers (Basel).* 2025;18:100.

CHAPTER EIGHT

Biofilms in specific cancers: *Helicobacter pylori, Fusobacterium nucleatum, Porphyromonas gingivalis, Escherichia coli*, and *Candida*

Sougata Rajak[a,*], Panteha Torabian[b], Naiera M. Helmy Mohamed[c], Neethu Asokan[d], Reham Hamed Abdelaziz[e,f], and Saanya Chaturvedi[g,*]

[a]Department of Biotechnology, Motilal Nehru National Institute of Technology Allahabad, Prayagraj, Uttar Pradesh, India
[b]Department of Biomedical and Chemical Engineering, Rochester Institute of Technology, Rochester, NY, United States
[c]Microbial Biotechnology Department, Biotechnology Research Institute, National Research Centre, Dokki, Cairo, Egypt
[d]Department of Lifesciences, Sri Sathya Sai University for Human Excellence, Kalaburagi, Karnataka, India
[e]Clinical and Chemical Pathology Department, Faculty of Medicine, Cairo University, Giza, Egypt
[f]Microbiology, Immunology and Infectious Disease Department, College of Medicine and Health Sciences (CMHS), Arabian Gulf University, Manama, Bahrain
[g]Department of Environmental Science and Engineering, Indian Institute of Technology (Indian School of Mines), Dhanbad, India
*Corresponding authors. e-mail address: sougatarajak08@gmail.com; chaturvedysaanya@gmail.com

Contents

Progress in Molecular Biology and Translational Science, Volume 224
ISSN 1877-1173, https://doi.org/10.1016/bs.pmbts.2026.07.017

Abstract

An organized microbial communities encapsulated in a protective matrix are called biofilms, that have a major impact on how different types of cancer form and spread. The effect of biofilms produced by *Helicobacter pylori*, *Fusobacterium nucleatum*, *Porphyromonas gingivalis*, *Escherichia coli*, and *Candida* spp on the development and spread of cancer is examined in this study. Because they cause cellular damage, genetic mutations, and persistent inflammation, *H. pylori* biofilms are strongly linked to stomach cancer, such as *F. nucleatum* biofilms have been connected to colorectal cancer, where they stimulate the formation of tumors by modifying the immune system and activating pathways associated to cancer. A common periodontal infection, *P. gingivalis*, has been linked to pancreatic and oral malignancies by increasing tumor invasiveness and facilitating immune evasion. Some of the enterotoxigenic pathovars of *E. coli* can cause chronic inflammation and have genotoxic effects, often are related to colorectal cancer. Moreover, *Candida* species, specifically *Candida albicans* can form biological films in the gastrointestinal and oral tracts, helping in epithelial invasion and synthesizing carcinogenic byproducts resulting in esophageal and other malignancies. Examining the mechanisms of cancer emergence and progression by biofilms can provide insights on the influence of microbes on the development of cancer. Targeting the biofilm development can provide novel approaches for the treatment regimen of cancer.

1. Introduction

Microbial biological films importantly contribute to the virulence of various diseases, including cancer.[1] These biological films comprise sessile microbial communities encapsulated in an extracellular polymeric substance (EPS) that attaches to substratum (such as skin or indwelling catheters) and interfaces (such as pus or mucus derived from the host).[2] These are a predominant form of microbial life in both natural and host environments.[3] The free-floating (planktonic) cells can sometimes not survive the hostilities of environment in contrast to biofilms that can survive tremendous adversities like acidic and immunologically strong region of the human GI tract.[4] Their development comprises of an array of events, initial reversible adsorption followed by irreversible attachment, micro-colonies development and EPS secretion, maturation into a three-dimensional structure and subsequent dispersion.[5] The EPS consists of polysaccharides, extracellular DNA (eDNA), proteins, and lipids that gives the biofilms their mechanical strength and stability, nutrient retention, and protection from hostilities, like antibiotics and host immune system.[6] These biofilms show spatial heterogeneity, difference in metabolism at different layers of EPS, and quorum sensing, that helps them in adaptation and persistence within host.[7] Earlier biological films were correlated with chronic infections, but now are known to be associated with pathogenesis of many cancers. Biofilms induce anti-inflammatory responses and metabolic changes in the host, which are comparable to those observed in cancer cells.[8] They stimulate the immune system by releasing proinflammatory cytokines and reactive oxygen species, contributing to DNA damage, immune suppression, and therapy resistance factors that support tumor progression.[8] Biofilms can also modulate the tumor microenvironment, facilitating oncogene expression and cancer progression.[2]

In recent years, the oncological relevance of biofilms has gained increasing attention. Several pathogenic species have been implicated in cancer development through biofilm-mediated mechanisms. *H. pylori* are the only bacterium classified as a Group I carcinogen by The International Agency for Research on Cancer (IARC), part of the World Health Organization,[9] due to its implication in gastric adenocarcinoma and mucosa-associated lymphoid tissue (MALT) lymphoma.[10] Similarly, *Fusobacterium nucleatum*, known for forming invasive biofilms in the colon, promotes colorectal cancer progression via multiple interrelated pathways, including adhesion, immune modulation, inflammation, oncogenic signaling and epithelial invasion.[11] *Porphyromonas gingivalis*, a Gram-negative anaerobic oral pathogen, has been associated with oral and esophageal cancers. It is known to invade epithelial cells, mediate immune evasion and increase inflammation in cancer cells. It also suppresses apoptosis.[12] *Escherichia coli*, especially genotoxic strains harboring the *pks* island, forms mucosal biofilms in the colon and secretes colibactin, a compound linked to DNA damage.[13] *Candida albicans*, a fungal species capable of forming robust biofilms on mucosal surfaces, causing oral microbial dysbiosis.[14] It is known to cause squamous cell carcinoma of oral cavity by chronic inflammation and dysplasia.[15] This chapter dives into the organization, immune evasion strategies and molecular contributions of biofilms and their role in cancer development from recent investigations Table 1.

Table 1 Biofilm-associated pathogens in cancer: key features, mechanisms, and cancer associations.

Microbe	Primary cancer association (s)	Biofilm niche & structure	Key virulence/ biofilm factors	Main oncogenic mechanisms	Evidence type
Helicobacter pylori	Gastric adenocarcinoma	Gastric mucosa, pits; exopolysaccharides, eDNA, proteins	CagA, VacA, BabA, SabA	NF-κB activation, IL-8 production, epithelial polarity loss, chronic inflammation	Clinical biopsies, in vivo & in vitro
Fusobacterium nucleatum	Colorectal cancer	Tumor crypt biofilms, co-aggregates with other gut/oral bacteria	FadA, Fap2	β-catenin activation, TIGIT-mediated immune suppression, inflammatory cytokines	CRC patient tissues, murine models
Porphyromonas gingivalis	Oral SCC, pancreatic, esophageal cancers	Oral biofilms, co-aggregation in polymicrobial consortia	Gingipains (Rgp, Kgp), PPAD, LPS	E-cadherin cleavage, TLR/NF-κB activation, systemic dissemination	Human tumors, animal models
Escherichia coli (pks^{+} strains)	Colorectal, bladder cancers	Mucus-associated colon biofilms, bladder urothelium	Colibactin, Type 1 pili (FimH)	DNA crosslinks/DSBs, chronic inflammation, biofilm persistence	Patient samples, in vivo/in vitro
Candida spp.	Oral, esophageal, GI cancers	Hyphal/yeast biofilms, mixed species with bacteria	β-glucan matrix, alcohol dehydrogenase, Saps	Acetaldehyde mutagenesis, chronic inflammation, immune evasion	Clinical, animal, ex vivo

2. Role of microbial biofilms in carcinogenesis

2.1 Chronic inflammation and immune evasion

Microbial biofilms act as reservoirs which trigger and sustain persistent inflammation by continuously stimulating host immune responses. The EPS matrix shields microbial antigens from full immune recognition while allowing controlled release of pathogen-associated molecular patterns (PAMPs) such as lipopolysaccharide (LPS), flagellin, and peptidoglycan.[16] These molecules activate pattern recognition receptors (PRRs) including Toll-like receptors (TLRs) and NOD-like receptors (NLRs), triggering NF-κB-mediated cytokine expression. A study by Di Russo et al. reported that NF-κB activation is shown by breast cancer cells with affinity to brain metastasis.[17] Biofilms facilitate chronic inflammation a hallmark of cancer by recruiting immune cells and promoting cytokine production, including IL-6, IL-8, and IL-1β.[18] This inflammatory environment, along with sustained oxidative stress and disrupted epithelial barriers, supports oncogenic transformation and tumor growth. For example, biofilm-associated *F. nucleatum* has been shown to activate IL-17 and STAT3 pathways in colorectal cancer (CRC), reduce E-cadherin expression, and promote epithelial-to-mesenchymal transition (EMT), thereby enhancing tumor cell invasiveness.[19] Concurrently, biofilms impair immune clearance through suppression of antigen presentation and recruitment of regulatory T cells (Tregs), shifting immune responses toward tolerance. *F. nucleatum*, for example, expresses Fap2 and RadD adhesins that inhibit natural killer (NK) cell activity and T-cell function, allowing tumor cells to evade immune surveillance.[20] Biofilm-enclosed *Staphylococcus aureus* and *Pseudomonas aeruginosa* also release immune-disruptive molecules, such as α-toxin and rhamnolipids, respectively, further contributing to an immunosuppressive tumor microenvironment.[21]

2.2 Genetic and epigenetic modulation

Biofilm-derived metabolites and toxins significantly modulate both tumor cells and the immune microenvironment through genetic and epigenetic mechanisms. Metabolites such as the polyamine N1, N12-diacetylspermine enhances CRC cell proliferation.[22] While deoxycholic acid produced by *Clostridium scindens*, can stimulate biofilm formation in *C. difficile*,[23] it also impairs $CD8^+$ T cell function by inhibiting the nuclear factor of activated T cells 2 (NFAT) signaling pathway.[24] Toxins like colibactin from *pks*$^+$ *E. coli* and BFT from enterotoxigenic *Bacteroides fragilis* (ETBF) induce DNA damage, activate oncogenic signaling pathways (e.g., β-catenin/NOTCH1), and promote cellular stemness and chemoresistance.[25]

Chronic inflammation associated with biofilm colonization contributes to tumorigenesis by inducing oxidative and nitrosative stress. Persistent oxidative stress caused by reactive oxygen and nitrogen species (ROS/RNS), leads to base modifications (e.g., 8-oxoG), double-stranded DNA breaks, and chromosomal instability.[26] Biofilms also influence the expression of DNA methyltransferases and histone deacetylases, resulting in silencing of tumor suppressor genes and activation of oncogenes.[27] In *H. pylori* infection, biofilm environments are linked to hypermethylation of key regulatory genes such as *CDH1*, *MLH1*, and *p16*, contributing to gastric epithelial dysplasia.[28] Similarly, *E. coli* biofilms harboring the *pks* genomic island produce colibactin, which induces interstrand DNA cross-links and senescence in epithelial cells factors contributing to colorectal tumor initiation.[13]

2.3 Impact on the tumor microenvironment

Biofilms modulate the tumor microenvironment (TME) by creating an immunosuppressive, hypoxic, and nutrient-depleted niche. The EPS matrix forms a physical barrier that limit immune cell infiltration, oxygen diffusion and nutrient flow, leading to hypoxia and nutrient depletion (metabolic stress) within the TME. These conditions stabilize hypoxia-inducible factors (HIFs) and promote angiogenesis, glycolytic metabolism, and matrix remodeling features of an aggressive tumor phenotype.[29] Biofilm-derived metabolites such as polyamines, short-chain fatty acids, and secondary bile acids can directly alter epithelial cell signaling and enhance carcinogenic transformation. In colorectal cancer, mucosa-associated biofilms have been shown to induce loss of E-cadherin-mediated adhesion, increase IL-6/STAT3 signaling, and promote epithelial-mesenchymal transition (EMT), all of which facilitate tumor invasion and metastasis.[30]

2.4 Biofilm-mediated resistance to therapy

Biological films show enhanced resilience towards antimicrobials and, even anticancer treatment regimen. The reduction in metabolism of dormant cells within biofilms, combined with obstruction from EPS, effluxes antibiotic by hindering its penetration and efficacy.[31] In *H. pylori*, biofilms are resilient to clarithromycin, metronidazole, and levofloxacin, being the prime reason for failure of treatment. The horizontal gene transfer (HGT) in biofilms is responsible for AMR dissemination through attainment of antibiotic-resistant genes.[32] In addition to this, biofilms can modulate immune system thus hindering chemotherapy and immunotherapy. For example, intratumoral microbe especially *F. nucleatum* minimizes the effectiveness of 5-fluorouracil in colorectal cancer by autophagy stimulation and activation and apoptosis hinderance in tumor cells.[33,34]

3. Helicobacter pylori and gastric cancer

3.1 Biofilm formation by H. pylori

H. pylori is a Gram-negative, with a microaerophilic nature that is responsible for stomach and 80 % of gastric ulcer in stomach.[35] It forms biofilms on both natural and artificial surfaces, like epithelial cells of GI tract and its mucosal lining. The development of biological film begins with flagella-mediated motility and attachment through outer membrane proteins (OMPs) like BabA and AlpB. Mature *H. pylori* biofilms are dome-shaped structures consisting of spiral and coccoid configuration.[36] Proteomics show expression of genes for adhesion (e.g., *homB*), stress responses (*spoT*), and efflux pumps (*hefA*, *gluP*) during biofilm growth.[37] The coccoid structured biofilms resist antibiotics and immune defenses, serving as dormant reservoirs being reactivated under favorable conditions.[38]

3.2 Virulence factors: CagA and VacA

The two prime virulence determinants of *H. pylori* CagA and VacA are pivotal in biological film mediated cancer development. CagA, is coded by the cytotoxin-related gene pathogenicity island (cagPAI), is diffused into host cells by type IV secretion system (T4SS), disturbing cell polarity, enhancing differentiation, and triggers proinflammatory signaling.[39] VacA, a secreted pore-forming toxin, disrupts the activity of mitochondria apoptosis, and lead to immune suppression. Both proteins are differentially expressed in biofilm conditions, with CagA promoting interleukin production and VacA facilitating immune evasion. Their combined actions maintain inflammation, damage DNA, and promote oncogenic transformation of gastric epithelial cells.[40]

3.3 Mechanisms promoting gastric carcinogenesis

Chronic *H. pylori* infection, particularly under biofilm conditions, drives the gastric carcinogenesis cascade described by Correa: from chronic gastritis to atrophy, intestinal metaplasia, dysplasia, and carcinoma.[41,42] Persistent colonization sustains mucosal inflammation, ROS production, and epithelial turnover. Biofilm-associated *H. pylori* increase mutation rates and promotes methylation of tumor suppressor genes such as *p53* and *CDKN2A*.[43] Additionally, the biofilm microenvironment allows escape from immune clearance, enhances bacterial adaptation, and perpetuates mucosal injury all of which contribute to oncogenesis.[44]

3.4 Clinical implications and eradication challenges

Biofilm formation by *H. pylori* complicates eradication efforts and increases the risk of gastric cancer. Conventional triple and quadruple therapies often fail due to the reduced susceptibility of biofilm-embedded bacteria.[45] The Minimum Biofilm Eradication Concentration (MBEC) for antibiotics is significantly higher than planktonic MIC values, leading to persistent infection despite apparent susceptibility. Novel strategies targeting biofilm integrity such as use of N-acetylcysteine (NAC), antimicrobial peptides (e.g., MSI-78A), and nanoparticle-based drug delivery are under investigation.[46,47] Chitosan-based micro- and nanoparticles, lipidic nanocarriers, and metallic systems have been developed to target *H. pylori* via mechanisms such as bacterial binding and membrane disruption. Targeted delivery using glycan or lectin ligands further increases therapeutic precision. Although these innovative approaches show great promise, their transition to clinical use is still limited.[47] Early detection and biofilm-directed therapy may improve treatment outcomes and reduce cancer incidence in high-risk populations.

4. Fusobacterium nucleatum and colorectal cancer

Colorectal cancer (CRC) is one of the most prevalent types of cancer spreading worldwide.[48] *Fusobacterium nucleatum* is Gram-negative anaerobic oral commensal and associated with periodontal pathogen.[49] The association between *Fusobacterium nucleatum* infection and periodontal pathogens suggests a potential role in cancer risk, highlighting its adaptability as a microbial factor contributing to cancer development.[50]

Both symbiotic and pathogenic microbiota can cause induction to local inflammation through invasion of the normal colon tissue, thus accelerating tumorigenesis through promotion of colonic epithelial cells genotoxicity then stimulation of CRC development.[51] Also, Metabolic products released from microbe's prompt DNA damage through induction of reactive oxidative species production in colon cells host cells thus induce tumor development.[51] There are many amino acid metabolites produced by *Fn* act as myeloid cell stimulators, comprising formyl methoxy leucine phenylalanine and short-chain fatty acids resulting in myeloid cell expansion in tumors. Also, The basic electron transport chains by *F.n* cause limitation to their capacity to inhale oxygen thus creating hypoxic tumor microenvironment.[52] There are mechanistic, biological, and clinical association between *F. nucleatum* and glucose metabolism and CRC.[53]

Fusobacterium adhesin A (FadA) and fibroblast activation protein 2 (Fap2) are produced virulence factors which enable adhesion or colonization.[54] The expression of such adhesins enables *F. nucleatum* binding and/or invasion of various cell types comprising oral, colonic and placental epithelial cells, keratinocytes, T-cells, and macrophages.[55] Epithelial cells infection with *F. nucleatum* cause increased angiogenesis through upregulation of vascular endothelial growth factor (VEGF) and its receptors VEGFR1 and 2.[56] A galactose adhesion hemagglutinin, Fusobacterial Fap2, is important factor in meditation of *F. nucleatum* colonization through binding to galactose-*N*-acetyl-D-galactosamine (Gal-GalNAc) as a tumor-overexpressed host factor resulting in invasion of CRC cells. Fap2 was shown to be vital for production of amyloid-like FadA, which converse acid tolerance and maintain gastrointestinal translocation and *F. nucleatum* colonization.[57] The binding of FadA adhesion protein to E-cadherin increases the inflammatory level and cancer genes by activating the Wnt/β-catenin signaling in epithelial cells of the intestine.[58] *F. nucleatum* caused enhancement of the CRC cells adhesion to endothelial cells, promotion of extravasation and metastasis through upregulating expression of intercellular adhesion molecule 1 (ICAM1).[59] *F. nucleatum* attachment to colorectal cancer (CRC) cells enabled by RadD in vitro. The direct binding between RadD and CD147, which is an overexpressed receptor on CRC cell surfaces, started a PI3K-AKT-NF-κB-MMP9 cascade, consequently enhanced tumorigenesis.[60] Succinic acid produced from *F. nucleatum* inhibited the cGAS-interferon-β pathway, thus diminishing the antitumor response through limitation in trafficking of $CD8^+$ T cell to the tumor microenvironment as detected in vivo.[61]

The link between microsatellite instability (MSI) in colon cancer and ability of *F. nucleatum* infected tumor cells to escape the immune response is one of the mechanisms resulting in tumor's protection from attacking through immune cells.[62] The association between *Fusobacterium* and distant metastases from primary human colorectal cancers was indicated by the presence of viable *Fusobacterium* strains through the metastatic process representing the migration of *Fusobacterium* with the colorectal cancer cells to the metastatic site.[63] The mechanisms of *Fusobacterium nucleatum*-associated tumorigenesis are illustrated in Fig. 1.

The process of biofilm- organized behaviour of *F. nucleatum* permits interactions with host cells via the expression of various adhesins. RadD is most well-defined fusobacterial adhesin, which allow binding the *Streptococcus mutans* adhesin SpaP resulting in mediating the two bacterial co-aggregation and progressive biofilm development. The introduction of physical and metabolic interactions in biofilms of *F. nucleatum*, comprising amino acid fermentation and glycolysis, could contributes to cancer development.[64] The direct contact between mucosal epithelial cells and formed bacterial biofilms may enhance permeability of colonic epithelial cells, which enables translocation of bacterial antigen and stimulates inflammation of pro-carcinogenic tissue.[65,66] During stress and disease conditions, *F. nucleatum* produces amyloid-like FadA acts as a scaffold for formation of biofilm, maintain acid tolerance, and mediates binding of *F. nucleatum* to the host cells.[67] The association between increased bacterial biomass of *Fusobacterium spp.* and inflammatory response in CRC samples

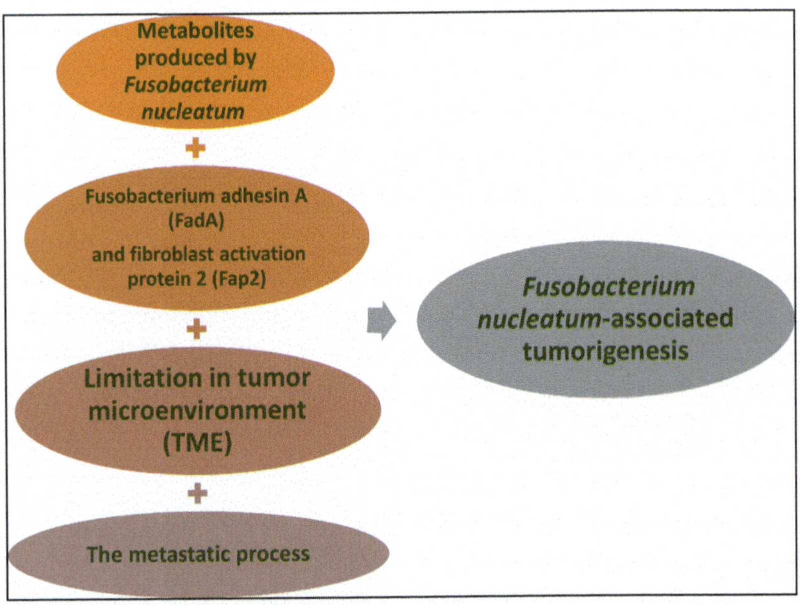

Fig. 1 The mechanisms of *Fusobacterium nucleatum*-associated tumorigenesis.

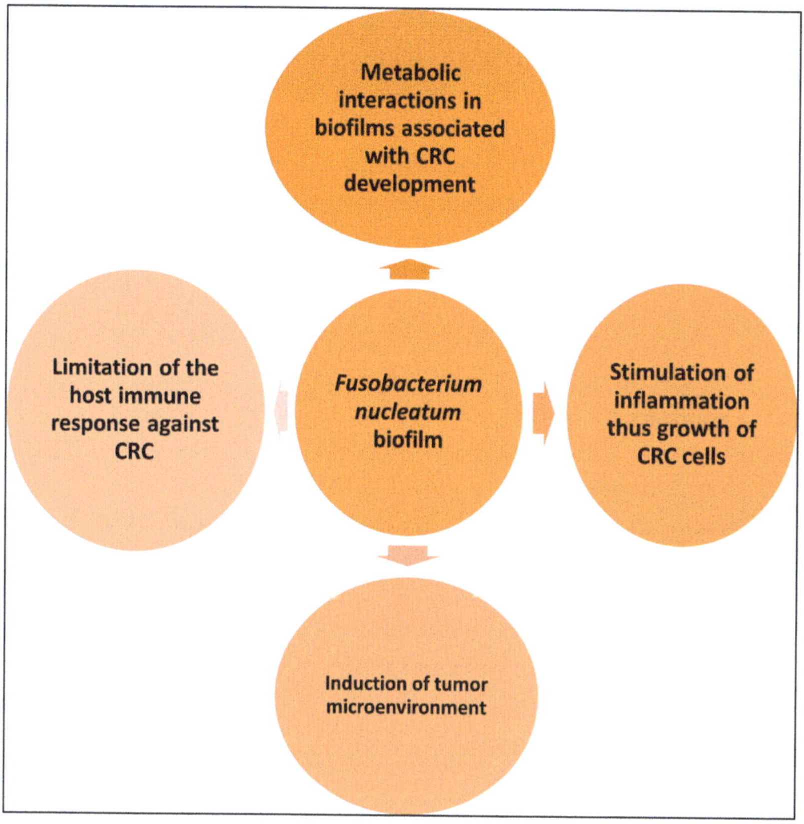

Fig. 2 The association between *Fusobacterium nucleatum* biofilm and colorectal cancer.

was indicative to the role of biofilms and core pathogens in characterization of the immune phenotype and the tumor microenvironment.[68] The association between *Fusobacterium nucleatum* biofilm and colorectal cancer illustrated in Fig. 2.

5. Porphyromonas gingivalis and oral/esophageal cancers

Oral cancer is a widespread cancer diagnosed in over 377,000 cases and causing death to 177,000 per year.[69] Oesophageal cancer grades as the seventh leading cause of cancer-related deaths worldwide. There are two major subtypes, the first is oesophageal squamous cell carcinoma and the second is oesophageal adenocarcinoma. According to epidemiological investigations, there is a gradual increase in the global occurrence of oesophageal adenocarcinoma in the last decade.[70] *Porphyromonas gingivalis* (*P. gingivalis*) is a gram-negative bacterium acts as the principal cause periodontitis, a closely related disease to periodontal disease development. Periodontitis progress to chronic infectious disease, is directly associated with the inflammatory immune response.[71]

Clinical studies investigated a great association between *P. gingivalis* and oral squamous cell carcinoma (OSCC). *P. gingivalis* stimulates the progression of OSCC through inhibition of the Cal-27 cells phagocytosis via macrophages, causing induction of macrophages functional polarization into macrophages tumor-associated macrophages (M2 TAMs), resulting in upregulation of expressed genes encoding for protumor molecules thus establishing immunosuppressive tumor microenvironment, which trigger OSCC immunoevasion and cause promotion of OSCC growth.[72]

The virulence factors of *P. gingivalis* such as gingipains (cysteine proteases), lipopolysaccharides (called *Pg*-LPS), capsule, and fimbriae are mechanisms contributed to progression of gastric carcinogenesis/GC through dysbiosis, systemic inflammation and tumorigenesis also innate immune response evasion. Pg-gingipains causes degradation of constituents of the tight junctions in gingival and gastrointestinal epithelial layer leading to disruption of their integrity.[73] In a study to investigate the relation between *P. gingivalis* and OSCC. *P. gingivalis*-infected OSCC cells was investigated to form metastatic foci in the lung more than uninfected cells.[74]

Bacterial adhesins and aggregation with colonizers from early- and late-stage such as *P. gingivalis* assigned with production of virulence factors. These factors influence the structure of biofilm communities, load of bacteria, and expression of other species gene affecting the host immune responses. Adhesins, proteolytic enzymes, and proinflammatory surface ligands in *F. nucleatum* permit its invasion of the oral epithelial cells and formation of a toxic environment. The progress of this environment develops the key to colonization of *P. gingivalis* in periodontal biofilms, resulting in proinflammatory tissue destruction.[75] Both *P. gingivalis* and *F. nucleatum* are associated with the sustained irritation and progression of tumor by NF-κB and NOD-like receptor (NLRP3) inflammasome pathways in esophageal cancer.[76,77] The mutual correlation between *P. gingivalis* and *F. nucleatum* in progress of toxic environment within tumor progression highlighted in Fig. 3.

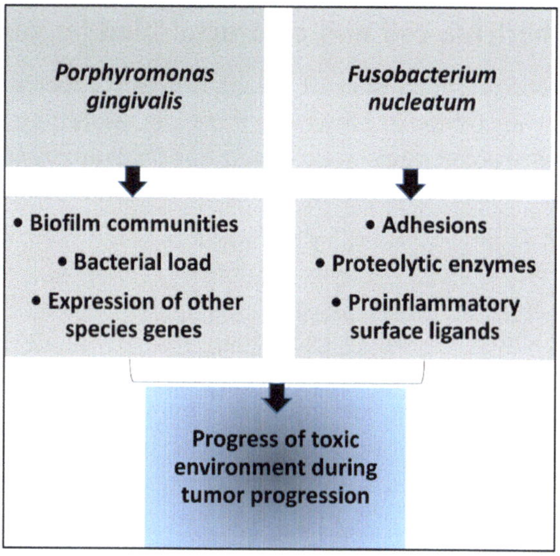

Fig. 3 The mutual correlation between *P. gingivalis* and *F. nucleatum* in progress of toxic environment within tumor progression.

The over expression of matrix metalloproteinases by *P. gingivalis* induce epithelial–mesenchymal transition (EMT) through enhancing the production of MMP-1 and MMP-10. The repeated infection of OSCC cells with *P. gingivalis* causes the transformation into a mesenchymal-like phenotype with epithelial markers decreased expression (such as cytokeratin 13) and mesenchymal biomarkers increased expression (such as N-cadherin and α-SMA).[78]

P. gingivalis shows anti-apoptotic activity through activation of signaling pathways. In the gingival epithelium, the P2×7 surface purinergic receptor has pro-apoptotic ability after binding to ATP. *P. gingivalis* produce the nucleoside diphosphate kinase that cleaves ATP consequently inhibits tumor cells apoptosis and promotes tumorigenesis. Other anti-apoptotic mechanisms stimulated by *P. gingivalis* such as Protein kinase B, Janus kinase 1 and STAT3, thus regulates the intrinsic mitochondrial apoptosis pathway.[79]

The accelerating role of *P. gingivalis* in progression of OSCC through enhancement of the autophagy pathway towards carcinoma apoptosis reduction.[80] ER-derived autophagosomal vesicles formed by *P. gingivalis* then incorporate and defend the bacterium against cytosolic ubiquitin-labeling and degradation by lysosomes. These double-membrane autophagosomal vesicles (also known as replicative autophagosomes) offer environment rich with nutrients to continue and replicate safely in the host cell.[81]

6. Escherichia coli and colorectal/bladder cancer

There is natural abundance of the gram-negative bacterium *E. coli* in the human gut microbiome. However, there was association between the high incidence of colon mucosa-associated *E. coli* colonization and patients with colorectal cancer. The produced toxins from the pathogenic *E. coli* strains such as cytotoxic necrosis factor (CNF), cyclomodulin, circulation inhibitory factor, colibactin, and cytolethal distending toxins (CDT) resulting in disruption of the cell cycle and/or promoting DNA damage, thus affecting differentiation of cells, apoptosis, in addition to cell proliferation.[50] Cytolethal distending toxin (CDT) and Colibactin carcinogenic activity results from induction of double-strand DNA breaking through its deoxyribonuclease activity.[82] The increase in *Escherichia coli* (*E. coli*) has been associated with inflammation of chronic tissue and the release of pro-inflammatory in addition to carcinogenic mediators.[83] The effect of produced toxins from the pathogenic *E. coli* on cancer progression highlighted in Fig. 4.

The prevalence of gallbladder cancer as the most widespread biliary tract malignancy reflecting one of the disease's risk factors as gallbladder disease history.[84] E-coli caused alteration to the heterogonous tumor cell population via epithelial mesenchymal transition (EMT), cancer stem cells (CSCs) and phenotypic and metabolic reprogramming, resulting in improvement of cancer cell ability in migration and self-renewal in vitro.[85]

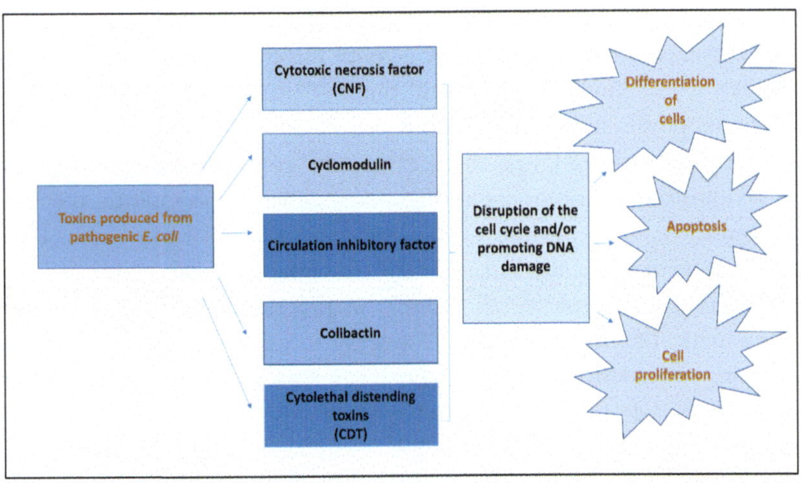

Fig. 4 The effect of produced toxins from the pathogenic *E. coli* on cancer progression.

The proteinaceous surface projections hairlike of uropathogenic *Escherichia coli* recognized as chaperone-usher pathway (CUP) pili. UTI progression mediated through type 1 pilus fibers by attachment to the urothelial surface via interacting with the mannose moieties of the bladder uroplakin surface in addition to biotic and abiotic surface and intracellular mediation of biofilm formation. The role of bacterial surface adhesion molecules contributes to aggregation of cancer cell.[86,87]

The direct degradation of c-MYC important in oncogenic transformation in malignancies and inhibition of *MYC* expression essential in cellular growth, metabolism and tissue development by uropathogenic *Escherichia coli* (UPEC) bacteria in human cells and murine tissues was investigated through identification of MYC inhibitors in the supernatants of bacterial culture, purification and characterization of a c-MYC-specific protease.[88]

7. Candida species and mucosal cancers

7.1 Biofilm formation in candida albicans

Biofilm formation of microbes on inert surfaces has become a matter of deep study in the past ten years and there seems to be a certain correlation between the capacity of the organism to form a biofilm and its pathogenicity.[89] In addition, it is believed that the architecture of biofilms favors the least oxygen availability around the central core region although how such an environment contributes towards altering the biofilm growth is still unknown, especially in Candida species. There is some conflicting evidence of the growth of Candida in anaerobic conditions by a small number of researchers.[89] Samarinayake et al., in earlier works, identified minute, but visible, surface growth of Candida albicans in Sabourauds dextrose deoxygenated slopes in rigorous anaerobic incubations.[90] In a subsequent report Webster and Odds, showed that anaerobic growth of seven *Candida* species which included *C. albicans* but not *C. guilliermondii* and *C. parapsilosis* were dormant, varied in anaerobiosis.[91] Biswas and Chaffin recently have not tried to grow *C. albicans* biofilms anaerobically, and have shown that in static conditions nor in mild shaking was there any growth of fungal biofilms on both plastic and denture acrylic.[89] Thus, the primary purpose of the carried-out research was to compare quantitatively the influence of the aerobic/anaerobic, and the static/dynamic conditions on the biofilm formation of *C. albicans* and the six non-albicans species of *Candida*, i.e., *C. krusei, C. glabrata, C. tropicalis, C. parapsilosis,*

C. guilliermondii, and *C. kefyr.*[89] The possibility was also exploited to determine the ultrastructure of the biofilm of *C. albicans* and *C. glabrata* (one strain each) in the above-mentioned environmental conditions through the application of scanning electronic microscopy.[89]

The presence of biofilms in surfaces of hosts or medical equipment links most infections caused by this pathogen to high morbidity and mortality. Notably, *C. albicans* biofilms are known to be naturally resistant to drug treatment, and hence the sensitivity of *C. albicans* biofilms to existing treatment options is very low.[92] Hence, the future unpredictability of any new molecules being amplified as new clinical antifungals is uncertain. Biofilm formation of *C. albicans* has been known to produce effects on antifungals susceptibility resulting in resistance also illustrated the need to focus on studies that look at prevention and control of these clinical microbial communities.[92]

The formation of a biofilm is an important determinant source of virulence. This begins by the *C. albicans* cells binding to the surfaces of a foreign body and creating a layer of cells and this enhances the formation and generation of the hyphal wall proteins.[93] Next, it is the actual growth of cells, and production of extracellular matrix, and at the end, there is hyphae formation that makes the biofilm permanent. The biofilm shields the *Candida* against the immune system and against antifungal treatment.[93] C. *albicans* also synthesizes an essential virulence factor, the extracellular hyphal-linked peptide toxin candidalysin, which is critical to triggering mucosal and systemic infections. It has been established over the years that candidalysin is a drug lytic component of *C. albicans*, which contributes to the acquisition of nutrients with a view to sustaining life and reproduction.[93] Moreover, candidalysin plays a crucial role in initiating the innate antifungal immunity in the course of infection that is mostly neutrophil and interleukin 17 (IL-17)-based. It is notable that candidalysin can indirectly stimulate epidermal growth factor receptor (EGFR), a complex process that includes EGFR and matrix metalloproteinase (MMP) ligands to be involved in most forms of cancer.[93]

Phenotypic transformation helps *C. albicans* to adapt to its environment (in this case, the host tissue).[94] When *C. albicans* invades epithelial mucosa cells, it can destroy the defensive barrier of the host epithelium, cause apoptosis or necrosis, and finally lead to structural changes in the epithelial cell.[95] This damages the epithelial cells and alters the normal structure, leading to abnormal proliferation associated with the development of cancer.[96] In addition, the ability of *Candida* to form biofilms can lead to a long-term exposure of host tissues to fungal carcinogens, such as acetaldehyde and an increased production of hydrolases, which can impair the innate immune system and cause chronic inflammation.[93] Risk factors for cancer development include chronic inflammation caused by microbial infections.[97]

7.2 Mycobiome and epithelial cell transformation

The interrelation of the gut microbiome with the human health has been within the focus since time immemorial and this composition of the gut microbiome consists not only of the bacteria but also various fungi which is named as the mycobiome. New studies have shown an important but perhaps underrated role played by fungal species, notably Candida albicans, in the maintenance of gut homeostasis, as well as the propagation of pathology in the case of the dysregulated state.[98] Allert et al.[99] concentrate on a particular fungus Candida albicans and the effect of translocation across intestinal epithelial barriers that can trigger systemic infections like candidemia. The authors provide evidence in vitro and through a high-throughput mutant screen that *C. albicans* translocation uses multiple steps to occur, including adhesion, invasion, damage of epithelia, and ultimate penetration of an epithelial barrier.[99] An important intermediary step in the process is the peptide toxin, candidalysin, which is encoded by ECE1 gene and triggers necrotic (but not apoptotic) cell death and results in the disruption of the barrier integrity.[99] Striking reduction in host cell debilitating capability and translocation in mutant strains uncursed in candidalysin production showed a direct mechanistic connection between harm to epithelium and spread to extra intestine, fundamentally explaining the connection between epithelial harm and systemic spreading. The results of this work show that *C. albicans* translocation process is neither entirely passive nor opportunistic but it is active and uses mechanisms derived by the fungus.[99]

Notably, epithelial damage that is germinal in translocation does not exclusively depend on exogenous influences such as surgery, or responses to immunosuppression but may be caused by the fungus itself. The most severe defect in translocation affecting deletion mutant incapability of forming hyphae, adherence or causing damage (e.g., ECE1Δ) shows that these processes are key virulence activities. The fact that physiological relevance was given to the results by the differentiated C2BBe1 intestinal epithelial cell line being used to represent the human gut barrier in a trans well model should be noted here. Corroborated by these mechanistic ideas, the general ecological and clinical background is offered in the review article about the gut mycobiome. Less diverse and abundant than the bacteriome, the human gut mycobiome has profound effects on the metabolism of the host, immune system, and the intestinal barrier. *C. albicans* (along with other fungi including Saccharomyces and Malassezia) are members of a dynamic fungal community that interacts complexly with bacteria and the host tissues.[98]

The review by Huang et al., highlights the point that fungal dysbiosis i.e., pathogen overrepresentation such as C. albicans has been associated with a wide range of conditions and diseases such as inflammatory bowel disease (IBD), obesity, type 2 diabetes, and colorectal cancer. In healthy humans, the overgrowths of fungi are restrained by the immune surveillance (e.g., through dectin-1 receptors signaling), and microbiota-mediated competition.[98] Activities like chemotherapeutics, immune-compromised condition or alterations in fooding pattern may lead to fungal proliferation and greater colonization. *C. albicans* has dual nature being commensals aiding immune priming and epithelial maturation and pathogen that causing inflammation and tissue infiltration. Such transition between the modes of commensalism and pathogenicity is regulated not only by the fungus (e.g., morphogenesis, candida lysin) but also by the host immune system condition or gut-microbiota balance). Another investigation by Huang et al., says that fungal cell wall (i.e beta glucans and Mannans) are powerful immune modulators that activate and down regulate inflammation, depending on their interaction with the pathways. These studies emphasizes on the need to think about mycobiome beyond the notion of a passive population of microbes but more so as a factor of host physiology and pathology.[98] Translocation model according to Allert et al. gives us molecular clearness on how *C. albicans* invades the gut barrier, whereas the mycobiome review places these findings in a more general context of host and microbe interaction, systemic disease risk, and therapeutic implications.[99] As the cases of fungal infections increase, especially among the immunocompromised, and as people grow to value the gut-brain-immune axis, in future, the microbiome needs to use fungal profiling. Precise guidance of host immunity, probiotic and antifungal interventions are potential solutions that can help treat diseases that are fungally related. Specifically, developing an awareness of what leads to candidalysin expression and host tissue penetration may enable methods to avoid spread throughout the body via the intestinal tract.[98]

7.3 Role in oral and esophageal cancer

C. albicans colonization and its invasion begins with the adhesion of the mucosal epithelial cells (which can be characterized as the first line of defense against microorganisms), weakening, in its turn, the host defense mechanisms.[93] A myriad of surface proteins, most of which are adhesins and facilitate adhesion of the C. albicans cells to the mucosal wall, have an important role here. Als agglutinin-like glycoproteins, especially the Als3 protein, form the most important set of adhesins because Als3 protein has been demonstrated to be involved during adhesion in the oral and vaginal mucosa as well as formation of the biofilm and induction of endocytosis.[93] Some of the Hwp hyphal wall proteins given are adhesins such as Hwp1 which facilitate covalent binding as they become the substrate of transglutaminase on the human cell surface. Therefore, the adhesins Als3 and Hwp1 are primary features of the process of hyphae generation of *C. albicans* and are central in their adhesion to a host cell. There are various factors that determine the adhesion process; this includes the nature of the proteins in the structure of the cell wall, and the physicochemical nature of cell surface.[93] When the integrity of the epithelial cells has been compromised, there is induced endocytosis and an active infiltration process which is regarded as a key event in the process of pathogenesis. Correlation between the decreased host defence mechanisms through the colonization and subsequent invasion by the fungus and the nature of the structural changes with the likelihood of developing an individual to cancer is directly related.[93]

In addition, the properties of invasive enzymes present in *C. albicans* are of utmost significance to the pathogenesis process, permitting it to disrupt the integrity of the mucosal tissue structure and thus contribute to virulence of the strain and subsequent effects on the cells.[93] *C. albicans* adhesins bind to ligands, including proteins, fibrinogen, and fibronectin. Human cells have an outer membrane that is fluid phospholipid bilayers and they contain proteins incorporated in them. Through the release of proteolytic enzymes, which break the peptide bonds and the phospholipases i.e. enzymes that break down phospholipids, *C. albicans* breaks through such a compromised membrane and is then able to enter the cell thereafter. This set of enzymes also can lyse microflora and, therefore, suppress the host defences.[93]

Planocellular carcinoma, also referred to as oral squamous cell carcinoma, is the commonest type of cancer superimposed on the oral cavity and it constitutes more than ninety percent of the total oral cancer cases.[93] It may occur in different locations of the mouth such as tongue, buccal mucosa, lips or gingiva or palate. Oral planocellular carcinoma occurs in over half of the oral diseases and it appears as the result of premalignant lesions that can be expressed through the presence of white filling or red masses on the mucosa (leukoplakia/erythroplakia).[93]

The association between oral carcinogenesis and *Candida spp.* has long been the subject of discussion. There have been different suggestions on the various ways in which this fungal genus might have been involved in the development of oral cancer.[93] Some of the suggested mechanisms include causing genetic instability in oral cells, encouraging the transformation of oral epithelial cells, carcinogenesis byproducts, immune regulation, and inflammation.[93] Recent systematic review established that the presence of hallmark species of the genus in the oral cavity, *C. albicans*, may be associated with the onset of oral cancer due to the alterations it induces in the phenotypic structure of the cell, but also because of genotypic changes. Also, it was emphasized that *C. albicans* releases carcinogenic materials that can trigger the progression of oral cancer.[93]

7.4 Fungal-bacterial interactions in tumorigenesis

The tumor microenvironment (TME) is a complex ecosystem comprising cancer cells, immune cells, stromal cells, blood vessels, along with a diverse community of microorganisms, including bacteria and fungi. While the influence of bacteria in cancer development is widely documented, the role of fungi and their associations with bacteria is relatively sparse. Recent studies say that fungi are not merely opportunists but instrumental in shaping TME and influencing the results of cancer progression. This paper will focus on the association between fungi and bacteria within the TME influencing the cancer development. The advancements in sequencing and robustness have led to the discovery of a plethora of different communities of microbes with respect to different forms of tumors. Tumors of the breast, lung, pancreas, colon, as well as other organs have known to harbor both the bacterium and fungi. The presence of bacterial communities in tumors differs based on the type, stage, and location of the tumor. Some of microbial genera within tumors are *Escherichia, Fusobacterium, Bacteroides, and Streptococcus*. Such bacteria may alter cancer progression by diverse methods, including regulating inflammation, changing metabolism, and contacting cancer cells themselves. This is also very diverse, with *Candida, Malassezia, Aspergillus and Saccharomyces* amongst the most identified genera in tumors. Fungal species can also be affected by the host immune status, diet, and exposure to antifungal agents since some can be present and abundant whereas others can be absent. Based on previous reports, fungi and bacteria that are found in the TME can interact in multiple ways:

Nutrient Competition: There are competitions with fungi and bacteria over essential nutrients of the TME like glucose, amino acids, and trace elements. Such competition can affect the growth and metabolism of fungal and bacteria population, which may affect cancer cell survival and proliferation.

Metabolites Generation: Fungi and bacteria have a large repertoire of metabolites potentially influencing the TME. As an example, certain bacteria synthesize short-chain fatty acids (SCFAs) which may interfere with the immune responses and cancer cell proliferation. Some fungi can make mycotoxins and other secondary metabolites with potential cytotoxic or immunomodulatory activity. Interaction of these metabolites may result in a complicated chemical environment with the possibility of affecting cancer progression.

Quorum Sensing: Quorum sensing (QS) or cell-to- cell communication system where bacteria coordinate gene expression with response to cell population density. The response to signals and production of QS molecules by some fungi are also possible. This inter-kingdom signalling has been shown to manipulate the actions of both fungal and bacterial populations with the possibility of having an effect of biofilm creation, pathogenicity, and antibiotic tolerance.

Biofilm Formation: Another effect is a potential ability of fungi and bacteria to constitute biofilms in the TME, and biofilms are communities of microorganisms embedded within a self-generated matrix. Microbial biofilms may effectively shield the microorganism against immune rejection and antibiotic therapy to support persistent infection and increase the spread of cancer. Synergy in the microbial consortia in biofilms may increase fungal and bacterial resistance to the adversities of the TME and the capacity of the microorganisms to sustain themselves in biofilms. Fig. 5.

Immune Modulation: Fungi and bacteria can interact with the immune system of the host, affecting inflammatory response and antitumor immunity. Certain fungi and bacteria have the ability of stimulating immune cells e.g., macrophages, dendritic cells resulting in inflammatory pro-cytokines. Immune suppression can be achieved by other fungi and

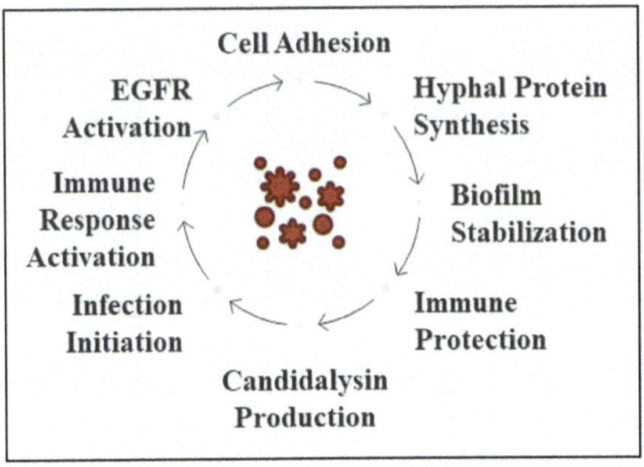

Fig. 5 Candida biofilm formation cycle.

bacteria that aid cancer cells to evade their immune systems. The relative pro-inflammatory to immunosuppressive action can play a role in the growth of cancer and reaction to immunotherapy.[93]

Fungi-bacterial interactions in the TME can have great potential effects on tumorigenesis in the following ways:

Cancer Initiation: Some species of fungi and bacteria are known to cause cancer initiation. As an example, there are bacteria that may form genotoxic metabolites dangerous to DNA and stimulating mutations. There is also an uprising risk of cancer in some communities due to fungal infections. Fungal and bacterial infections may co-operate in the process of cancer promotion through their synergistic action.

Cancer Progression: The interaction between fungus and bacteria may have a significant role in the progression of cancer cells and their proliferation, invasion, and metastasis. Bacteria can generate growth factors augmenting the cancer cells development. Cancer cells can be attacked by the extracellular matrix of fungi, which is actively secreted by the extracellular matrix. The cancer cell growth can also lead to the formation of biofilms, thus promoting cancer cell dispersal and metastasis.[93]

Therapeutic Resistance: There may be fungal-bacterial interactions that increase therapeutic resistance. Cancer and microorganisms resistant to traditional forms of chemotherapy and radiation may be shielded by biofilms. Certain bacteria may break down chemotherapy medicines, making their effects diminished. Another way by which fungi infections influence the tumor microenvironment altering it in a way that does not respond to treatment Fig. 6.

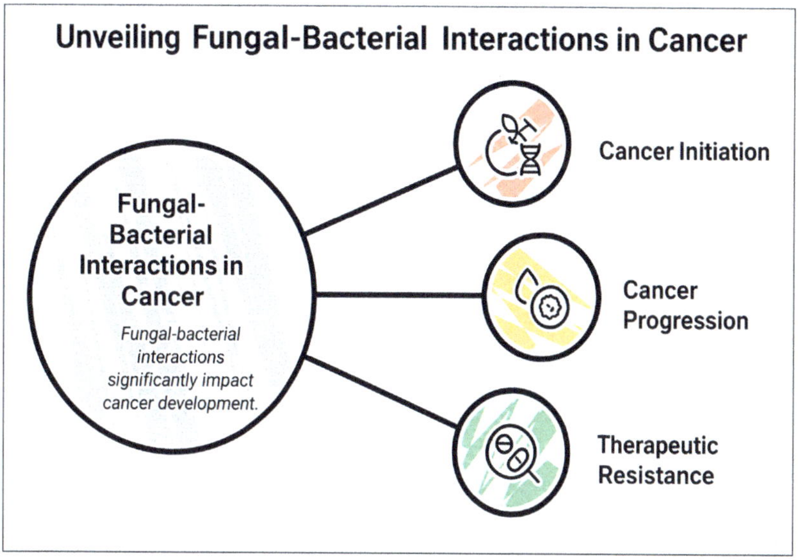

Fig. 6 Unveiling fungal-bacterial interactions in cancer.

The key is to gain an understanding of this intricate fungi-bacteria relationship within a TME, which creates the opportunity to develop new approaches to cancer treatment:

Targeting Polymicrobial Communities: Factors to weaken fungal-bacterial interactions, like quorum sensing inhibition or biofilm destabilization have the potential to increase the effectiveness of the traditional cancer treatments.

DCs and Tumor Microenvironment: Some techniques to manipulate the TME through modulation of the fungal and bacterial microbiome may trigger a better anti-tumour immunity and improve immunotherapy response.

Emerging antimicrobial agents: Newly developed antimicrobial agents targeting specifically fungal-bacterial biofilms or interfering with their interactions can be a new therapeutic paradigm in the treatment of patients with cancer.

Fungal-Bacteria interaction in the tumor microenvironment constitutes an extremely complex and dynamic ecosystem that can have serious consequences on tumorigenesis. There is a need to further study the mechanisms of these interactions and develop therapeutic intervention that can selectively target these interactions to facilitate positive outcome in cancer. Through the complex interaction of fungi, bacteria, and cancer cells, we can make sure that stronger and more precision-driven cancer treatment can be achieved Fig. 7.

The correlation between infections by microorganisms and cancer is also a matter of concern at present. The number of studies that favor the theory that there exists an association is increasing especially when the causative agents are the bacteria or virus. *Candida* is a series of opportunistic pathogenic fungi that are widespread in skin, mouth, and the gastrointestinal tract of the host. Reduction of the immunity of the host facilitates occurrence of yeast infection by *Candida*. The most common pathogenic fungus is *Candida albicans* among them.[100] Over the recent years, epidemiological and pathological data have been continuously indicating the influence of pathogenic microorganisms on the rates of occurrence of cancers in the world as of significant contribution. Nevertheless, cancers have only been associated with fungal infection in a few studies. The interconnection of carcinogenesis and fungal microbiota disproportion is mostly unfamiliar. It stems from the fact that the number of fungi is relatively scarce and there is no such clearly defined reference genome.[100] Moreover, the research techniques are a demanding one, and thus the fungal microbiota is typically examined comparatively little compared to different microorganisms. To date, research has revealed that *C. albicans* in the fungal microorganism is highly related to the development of cancer.

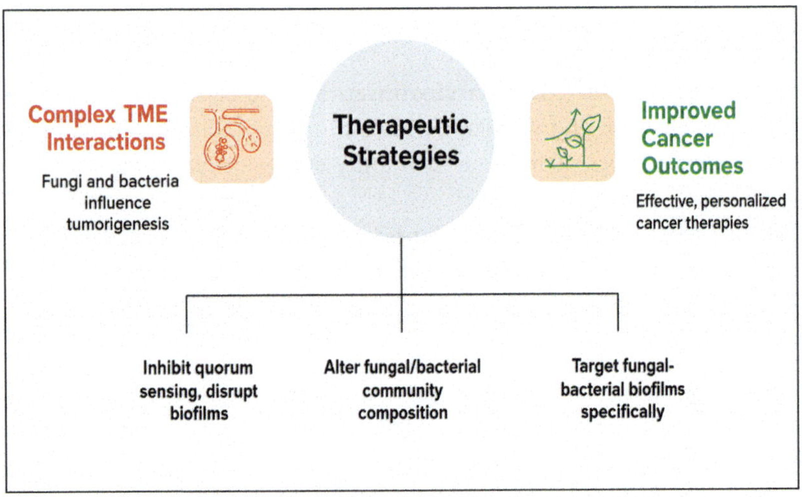

Fig. 7 Targeting fungal-bacterial interactions for cancer therapy.

Candida albicans is an opportunistic fungus associated with a high level of carcinogenesis and metastasis. *Candida spp.* has traditionally been associated with cancerous activity though that was primarily since *Candida* is an opportunist pathogen that exploits the immunosuppressed nature of patients especially courtesy of chemotherapy.[100] Conversely, the latest evidence shows that *C. albicans* can promote cancer in multiple ways, as discussed in the review; creation of cancerous byproducts, inflammatory trigger, Th17 response activation and molecular mimicry. The necessity to prevent not only this type of infection during cancer treatment but also to develop new ways of therapy to prevent pro-tumor effect of this yeast species was emphasized as well since the role of this species of yeasts is widely discussed in nosocomial infections as well. Induced oral leukoplakia and malignant tumors is reported.[100] Based on the available epidemiological data, the broadening of *Candida* identifying tendency in the patients with oral cancer is observed, and the primary one is *C. albicans*. Sadly, these are merely epidemiological and descriptive studies that have no effective experimental evidence. The production of IL-1 beta that was induced by NLRP3 inflammasome resulted in vitro and in vivo 5-FU resistance in OSCC. This could be deduced that the availability of *C. albicans* in oral cancer could have contributed to the impact of chemotherapy in producing IL-1 b which was also a possible target in treating oral cancer.[100]

The outcomes of some studies also revealed that *Candida albicans* strains in oral precancer lesions had the highest nitrosation potential with *Candida tropicalis, Candida parapsilosis, and Torulopsis glabrata* depicting lesser nitrosation potential, which can cause epithelial carcinogenesis inducing tumor development.[100]

8. Conclusion

Microbial biofilms play a crucial role in the initiation, progression, and persistence of several cancers. The ability of species such as *Helicobacter pylori, Fusobacterium nucleatum, Porphyromonas gingivalis, Escherichia coli*, and *Candida albicans* to establish biofilms enhances their survival, promotes chronic inflammation, and supports a tumor-friendly microenvironment. These biofilm-mediated processes are not only responsible for genetic alterations but also help in immune evasion and resilience to chemotherapeutics, that can escalate cancer development.

Future research should stress on novel approaches for mitigation of biofilms, rapid innovations for biofilm detection, and diving into the interplay between microbial consortia and the tumor microenvironment, as a result of which, treatment regimens would be rendered convenient and effective.

CRediT authorship contribution statement

Sougata Rajak: Project administration, writing original draft, conceptualization, methodology, data curation, resources, validation, review and editing original draft, formal, formal analysis. **Panteha Torabian:** Writing original draft, conceptualization. **Naiera M. Helmy Mohamed:** Writing original draft. **Neethu Asokan:** Writing original draft. **Reham Hamed Abdelaziz:** Writing original draft. **Saanya Chaturvedi:** Project administration, writing original draft, review and editing original draft, formal analysis, provided critical feedback.

Declaration of competing interest

The authors declare that they have no known competing financial interests or personal relationships that could have appeared to influence the work reported in this paper.

References

1. Vestby LK, Grønseth T, Simm R, Nesse LL. Bacterial biofilm and its role in the pathogenesis of disease. *Antibiotics.* 2020;9(2):59. https://doi.org/10.3390/antibiotics9020059
2. Choi E, Murray B, Choi S. Biofilm and cancer: interactions and future directions for cancer therapy. *Int. J. Mol. Sci.* 2023;24(16):12836. https://doi.org/10.3390/ijms241612836
3. Zhao A, Sun J, Liu Y. Understanding bacterial biofilms: from definition to treatment strategies. *Front Cell Infect. Microbiol.* 2023;13:1137947. https://doi.org/10.3389/fcimb.2023.1137947
4. Sharma S, Mohler J, Mahajan SD, Schwartz SA, Bruggemann L, Aalinkeel R. Microbial biofilm: a review on formation, infection, antibiotic resistance, control measures, and innovative treatment. *Microorganisms.* 2023;11(6):1614. https://doi.org/10.3390/microorganisms11061614
5. Ahmed ZHT, Singh P, Saraswathi A, et al. Introduction to microbial biofilms. In: Busi S, Pattnaik S, Prasad R, eds. *Omics Approaches in Biofilm Research.* Cham: Springer; 2025. https://doi.org/10.1007/978-3-031-91863-6_1.
6. Elshenawi Y, Hu S, Hathroubi S. Biofilm of *Helicobacter pylori*: life cycle, features, and treatment options. *Antibiotics.* 2023;12(8):1260. https://doi.org/10.3390/antibiotics12081260
7. Jo J, Price-Whelan A, Dietrich LEP. Gradients and consequences of heterogeneity in biofilms. *Nat. Rev. Microbiol.* 2022;20(10):593–607. https://doi.org/10.1038/s41579-022-00692-2
8. Gupta A, Bhattacharya D, Elayaperumal S, Sivamani Y, Lahiri D, Nag M. Biofilms: a cause for the development of cancer: a review. *Microbe.* 2025;6:100236. https://doi.org/10.1016/j.microb.2025.100236
9. Reyes VE. *Helicobacter pylori* and its role in gastric cancer. *Microorganisms.* 2023;11(5):1312. https://doi.org/10.3390/microorganisms11051312
10. Duan Y, Xu Y, Dou Y, Xu D. Helicobacter pylori and gastric cancer: mechanisms and new perspectives. *J. Hematol. & Oncol.* 2025;18(1):10. https://doi.org/10.1186/s13045-024-01654-2

11. Ou S, Wang H, Tao Y, et al. Fusobacterium nucleatum and colorectal cancer: from phenomenon to mechanism. *Front Cell Infect. Microbiol.* 2022;12:1020583. https://doi.org/10.3389/fcimb.2022.1020583
12. Muñoz-Medel M, Pinto MP, Goralsky L, et al. Porphyromonas gingivalis, a bridge between oral health and immune evasion in gastric cancer. *Front Oncol.* 2024;14:1403089. https://doi.org/10.3389/fonc.2024.1403089
13. Sadeghi M, Mestivier D, Sobhani I. Contribution of pks+ *Escherichia coli* (*E. coli*) to colon carcinogenesis. *Microorganisms.* 2024;12(6):1111. https://doi.org/10.3390/microorganisms12061111
14. Kashyap B, Padala SR, Kaur G, Kullaa A. *Candida albicans* induces oral microbial dysbiosis and promotes oral diseases. *Microorganisms.* 2024;12(11):2138. https://doi.org/10.3390/microorganisms12112138
15. Malavika G, Ravi S, Maheswary D, Leela K, Lathakumari R, Priya L. Role of Candida albicans in chronic inflammation and the development of oral squamous cell carcinoma. *Cancer Pathogenesis Ther.* 2025. https://doi.org/10.1016/j.cpt.2025.03.002
16. Montanari E, Bernardo G, Le Noci V, et al. Biofilm formation by the host microbiota: a protective shield against immunity and its implication in cancer. *Mol. cancer.* 2025;24(1):148. https://doi.org/10.1186/s12943-025-02348-0
17. Di Russo S, Borsatti G, Bouzidi A, et al. NF-κB-mediated cytokine secretion and glutamate metabolic reprogramming converge in breast cancer brain tropism. *Cancer Lett.* 2025;630:217907. https://doi.org/10.1016/j.canlet.2025.217907
18. Kvich L, Fritz BG, Zschach H, et al. Biofilms and core pathogens shape the tumor microenvironment and immune phenotype in colorectal cancer. *Gut Microbes.* 2024;16(1):2350156. https://doi.org/10.1080/19490976.2024.2350156
19. Binder GA, Fischman S, Revach B, et al. Periodontal pathogens Porphyromonas gingivalis and Fusobacterium nucleatum promote tumor progression in an oral-specific chemical carcinogenesis model. *Oncotarget.* 2015;6:22613 23 https://doi.org/10.18632/oncotarget.4209.
20. Galasso L, Termite F, Mignini I, et al. Unraveling the role of *Fusobacterium nucleatum* in colorectal cancer: molecular mechanisms and pathogenic insights. *Cancers.* 2025;17(3):368. https://doi.org/10.3390/cancers17030368
21. Scherr TD, Hanke MLHuang O, James DBA, et al. Staphylococcus aureus biofilms induce macrophage dysfunction through leukocidin AB and alpha-toxin. *mBio.* 2015;6(10):1128. https://doi.org/10.1128/mbio.01021-15
22. Holbert CE, Cullen MT, Casero Jr RA, Stewart TM. Polyamines in cancer: integrating organismal metabolism and antitumour immunity. *Nat. Rev. Cancer.* 2022;22(8):467–480. https://doi.org/10.1038/s41568-022-00473-2
23. Dubois T, Tremblay YDN, Hamiot A, et al. A microbiota-generated bile salt induces biofilm formation in *Clostridium difficile. NPJ Biofilms Microbiomes.* 2019;5(1):14. https://doi.org/10.1038/s41522-019-0087-4
24. Cong J, Liu P, Han Z, et al. Bile acids modified by the intestinal microbiota promote colorectal cancer growth by suppressing CD8 + T cell effector functions. *Immunity.* 2024;57:876–e88911. https://doi.org/10.1016/j.immuni.2024.02.014
25. Zhang S, Huang J, Jiang Z, Tong H, Ma X, Liu Y. Tumor microbiome: roles in tumor initiation, progression, and therapy. *Mol. biomedicine.* 2025;6(1):9. https://doi.org/10.1186/s43556-025-00248-9
26. Đikić D, Bogdanović A, Marković D, et al. Inflammation promotes oxidative and nitrosative stress in chronic myelogenous leukemia. *Biomolecules.* 2022;12(2):247. https://doi.org/10.3390/biom12020247
27. Xie X, Liu W, Yuan Z, Chen H, Mao W. Bridging epigenomics and tumor immunometabolism: molecular mechanisms and therapeutic implications. *Mol. cancer.* 2025;24(1):71. https://doi.org/10.1186/s12943-025-02269-y

28. Otálora-Otálora BA, López-Rivera JJ, Aristizábal-Guzmán C, Isaza-Ruget MA, Álvarez-Moreno CA. Host transcriptional regulatory genes and microbiome networks crosstalk through immune receptors establishing normal and tumor multiomics metafirm of the oral-gut-lung axis. *Int. J. Mol. Sci.* 2023;24(23):16638. https://doi.org/10.3390/ijms242316638
29. Chen G, Wu K, Li H, Xia D, He T. Role of hypoxia in the tumor microenvironment and targeted therapy. *Front Oncol.* 2022;12:961637. https://doi.org/10.3389/fonc.2022.961637
30. Li J, Zhang AH, Wu FF, Wang XJ. Alterations in the gut microbiota and their metabolites in colorectal cancer: recent progress and future prospects. *Front Oncol.* 2022;12:841552. https://doi.org/10.3389/fonc.2022.841552
31. Asma ST, Imre K, Morar A, et al. An overview of biofilm formation-combating strategies and mechanisms of action of antibiofilm agents. *Life (Basel, Switz).* 2022;12(8):1110. https://doi.org/10.3390/life12081110
32. Wang S, Ding H, Li L, Zhao R, Chai N. Targeted nanomedicines for the treatment of Helicobacter pylori infection. *Mater Today Bio.* 2025;32:101820. https://doi.org/10.1016/j.mtbio.2025.101820
33. Modeel S, Siwach S, Dolkar P, et al. Emerging risk factors and the role of gut microbiota in immunomodulation and therapeutic implications in colorectal cancer. *Cancer Pathogenesis Ther.* 2025. https://doi.org/10.1016/j.cpt.2025.06.007
34. LaCourse KD, Zepeda-Rivera M, Kempchinsky AG, et al. The cancer chemotherapeutic 5-fluorouracil is a potent Fusobacterium nucleatum inhibitor and its activity is modified by intratumoral microbiota. *Cell Rep.* 2022;41:111625. https://doi.org/10.1016/j.celrep.2022.111625
35. Gu Y, Xu Y, Wang P, Zhao Y, Wan C. Research progress on molecular mechanism of pyroptosis caused by *Helicobacter pylori* in gastric cancer. *Ann. Med. Surg. (2012).* 2024;86(4):2016–2022. https://doi.org/10.1097/MS9.0000000000001802
36. Sedarat Z, Taylor-Robinson AW. *Helicobacter pylori* outer membrane proteins and virulence factors: potential targets for novel therapies and vaccines. *Pathogens.* 2024;13(5):392. https://doi.org/10.3390/pathogens13050392
37. Ge X, Cai Y, Chen Z, et al. Bifunctional enzyme SpoT is involved in biofilm formation of helicobacter pylori with multidrug resistance by upregulating efflux pump Hp1174 (*gluP*). *Antimicrobial Agents Chemother.* 2018;62(11):e00957-18. https://doi.org/10.1128/AAC.00957-18
38. Andersen L, Rasmussen L. *Helicobacter pylori*-coccoid forms and biofilm formation. *FEMS Immunol. & Med. Microbiol.* 2009;56(2):112–115. https://doi.org/10.1111/j.1574-695X.2009.00556.x
39. Sharndama HC, Mba IE. Helicobacter pylori: an up-to-date overview on the virulence and pathogenesis mechanisms. *Braz. J. Microbiol.: Publ. Braz. Soc. Microbiol.* 2022;53(1):33–50. https://doi.org/10.1007/s42770-021-00675-0
40. Baj J, Forma A, Sitarz M, et al. *Helicobacter pylori* virulence factors-mechanisms of bacterial pathogenicity in the gastric microenvironment. *Cells.* 2020;10(1):27. https://doi.org/10.3390/cells10010027
41. Correa P, Piazuelo MB. The gastric precancerous cascade. *J. Digestive Dis.* 2012;13(1):2–9. https://doi.org/10.1111/j.1751-2980.2011.00550.x
42. Cheng H, Yang Y, Yang H, et al. Evolution of the Correa's cascade steps: a long-term endoscopic surveillance among non-ulcer dyspepsia and gastric ulcer after H. pylori eradication. *J. Formos Med. Assoc.* 2023;122(5):400–410. https://doi.org/10.1016/j.jfma.2022.11.008
43. Valenzuela MA, Canales J, Corvalán AH, Quest AF. Helicobacter pylori-induced inflammation and epigenetic changes during gastric carcinogenesis. *World J. Gastroenterol.* 2015;21(45):12742–12756. https://doi.org/10.3748/wjg.v21.i45.12742

44. Muttiah B, Wahid W, Sukri A, Hanafiah A. Towards effective *Helicobacter pylori* eradication: emerging therapies in the wake of antibiotic resistance. *Int. J. Mol. Sci.* 2025;26(13):6064. https://doi.org/10.3390/ijms26136064
45. Rocha GR, Lemos FFB, Silva LGO, et al. Overcoming antibiotic-resistant *Helicobacter pylori* infection: current challenges and emerging approaches. *World J. Gastroenterol.* 2025;31(10):102289. https://doi.org/10.3748/wjg.v31.i10.102289
46. Guo Y, Mao Z, Ran F, et al. Nanotechnology-based drug delivery systems to control bacterial-biofilm-associated lung infections. *Pharmaceutics.* 2023;15(11):2582. https://doi.org/10.3390/pharmaceutics15112582
47. Fonseca D, Chitas R, Parreira P, Cristina M, Martins L. How to manage Helicobacter pylori infection beyond antibiotics: the bioengineering quest. *Appl. Mater Today.* 2024;37:102123. https://doi.org/10.1016/j.apmt.2024.102123
48. Feng L, Wang X, Guo X, et al. Identification of novel target DCTPP1 for colorectal cancer therapy with the natural small-molecule inhibitors regulating metabolic reprogramming. *Angew Chem. Int. Ed. Engl.* 2024;63(47):e202402543. https://doi.org/10.1002/anie.202402543
49. Han YW. Fusobacterium nucleatum: a commensal-turned pathogen. *Curr. Opin. Microbiol.* 2015;23:141–147. https://doi.org/10.1016/j.mib.2014.11.013
50. Yusuf K, Sampath V, Umar S. Bacterial infections and cancer: exploring this association and its implications for cancer patients. *Int. J. Mol. Sci.* 2023;24(4). https://doi.org/10.3390/ijms24043110
51. Dai Z, Zhang J, Wu Q, et al. The role of microbiota in the development of colorectal cancer. *Int. J. Cancer.* 2019;145(8):2032–2041. https://doi.org/10.1002/ijc.32017
52. Li R, Shen J, Xu Y. Fusobacterium nucleatum and colorectal cancer. *Infect Drug Resist.* 2022;15:1115–1120. https://doi.org/10.2147/idr.S357922
53. Hong J, Guo F, Lu SY, et al. F. nucleatum targets lncRNA ENO1-IT1 to promote glycolysis and oncogenesis in colorectal cancer. *Gut.* 2021;70(11):2123–2137. https://doi.org/10.1136/gutjnl-2020-322780
54. Dougherty MW, Jobin C. Intestinal bacteria and colorectal cancer: etiology and treatment. *Gut. Microbes.* 2023;15(1):2185028. https://doi.org/10.1080/19490976.2023.2185028
55. McIlvanna E, Linden GJ, Craig SG, Lundy FT, James JA. Fusobacterium nucleatum and oral cancer: a critical review. *BMC Cancer.* 2021;21(1):1212. https://doi.org/10.1186/s12885-021-08903-4
56. Hashemi Goradel N, Heidarzadeh S, Jahangiri S, et al. Fusobacterium nucleatum and colorectal cancer: a mechanistic overview. *J. Cell Physiol.* 2019;234(3):2337–2344. https://doi.org/10.1002/jcp.27250
57. Wang N, Fang JY. Fusobacterium nucleatum, a key pathogenic factor and microbial biomarker for colorectal cancer. *Trends Microbiol.* 2023;31(2):159–172. https://doi.org/10.1016/j.tim.2022.08.010
58. Qu R, Zhang Y, Ma Y, et al. Role of the gut microbiota and its metabolites in tumorigenesis or development of colorectal cancer. *Adv. Sci. (Weinh).* 2023;10(23):e2205563. https://doi.org/10.1002/advs.202205563
59. Zhang Y, Zhang L, Zheng S, et al. Fusobacterium nucleatum promotes colorectal cancer cells adhesion to endothelial cells and facilitates extravasation and metastasis by inducing ALPK1/NF-κB/ICAM1 axis. *Gut. Microbes.* 2022;14(1):2038852. https://doi.org/10.1080/19490976.2022.2038852
60. Zhang L, Leng XX, Qi J, et al. The adhesin RadD enhances Fusobacterium nucleatum tumour colonization and colorectal carcinogenesis. *Nat. Microbiol.* 2024;9(9):2292–2307. https://doi.org/10.1038/s41564-024-01784-w
61. Jiang SS, Xie YL, Xiao XY, et al. Fusobacterium nucleatum-derived succinic acid induces tumor resistance to immunotherapy in colorectal cancer. *Cell Host Microbe.* 2023;31(5):781–797.e789. https://doi.org/10.1016/j.chom.2023.04.010

62. Hernández-Luna MA, López-Briones S, Luria-Pérez R. The four horsemen in colon cancer. *J. Oncol.* 2019;2019:5636272. https://doi.org/10.1155/2019/5636272
63. Bullman S, Pedamallu CS, Sicinska E, et al. Analysis of Fusobacterium persistence and antibiotic response in colorectal cancer. *Science.* 2017;358(6369):1443–1448. https://doi.org/10.1126/science.aal5240
64. Brennan CA, Garrett WS. Fusobacterium nucleatum - symbiont, opportunist and oncobacterium. *Nat. Rev. Microbiol.* 2019;17(3):156–166. https://doi.org/10.1038/s41579-018-0129-6
65. Cheng Y, Ling Z, Li L. The intestinal microbiota and colorectal cancer. *Front Immunol.* 2020;11:615056. https://doi.org/10.3389/fimmu.2020.615056
66. Hold GL, Garrett WS. Gut microbiota. Microbiota organization–a key to understanding CRC development. *Nat. Rev. Gastroenterol Hepatol.* 2015;12(3):128–129. https://doi.org/10.1038/nrgastro.2015.25
67. Meng Q, Gao Q, Mehrazarin S, et al. Fusobacterium nucleatum secretes amyloid-like FadA to enhance pathogenicity. *EMBO Rep.* 2021;22(7):e52891. https://doi.org/10.15252/embr.202152891
68. Kvich L, Fritz BG, Zschach H, et al. Biofilms and core pathogens shape the tumor microenvironment and immune phenotype in colorectal cancer. *Gut. Microbes.* 2024;16:2350156. https://doi.org/10.1080/19490976.2024.2350156
69. Li J, Kot WY, McGrath CP, Chan BWA, Ho JWK, Zheng LW. Diagnostic accuracy of artificial intelligence assisted clinical imaging in the detection of oral potentially malignant disorders and oral cancer: a systematic review and meta-analysis. *Int. J. Surg.* 2024;110(8):5034–5046. https://doi.org/10.1097/js9.0000000000001469
70. Yang H, Wang F, Hallemeier CL, Lerut T, Fu J. Oesophageal cancer. *Lancet.* 2024;404(10466):1991–2005. https://doi.org/10.1016/s0140-6736(24)02226-8
71. Huang Z, Hao M, Shi N, et al. Porphyromonas gingivalis: a potential trigger of neurodegenerative disease. *Front Immunol.* 2025;16:1482033. https://doi.org/10.3389/fimmu.2025.1482033
72. Liu S, Zhou X, Peng X, et al. Porphyromonas gingivalis promotes immunoevasion of oral cancer by protecting cancer from macrophage attack. *J. Immunol.* 2020;205(1):282–289. https://doi.org/10.4049/jimmunol.1901138
73. Muñoz-Medel M, Pinto MP, Goralsky L, et al. *Porphyromonas gingivalis*, a bridge between oral health and immune evasion in gastric cancer. *Front Oncol.* 2024;14:1403089. https://doi.org/10.3389/fonc.2024.1403089
74. Woo BH, Kim DJ, Choi JI, et al. Oral cancer cells sustainedly infected with Porphyromonas gingivalis exhibit resistance to Taxol and have higher metastatic potential. *Oncotarget.* 2017;8(29):46981–46992. https://doi.org/10.18632/oncotarget.16550
75. Zhu J, Li M, Li J, Wu J. Sialic acid metabolism of oral bacteria and its potential role in colorectal cancer and Alzheimer's disease. *Carbohydr Res.* 2024;541:109172. https://doi.org/10.1016/j.carres.2024.109172
76. Dejea CM, Wick EC, Hechenbleikner EM, et al. Microbiota organization is a distinct feature of proximal colorectal cancers. *Proc. Natl. Acad. Sci. U. S. A.* 2014;111(51):18321–18326. https://doi.org/10.1073/pnas.1406199111
77. Moe KT, Tan KS. Mechanistic insights on microbiota-mediated development and progression of esophageal cancer. *Cancers (Basel).* 2024;16(19). https://doi.org/10.3390/cancers16193305
78. Wang X, He X, Zhong B. Oral microbiota: the overlooked catalyst in cancer initiation and progression. *Front Cell Dev Biol.* 2024;12:1479720. https://doi.org/10.3389/fcell.2024.1479720
79. Zhang X, Chen Y, Xia Y, et al. Oral microbiota in colorectal cancer: unraveling mechanisms and application potential. *Life Sci.* 2025;365:123462. https://doi.org/10.1016/j.lfs.2025.123462

80. Yuan K, Xu S, Liu G, et al. Porphyromonas gingivalis promotes oral squamous cell carcinoma progression by modulating autophagy. *Oral Dis.* 2025;31(2):492–502. https://doi.org/10.1111/odi.15157
81. Sheridan M, Chowdhury N, Wellslager B, et al. Opportunistic pathogen Porphyromonas gingivalis targets the LC3B-ceramide complex and mediates lethal mitophagy resistance in oral tumors. *iScience.* 2024;27(6):109860. https://doi.org/10.1016/j.isci.2024.109860
82. Wong SH, Yu J. Gut microbiota in colorectal cancer: mechanisms of action and clinical applications. *Nat. Rev. Gastroenterol. & Hepatol.* 2019;16(11):690–704. https://doi.org/10.1038/s41575-019-0209-8
83. Quaglio AEV, Grillo TG, De Oliveira ECS, Di Stasi LC, Sassaki LY. Gut microbiota, inflammatory bowel disease and colorectal cancer. *World J. Gastroenterol.* 2022;28(30):4053–4060. https://doi.org/10.3748/wjg.v28.i30.4053
84. Waller GC, Sarpel U. Gallbladder cancer. *Surg. Clin. North Am.* 2024;104(6):1263–1280. https://doi.org/10.1016/j.suc.2024.03.006
85. Abd-El-Raouf R, Ouf SA, Gabr MM, Zakaria MM, El-Yasergy KF, Ali-El-Dein B. Escherichia coli foster bladder cancer cell line progression via epithelial mesenchymal transition, stemness and metabolic reprogramming. *Sci Rep.* 2020;10(1):18024. https://doi.org/10.1038/s41598-020-74390-5
86. Maalouf N, Gur C, Yutkin V, Scaiewicz V, Mandelboim O, Bachrach G. High mannose level in bladder cancer enhances type 1 fimbria-mediated attachment of uropathogenic E. coli. *Front Cell Infect. Microbiol.* 2022;12:968739. https://doi.org/10.3389/fcimb.2022.968739
87. Van Eyssen SR, Samarkina A, Isbilen O, Zeden MS, Volkan E. FimH and Type 1 Pili mediated tumor cell cytotoxicity by uropathogenic Escherichia coli *in vitro*. *Pathogens.* 2023;12(6) https://doi.org/10.3390/pathogens12060751
88. Butler DSC, Cafaro C, Putze J, et al. A bacterial protease depletes c-MYC and increases survival in mouse models of bladder and colon cancer. *Nat. Biotechnol.* 2021;39(6):754–764. https://doi.org/10.1038/s41587-020-00805-3
89. Thein ZM, Samaranayake YH, Samaranayake LP. In vitro biofilm formation of Candida albicans and non-albicans Candida species under dynamic and anaerobic conditions. *Arch. Oral Biol.* 2007;52(8):761–767.
90. Samaranayake LP, Geddes DA, Weetman DA, MacFarlane TW. Growth and acid production of Candida albicans in carbohydrate supplemented media. *Microbios.* 1983;37(148):105–115.
91. Webster CE, Odds FC. Growth of pathogenic Candida isolates anaerobically and under elevated concentrations of CO_2 in air. *J Med Veterinary Mycology.* 1987;25(1):47–53.
92. Pereira R, dos Santos Fontenelle RO, De Brito EHS, De Morais SM. Biofilm of Candida albicans: formation, regulation and resistance. *J. Appl. Microbiol.* 2021;131(1):11–22.
93. Talapko J, Meštrović T, Dmitrović B, et al. A putative role of Candida albicans in promoting cancer development: a current state of evidence and proposed mechanisms. *Microorganisms.* 2023;11(6):1476.
94. Alves R, Barata-Antunes C, Casal M, Brown AJ, Van Dijck P, Paiva S. Adapting to survive: how Candida overcomes host-imposed constraints during human colonization. *PLoS Pathog.* 2020;16(5):e1008478.
95. Lapaquette P, Ducreux A, Basmaciyan L, et al. Membrane protective role of autophagic machinery during infection of epithelial cells by Candida albicans. *Gut Microbes.* 2022;14(1):2004798.
96. Hinck L, Näthke I. Changes in cell and tissue organization in cancer of the breast and colon. *Curr. OpCell Biol.* 2014;26:87–95.

97. Kouzu K, Tsujimoto H, Kishi Y, Ueno H, Shinomiya N. Role of microbial infection-induced inflammation in the development of gastrointestinal cancers. *Medicines*. 2021;8(8):45.
98. Huang H, Wang Q, Yang Y, Zhong W, He F, Li J. The mycobiome as integral part of the gut microbiome: crucial role of symbiotic fungi in health and disease. *Gut Microbes*. 2024;16(1):2440111.
99. Allert S, Förster TM, Svensson CM, et al. Candida albicans-induced epithelial damage mediates translocation through intestinal barriers. *MBio*. 2018;9(3):10–1128.
100. Yu D, Liu Z. The research progress in the interaction between Candida albicans and cancers. *Front Microbiol*. 2022;13:988734.

CHAPTER NINE

Role of biofilm in cancer invasion

Pritam Dey Roy[a], Arpita Ghosh[a], Subhajit Nath[b], and Sahana Ghosh[a,*]
[a]Department of Microbiology, Vijaygarh Jyotish Ray College, Kolkata, West Bengal, India
[b]Department of Medical & Molecular Virology, University of Manchester, Manchester, United Kingdom
*Corresponding author. e-mail address: sahanaghosh2102@gmail.com

Contents

Abstract

Biofilms produced by bacteria are one of the recognized causes in cancer development and progression. These microbial communities with specific structures are embedded in extracellular matrices. They facilitate chronic infections by enhancing the resistance to various antibiotics and immune defense system of the host. An underlying key mechanism for this process is called quorum sensing (QS). It's basically a cell – density dependent signaling system that regulates bacterial behavior and biofilm formation. Different QS molecules such as N-acyl homoserine lactones have been seen to modulate immune responses and promote different processes like epithelial-mesenchymal transition (EMT), thereby contributing to tumor invasion and metastasis. Also, biofilm associated bacteria can directly reprogram host cells too. Certain bacterial toxins such as

Progress in Molecular Biology and Translational Science, Volume 224
ISSN 1877-1173, https://doi.org/10.1016/bs.pmbts.2026.07.006

colibactin from genotoxic strains of *E.coli* induce DNA damage and somewhat of genetic instability, both of which are hallmarks of cancer initiation. Similarly, adhesins like FadA activate signaling cascades in host epithelial cells, which are considered as genetic expression altering molecules. Therefore, disrupting the QS pathways and targeting microbial effectors that hijack host signaling, are emerging as very promising therapeutic strategies. Further future research, aimed mainly at decoding the biofilm and its host interactions, may unlock novel treatments to cancer progression and eliminate them.

1. Introduction

Biofilms develop on both animate and inanimate material and are formed when microorganisms band together to create large numbers of cells and stay in one location. The structured microbial communities embedded within an extracellular polymeric substance (EPS) matrix which they secrete to protect themselves is a combination of proteins, lipids, polysaccharides, and eDNA. The EPS protects the individual cells (and therefore bacteria) from environmental stress, antibiotic action, and other ways that impede or prevent long-term colonization.[1] Whereas conventionally biofilms have been associated with chronic infections, it is increasingly evident from recent scientific evidence that the process of biofilm formation is related to the establishment and progress of several cancers.[2] Biofilms consist of several functional stages: their dynamic process involves attachment, micro-colony development, maturation, and dispersion. They often occur as multi-layered structures on mucosal surfaces, such as the gastrointestinal tract, oral cavity, and respiratory passages, in the human body.[3] Chronic inflammation and the growth of microbes are both associated with the formation of tumors in these regions.[4] Biofilm activity in these locations leads to chronic inflammation, which damages the tissue and causes DNA mutations, and increases the number of cells that are growing abnormally.[5] Over time, transformative changes can occur in healthy cells to create malignant cells and thus demonstrate a relationship between biofilm persistence and the initiation/invasion of malignancy/cancer.[6] Quorum sensing is a primary mechanism to explain communication and organization within biofilms. Quorum sensing is a density-dependent system used by bacteria to communicate with each other by sensing their own "population size", based on the release and detection of small molecules called autoinducers. When the concentration of autoinducers in the environment has reached a certain threshold, bacteria will activate genes associated with biofilm maturation, virulence, and modulating the immune response.[7] Cancer has been linked to changes in the host cell's signaling pathways due to the action of quorum sensing molecules (such as N-acyl homoserine lactones, self-induced molecules and autoinducer-2) activating pathways that promote inflammation (such as NF-κB and STAT3). By activating these pathways, quorum sensing molecules can promote cell survival, angiogenesis and resistance to the cellular death (apoptosis) that normally occurs in cancerous cells.[8] In addition, through promoting the transition from being epithelial to mesenchymal, quorum sensing is involved in the invasion of cancer via allowing cancer cells to infiltrate the local tissue and then spread to distant sites.[9] Bacteria that form biofilm can also secrete substances that are genotoxic, such as colibactin from E. coli of some strains, which cause double-strand breaks in DNA and increased genome instability, which is a fundamental aspect of the initiation of cancer.[10] Bacterial adhesion molecules such as FadA can bind to epithelial cells and activate cellular signal transduction pathways to promote the growth of tumors and metastasis.[11]

2. Biofilm: A survival strategy of human-associated bacteria

A biofilm is considered to be one of nature's most successful lifestyles for microbes and has a significant impact on the human infectious process. A biofilm is a community of microbial organisms – including viruses, bacteria, fungi, protozoa, and even algae – that attach to surfaces, most commonly to liquid-solid interfaces.[12] In contrast to organisms that are free living, a biofilm has emergent properties and unique characteristics that are not predictable based on a study of planktonic bacteria. Therefore, it is also one of the most broadly successful and widely distributed lifeforms. An important characteristic of this complex lifestyle means that the organisms are able to use the biofilm to protect themselves from adverse effects, including a bacterial biofilm that can be highly resistant to antibacterial agents and the human immune system.[13] As a result, up to 80 % of the bacterial infections in humans have been reported to be associated with biofilms.[14]

3. Adhesion and biofilm development cycle: Link to cancer progression

The development of biofilm is not an accident but a controlled, cyclic event that helps pathogenic organisms survive within host organisms. This stepwise process has been well studied in the disease-causing bacterium *Pseudomonas aeruginosa*, which is used as a model organism to study the characteristics of biofilm-associated infections. As illustrated in Fig. 1, biofilm development in *Pseudomonas aeruginosa* proceeds through five sequential stages: reversible attachment, irreversible attachment, maturation-I, maturation-II, and dispersion.[16,17] These stages collectively contribute to the establishment, maintenance, and dissemination of biofilm-associated bacterial communities.

In the initial (early) development of bacterial cells, they will travel to and interact with their environments either from natural (organic-) based to synthetic (man-made) substrates. In this case, P. aeruginosa connects to their surfaces (substrata) using their appendages (e.g. flagella and pili) which enable them to move through the water/swim and move over the surface of an object (twitch) without permanently adhering to the surface.[18,19] The temporary interaction enables bacteria to measure the outside surface of their environment to make a decision about continuing to attach based on previously established criteria. After determining the surface satisfies the required conditions to allow further colonization, the type IV pilus of Pseudomonas aeruginosa will elongate, leading to attachment of the cell to the surface permanently. [20]

The molecular level provides further stability to this connection and strengthens it through both irreversibly covalent bonding interactions and

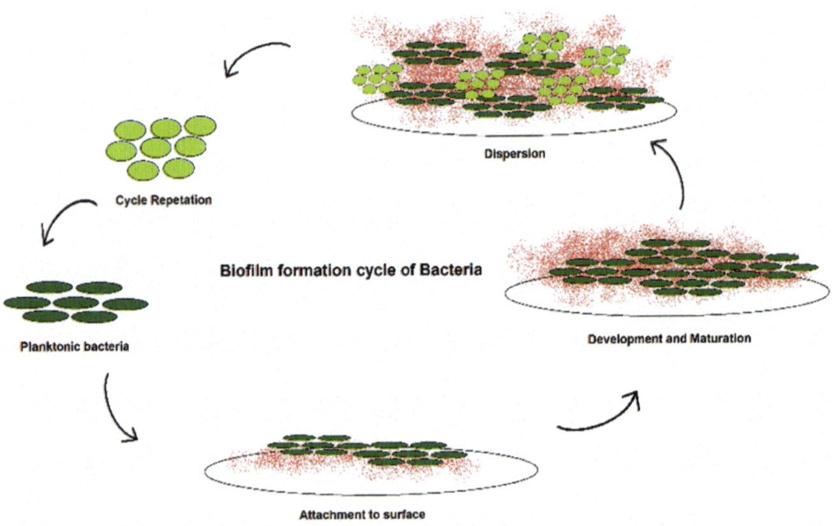

Fig. 1 Biofilm formation cycle of bacteria.[15]

hydrogen bonding interactions to promote irreversible attachment. In contrast, cellular attachment to abiotic surfaces is primarily through non-specific interactions.[21] In addition, as more cells accumulate, structural cohesion is critical for supporting and maintaining the cohesiveness of the entire growing colony.[22] In addition to supporting the structural integrity of the biofilm by interacting with Pel and Psl, CdrA has also demonstrated the ability to promote cell-clustering independent of exopolysaccharide synthesis.[22]

Biofilm structure becomes more stable through the production of matrix material. At this point in the development of the biofilm, it is at its highest level of expression of extracellular polymers/exopolysaccharides (in particular, Pel, Psl, and alginate).[23] Continued development and accumulation lead to the establishment of spatial orientation within the biofilm. Finally, when the biofilm is formed into its final size and thickness, the biofilm as a three-dimensional structure will become visible to the naked eye and appear as a multicellular association that looks like a mushroom.[24]

Biofilms are highly hydrated, three-dimensional microbial communities embedded within EPS matrix. They often contain approximately 70–97 % water, although the exact hydration level varies depending on the species, EPS composition (including Pel, Psl, alginate, cellulose, and extracellular DNA), and surrounding environmental conditions. In addition to water, the biofilm matrix contains polysaccharides, proteins, lipids, nucleic acids, and insoluble structural components such as amyloids, fimbriae, pili, and flagella.[25] The composition and organization of these matrix contribute to the structural integrity, resilience, and functional properties of biofilms. Consequently, the interplay between genetic regulation, physical architecture, and collective microbial behaviour enables biofilms to survive, persist, and dominate across diverse biological habitats.

4. Microbial dynamics in biofilms: Cooperation, competition and communication

4.1 Cooperation within biofilms

In the cooperative mode, bacterial cells in a biofilm produce public goods such as extracellular polymeric substances (EPS), enzymes, and metabolites that provide benefits to other bacterial cells in the community.[26] Public goods in biofilms help in the attachment of bacteria to surfaces, the capture of nutrients, and the protection of bacteria from environmental stressors, antibiotics, and the host immune response.[27,28] Quorum sensing helps bacteria regulate gene expression to ensure that expensive social behaviors, such as the production of the biofilm matrix or the production of virulence factors, are only expressed when a sufficient number of bacterial cells are present.[29]

5. Competition, antagonism, and competition sensing in biofilms

Within biofilms, microbial interactions involve not only cooperation but also intense competition for nutrients and ecological dominance. Different species of bacteria or strains of the same species of bacteria employ antagonistic strategies by producing antimicrobial compounds, bacteriocins, and other toxic substances to inhibit the growth of competing microorganisms. The expression of these competitive traits is tightly regulated by quorum-sensing systems, which ensure that toxin production occurs only when the bacterial population reaches a sufficient density and environmental conditions permit the local accumulation of secreted molecules around target cells. In addition to quorum-sensing-mediated chemical communication, bacteria can detect the presence of competitors through a phenomenon known as competition sensing. This mechanism enables microorganisms to recognize biotic stresses arising from physical proximity, nutrient limitation, cell wall damage, or other competitor-induced insults, while distinguishing them from abiotic stresses such as heat or osmotic shock.[30] Consequently, stress-responsive pathways trigger the upregulation of defensive and offensive mechanisms only when a genuine biological threat is encountered, thereby optimizing energy expenditure and enhancing competitive fitness within the biofilm community.[30]

6. Relevance to cancer progression

The tumor microenvironment is heavily influenced by the cooperative and competitive dynamics of biofilms. Intercellular communication occurs over time in biofilms as a result of both quorum sensing and stress responses. Chronic inflammation, immune modulation, and remodelling of tissue are all major contributors to the establishment of a microenvironment that supports the survival, invasion, and eventual metastasis of cancer cells.[31,32] All of the events listed above are complex and interconnected to one another to create a unique microenvironment that promotes cancer cell growth in an unsuspected way.[33,34]

7. The dual role of biofilms in the host: From beneficial to pathogenic

The two-part relationship between biofilm communities and the host and adjacent microbial populations provides the primary definition of biofilms. Biofilms that reside within various anatomical locations of the human body (e.g., gastrointestinal tract and oral cavity) represent an essential component of the maintenance of normal physiology because they assist in maintaining the host's barrier function against invading organisms. According to one literature source, "Human oral and gastrointestinal biofilms contribute to host defense by acting as a physical barrier through their extracellular matrix that blocks the entry of potentially pathogenic organisms and compete for nutriments with pathogens, produce anti-microbial substances to impede the establishment of pathogens, and prevent the establishment of pathogens".[35]

In this positive light, biofilm communities help sustain equilibrium between communities and provide long-lasting support for microbiota stability. Biofilm formation has a multifaceted function in that its primary function is to serve as a natural interface between organisms and their environment and create inhospitable microenvironments for pathogen and invasive species establishment, thereby preserving equilibrium within a community.[36] The body's immune response has also learned to recognize and accept these communities as part of its normal physiology through receptor interactions with commensal biofilms, resulting in immune homeostasis instead of chronic inflammation. Once the microbial ecosystem becomes disrupted (for example, by antibiotic therapy, immune suppression, senescence, or epithelial damage), biofilms may become pathological in nature; pathological biofilms often persist despite the presence of the host immune system and play a huge role in creating and maintaining chronic conditions. This point is further emphasized in an important review, which states the following: biofilms possess both pro-healthy and pathogenic characteristics, depending on the situation, and therefore represent a dual dimension to their functionality.[36] Thus, the role of biofilms in maintaining homeostasis as well as their potential to contribute to the development of disease depends on specific situational factors.

Pathogenic biofilms are also linked to chronic inflammatory diseases and cancer, in addition to being linked to long-term infections. Biofilm microenvironments create conditions that can produce cellular stress, DNA damage, and abnormal signalling from low-grade inflammation over long periods of time. Another literature review highlights the effect of biofilms on disease pathology throughout the body's tissue; specifically, bacterial biofilm disease has been known for many years to cause acute long-term human disease and additional diseases such as inflammatory bowel disease and colorectal cancer.[37]

8. Microbial biofilms and the host immune system

The relationship between biofilms and the host immune system is intricate because microbial biofilms can simultaneously elicit an immunologic response and also evade the immune system. When biofilms are newly formed, the organized extracellular matrix of the biofilm serves as an excellent mechanical barrier for the bacteria against the host immune system. According to Montanari et al., biofilms produced from the host microbiota act as a protective barrier, providing less opportunity for the host immune system to recognize the microbe and aiding in the maintenance of a delicate balance between the microbe and the host.[36] This provides evidence that biofilms can aid in maintaining a healthy balance at first between microorganisms and their host.

Pathogenic species become dominant and no longer remain nonpathological by resisting elimination by the immune system. Biofilms display increased resistance to host immune responses, even in the presence of strong inflammatory responses as shown by Jensen et al. who reported that Pseudomonas aeruginosa biofilms resist neutrophil activity as well as antimicrobial peptides and persist in the presence of strong inflammation.[38] The resistance occurs via the EPS matrix (extracellular polymeric substance) that inhibits access of immune cells to the biofilm; this matrix masks bacterial antigens from detection by the immune system; and regulates the metabolism of microorganisms thus rendering them more difficult to detect.

The continual interaction of immune cells targeted by biofilms will lead to a state of chronic immune activation creating conditions for chronic biofilm-associated infection and creating significant and ongoing innate immune response leading to ongoing inflammation and damage to the surrounding tissue. Furthermore, a recent review indicated that these chronic infections caused by biofilms would cause prolonged activation of innate immune signalling pathways, such as NF-κB and STAT3 that affect immune system regulation and tissue damage.[39] Thus biofilms function not only as escape mechanisms from being resolved by the immune system but as strong modulators of host immunity that create prolonged inflammation and disease progression.

9. Inflammation acts as a bridge between biofilm and cancer

Chronic inflammation is considered a key mechanism underlying the association between microbial biofilms and cancer development (oncogenesis). Under physiological conditions, commensal biofilms contribute to tissue homeostasis through regulated interactions with the host immune system, promoting immune tolerance and preventing excessive inflammatory responses.[36] The effects of biofilms on host tissues are largely influenced by their microbial composition and ecological context. Disruption of this balanced host–microbiota relationship, termed microbial dysbiosis, facilitates the establishment of pathogenic biofilms that provoke persistent inflammatory responses. Biofilm-forming pathogens, such as *Pseudomonas aeruginosa*, induce sustained activation of innate immune cells, including neutrophils and macrophages, resulting in prolonged production of pro-inflammatory cytokines and other inflammatory mediators.[38] Due to their inherent resistance to immune clearance, pathogenic biofilms often evade complete eradication, leading to unresolved inflammation, continuous tissue injury, and prolonged cellular stress. This chronic inflammatory micro-environment promotes genomic instability, aberrant cell proliferation, and other oncogenic processes, thereby contributing to the initiation and progression of cancer.

Tumor-supportive microenvironments are generated as a result of chronic inflammation. Tumor-supportive microenvironments contain very high levels of reactive oxygen species (ROS), pro-inflammatory cytokines (especially IL-6, TNF-α, and IL-1β), chemokines, and growth factors. Reactive oxygen species produce damage to DNA by causing mutations and increasing instability of the genomic material. Pro-inflammatory cytokines initiate transcriptional signalling events, (such as through NFκB and STAT3) that enhance cell survival, proliferation and resistance to apoptosis. Growth factors, such as VEGF promote the formation of new blood vessels (angiogenesis) to supply both nutrients and oxygen to the growing tumours.

Over time, the action of these pro-inflammatory mediators causes EMT to occur in epithelial cells; EMT is a process whereby epithelial cells lose their normal structure and behaviour and acquire the ability to invade. Also occurring simultaneously is the development of immune suppression within the tissue microenvironment, which allows mutated cells to evade detection by the immune system and continue to proliferate. All of the above factors have been demonstrated to result in malignant tumor formation from chronic inflammation. The chronic inflammation produced by biofilms enhances the four characteristics of cancer progression (DNA damage, epithelial-to-mesenchymal transition, angiogenesis, and immune suppression).[39]

10. Biofilm in specific cancers

10.1 Colorectal cancer (CRC)

CRC is considered a biofilm-associated cancer in which microbial communities have colonized the surface of the tumour and are involved in promoting cancer growth. The most consistently enriched microbial species within CRC tumours are *Fusobacterium nucleatum*, an anaerobic oral bacterium that develops invasive, dense biofilms that establish direct contact with the epithelial cells and the surrounding tumour microenvironment. The enrichment of *F. nucleatum* in invasive CRC biofilms highlights the significance of complex community partnerships in tumour pathogenesis.[40] Fig. 2 illustrates the multistep progression of colorectal cancer from normal epithelium through adenomatous lesions to carcinoma, a process closely associated with microbial dysbiosis, chronic inflammation, and dysregulation of oncogenic

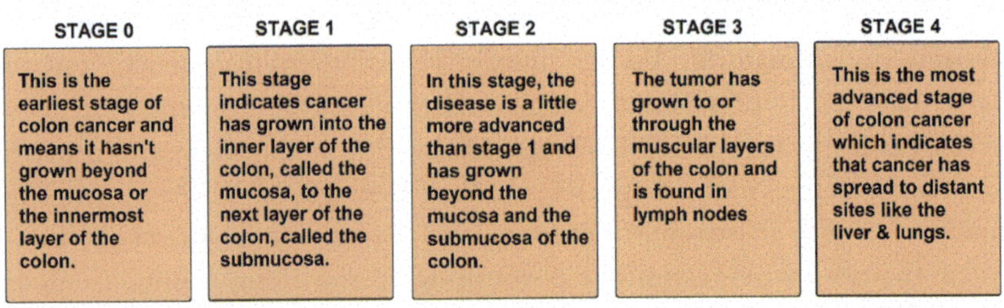

Fig. 2 Stages of colorectal cancer progression.[41]

signalling pathways. These biofilms should not be viewed as passive structures but rather as highly dynamic, evolving ecosystems that drive molecular signalling pathways that contribute to cancer development, promote chronic inflammation, and protect the bacteria within the biofilms from being cleared by the host immune system.

A major molecular mechanism by which *F. nucleatum* has been classified as an oncogenic factor is via its adhesin protein (FadA), which binds to cadherin molecules on the host cell surface, allowing the bacteria to attach and facilitating signal transduction in bacteria. In particular, FadA is secreted by the organism and binds to VE-cadherin on endothelial cells and E-cadherin epithelial cells. The binding of FadA to E-cadherin activates oncogenic signalling pathways (Wnt/β-catenin) as it relates to colorectal cancer and oral cancer.[42] The interactions of FadA and E-cadherin disrupt Junctions between Cells and activate Wnt/β-catenin signalling pathways, leading to nuclear accumulation of β-catenin and ultimately driving oncogenesis through the transcription of genes that drive unregulated cell proliferation, invasion and metastasis.

10.2 Oral cancer and biofilm-associated carcinogenesis

Numerous studies reveal that the OSCC tissues containing biofilms are heavily infected with several anaerobic periodontal pathogens, notably Porphyromonas gingivalis and Fusobacterium nucleatum. Both of these microorganisms have the capability to adhere to the epithelial cells situated within the oral cavity, and then penetrate into the surrounding deep tissue layers, where they manipulate the normal cellular functions of the host. In the review "Biofilms on Oral Squamous Cell Carcinomas", the authors discuss the various mechanisms in which biofilms can contribute to carcinogenesis, including promoting cell proliferation, promoting cellular invasion, inducing chronic inflammation, and allowing evasion of immunity.[41] This provides strong evidence that the species composition of oral tumors is strongly correlated with aggressive disease behaviour.

10.3 Gastric cancer and biofilm involvement

Helicobacter pylori is known to be the key bacterial risk factor for gastric cancer; evidence suggests that its ability to create biofilms makes those infections and disease progression much more likely, and those infections can persist over an extended period of time. *H. pylori* produces biofilms in the context of the gastric mucosa creating very structured communities that resist harsh acidic environments and immune attacks. Therefore, the ability of the bacteria to survive in the stomach (and constantly irritate and inflame the epithelium of the stomach) serves to increase the risk of the development of malignancy.

Biofilms also make it more difficult to use antibiotics in order to treat the infection because the biofilm creates a physical barrier that hinders the delivery of antibiotics and makes it much less effective. In other words, due to the biofilm mode of growth, *H. pylori* is much harder to eliminate, therefore there is persistent colonization. As discussed, *H. pylori* biofilm formation adds to persistent colonization and treatment resistance, likely causing chronic inflammation associated with gastric carcinogenesis[43] and therefore biofilms may significantly contribute to exacerbating inflammation-driven cancer pathways in the gastric mucosa.

As illustrated in Fig. 3, persistent *Helicobacter pylori* infection is a major etiological factor in gastric carcinogenesis and promotes tumour development through multiple interconnected mechanisms, including chronic inflammation, immune modulation, DNA damage, epithelial–mesenchymal transition (EMT), and activation of oncogenic signaling pathways. In addition to its direct effects on host cells, chronic *H. pylori* infection disrupts the composition

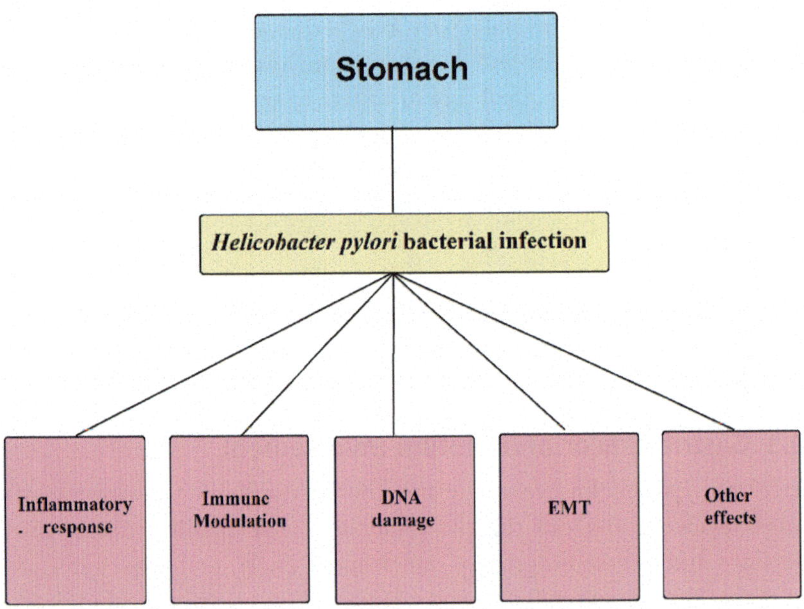

Fig. 3 Role of *Helicobacter pylori* infection and associated inflammatory pathways in gastric carcinogenesis.[44]

and homeostasis of the gastric microbiota, resulting in microbial dysbiosis. The altered microbial ecosystem, together with *H. pylori*, further amplifies inflammatory and tumorigenic signaling cascades in gastric epithelial cells. These complex interactions among the pathogen, host, and resident microbiome contribute to persistent tissue injury, aberrant cellular proliferation, and ultimately the initiation and progression of gastric cancer.[44]

10.4 Pancreatic and other cancers linked to biofilm-associated microbes

Although preliminary research has been published in this topic, there have been previous studies that show that biofilm-forming microorganisms could play a role in pancreatic ductal adenocarcinoma (PDAC) or, more generally, gastrointestinal cancers.[45,46] Bacterial populations (specifically Fusobacterium species) have been identified in the PDAC tissues using newer microbiome analysis and sequencing technologies.[47] High levels of Fusobacterium species have been found in PDAC tumors, and there is a correlation between increased quantities of Fusobacterium species within a PDAC tumor and aggressiveness of the tumor and unfavorable clinical outcome for the patient. Because Fusobacterium species could have been introduced into the PDAC tumor from an environmental source, further research must determine whether the species are actively modifying tumor behavior or just comprise part of the microbiota of PDAC tumors.[48]

As mentioned previously, genomic and clinical analyses have shown an increase in the abundance of Fusobacterium in multiple forms of gastrointestinal malignancies, thus establishing Fusobacterium's potential association with pancreatic and gastric cancers.[49] This extends the ability of biofilm-associated microorganisms to participate in the development of malignancies outside of the site of origin, and therefore shows how the presence of other microbiota in the gastrointestinal tract (GIT) may be important for tumour development and progression.

10.5 Breast cancer and biofilm-associated microbiome

Genotoxins produced by a subset of bacterial agents (particularly Escherichia coli & Staphylococcus epidermidis) that regularly exist in breast tumor samples damages the host's genetic material directly; the presence of these toxins has been shown to cause an organism to accumulate mutations at an accelerated rate which in turn leads to an increase in the speed at which the organism develops a tumor. This statement can be supported by data showing that strains of E. coli that have been identified in breast cancer specimens produce colibactin-like genotoxins that can induce double-stranded breaks in the DNA of the host, according to reference.[50]

In addition to inducing DNA damage, intratumoral biofilm-associated bacteria can compromise the efficacy of anticancer therapies. These microorganisms may metabolically or enzymatically modify chemotherapeutic agents, thereby reducing their bioavailability and therapeutic activity. Moreover, biofilm-associated bacteria can activate pro-survival signaling pathways in malignant cells, enhancing their resistance to treatment. Consequently, tumour-associated bacteria contribute to therapeutic failure through both drug inactivation and the promotion of cellular mechanisms that support cancer cell survival and chemoresistance.[51]

11. Impact of biofilm on anticancer therapies

The chemical alteration or breakdown of chemotherapeutic agents is a very direct way in which biofilms can reduce the effectiveness of chemotherapy treatments. Specific intratumoral bacteria contain enzymes that act on anticancer agents and render them metabolized before they are able to kill the tumor cells. This process is very well established: the ability of intratumoral bacteria to metabolically degrade and inactivate chemotherapeutic agents results in decreased efficacy of the chemotherapeutic agent and acceleration of tumor growth.[52] Therefore, although patients may be dosed appropriately with a chemotherapeutic agent, the amount of active drug delivered to the tumor cells is much less than would have been possible in the absence of bacterial interference.

The presence of the tumor microbiome impacts how well patients respond to both chemopreventive and therapeutic treatments with immunotherapeutic agents - especially immunological checkpoint inhibitors that depend on reactivating the body's own natural immunity via various mechanisms dependent on activation of different cell types within the immune system (both adaptive and innate). Due to the ability for certain types of bacteria to establish variability in cellular signalling pathways such as those mentioned above, it stands to reason that different types of microbial communities can produce different immunological responses within a given host (i.e., a person's own unique immunity). As such, different types of biofilm-associated microorganisms have been demonstrated to influence the magnitude of anti-tumor/anti-microbial immunity resulting from the use of immune checkpoint inhibitors depending upon particular characteristics of each specific microbe present in the tumor.[52]

The ability of bacteria products and metabolic products of bacteria to improve resistance through activation of cancer cell survival pathways is based on the molecular level. Microbial products trigger transcription factors (e.g., NF-κB and STAT3) that control the expression of genes associated with cell proliferation, inflammatory response, and resistance against apoptosis. Bacterial-derived products and metabolites derived from the microbiota, have been shown to activate NF-κB and STAT3 pathway in tumor promoting cell proliferation and longevity and developing drug-resistance.[53,54] Activation of this signalling reprogrammes tumor cells to become drug resistant, allowing them to continue to live after aggressive treatments.

Overall, these mechanisms indicate that tumor-related biofilms serve as a form of biological armor and act as biochemical modulators in decreasing the effectiveness of both chemotherapy and immunotherapy. Therefore, controlling microbial biofilms may be a future strategy to improve treatment efficacy.

12. Future perspectives

As evidence continues to grow for the involvement of microbes in initiating cancer, invading the body, and ultimately worse disease outcomes, much of the work done to date has been observational, with relatively little known about the intricacies of the ways in which different members of microbial communities affect cancer cells. To get to mechanistic (how they work) functional (what they do) action we need to do mechanistic and functional research will require moving outside of microbiome studies to determine the connections between many groups of microbes and how cancer cells are impacted by their presence.

In order to do this, detailed experimental methods are going to be necessary in order to replicate the tumors within their microenvironment including:

Organoids/3D co-culture methods/tumor on-a-chip (microfluidic-based technology). This will help us rebuild the complex nature of the interactions occurring between biofilm, tumor cell, and immune cells, looking at these interactions under physiologically relevant culture conditions so we can observe in real time the actions of quorum sensing, genotoxic production, immune modulation and extracellular matrix remodeling during the process of tumor development.

The high-level methods of studying the multiscience universe—metagenomics, transcriptomics, metabolomics, and proteomics—are necessary for developing a full picture of the intratumoral biofilm community as well as developing a better understanding of their virulence factors, particularly those with oncogenic signaling potential. This information could also be integrated with AI/machine learning tools to produce more accurate predictions about biofilm-based therapies and progression of disease.

Regarding the translation aspect, biofilms present additional opportunities for treating cancer with novel therapies. There are a number of methods that may lead to greater drug delivery to tumors and restoration of drug sensitivity, including quorum sensing inhibitors, biofilm breaking enzymes, bacteriophages, microbiome manipulation and the delivery of antibiotics to the patients' tumors. In predictive medicine, microbiome profiling can be used to generate a personalized medicine approach to treating a patient's cancer based on the specific microbiome characteristics of their tumor.

In summary, biofilms have great potential to be redefined as cancer therapies and for the advancement of personalized cancer medicine through the discovery and targeting of tumor biofilms.

13. Conclusion

This review provides insight into the transformation of biofilm-associated microbes from participating in cancer biology as peripheral players into the major players in cancer biology. The initiation, invasion, and metastasis of cancer; as well as the resistance of cancer to treatment; is all mediated by biofilm. Up until this point, microbes that produce biofilms have been viewed as benign commensal inhabitants of the host. However, today these microbes play an integral role in the tumor microenvironment by communicating with each other through quorum sensing; producing genotoxic compounds; modulating the immune response; and initiating signal transduction cascades for oncogenesis and biomodification of the biofilm niche to promote the development of cancer.

Bacterial biofilms are responsible for causing chronic inflammatory processes because they can lead to epithelial-mesenchymal transformation (EMT).

These processes aid in the growth of cancerous cells as they invade surrounding tissues and organs. In addition to the above, cancer cells have been shown to be impeded by extracellular matrix limitations for drug absorption due to the presence of biofilms. These findings indicate that cancer is not only a function of the host's genetics but also a product of an intricate biological system formed by the host's microbial population.

A better understanding of the biofilm-cancer relationship provides an alternative view of cancer heterogeneity and the causes of many failed cancer therapies. Targeting biofilms and signalling pathways with targeted treatment may become feasible treatments in addition to using other types of treatment for cancer. Future success in implementing these concepts into practical applications will require collaboration between disciplines such as microbiology, oncology, immunology, and systems biology. There is no doubt that disrupting biofilm-controlled processes creates opportunities for more effective cancer prevention, improved early diagnosis, and more effective therapies.

References

1. Flemming HC, Wingender J. The biofilm matrix. *Nat Rev Microbiol*. 2010;8:623–633. https://doi.org/10.1038/nrmicro2415
2. Garrett WS. Cancer and the microbiota. *Science*. 2015;348:80–86 10.1126/science.aaa4972.
3. Costerton JW, et al. Bacterial biofilms: a common cause of persistent infections. *Science*. 1999;284:1318–1322. https://doi.org/10.1126/science.284.5418.1318
4. Coussens L, Werb Z. Inflammation and cancer. *Nature*. 2002;420:860–867. https://doi.org/10.1038/nature01322
5. Yoshino T, Nakamura S, Matsuno Y, Ochiai A, Yokoi T, Kitadai Y, et al. Epstein–Barr virus involvement is a predictive factor for the resistance to chemoradiotherapy of gastric diffuse large B-cell lymphoma. *Cancer Sci*. 2006;97:163–166. https://doi.org/10.1111/j.1349-7006.2006.00155.x
6. Schwabe R, Jobin C. The microbiome and cancer. *Nat Rev Cancer*. 2013;13:800–812. https://doi.org/10.1038/nrc3610
7. Christopher M, et al. Quorum sensing: cell-to-cell communication in bacteria. *Annu Rev Cell Dev Biol*. 2005;21:319–346. https://doi.org/10.1146/annurev.cellbio.21.012704.131001
8. Hughes D, Sperandio V. Inter-kingdom signalling: communication between bacteria and their hosts. *Nat Rev Microbiol*. 2008;6:111–120. https://doi.org/10.1038/nrmicro1836
9. Maric G, et al. GPNMB augments Wnt-1 mediated breast tumor initiation and growth by enhancing PI3K/AKT/mTOR pathway signaling and β-catenin activity. *Oncogene*. 2019;38:5294–5307. https://doi.org/10.1038/s41388-019-0793-7
10. Nougayrède JP, et al. *Escherichia coli* Induces DNA double-strand breaks in eukaryotic. *Cells Science*. 2006;313:848–851. https://doi.org/10.1126/science.1127059
11. Kostic, A.D., et al., *Fusobacterium nucleatum* potentiates intestinal tumorigenesis and modulates the tumor-immune microenvironment. Cell Host & Microbe. no.date; 14(2), 207 – 215. https://doi.org/10.1016/j.chom.2013.07.007.

12. Kang X, et al. Strategies and materials for the prevention and treatment of biofilms. *Mater Today Bio*. 2023;23:100827. https://doi.org/10.1016/j.mtbio.2023100827
13. Stoodley P, et al. Biofilms as complex differentiated communities. *Annu Rev Microbiol.* 2002;56:187–209. https://doi.org/10.1146/annurev.micro.56.012302.160705
14. Costerton JW, Stewart PS, Greenberg EP. Bacterial biofilms: a common cause of persistent infections. *Science*. 1999;284(5418):1318–1322. https://doi.org/10.1126/science.284.5418.1318
15. Behura A, et al. *Mycobacterial biofilm: structure and its functional relevance in the pathogenesis, Editor(s): Surajit Das, Neelam Amit Kungwani. Understanding Microbial Biofilms*. Academic Press; 2023:461–474 ISBN 9780323999779. https://doi.org/10.1016/B978-0-323-99977-9.00018-1.
16. Jiang Z, et al. Searching for the secret of stickiness: how biofilms adhere to surfaces. *Front Microbiol*. 2021;12:686793. https://doi.org/10.3389/fmicb.2021.686793
17. de Sousa T, et al. Genomic and metabolic characteristics of the pathogenicity in pseudomonas aeruginosa. *Int J Mol Sci*. 2021;22(23):12892. https://doi.org/10.3390/ijms222312892
18. Chang CY. Surface sensing for biofilm formation in pseudomonas aeruginosa. *Front Microbiol*. 2017;8:2671. https://doi.org/10.3389/fmicb.2017.02671
19. Mikkelsen H, Sivaneson M, Filloux A, et al. Key two-component regulatory systems that control biofilm formation in Pseudomonas aeruginosa. *Env Microbiol*. 2011;13(7):1666–1681. https://doi.org/10.1111/j.1462-2920.2011.02495.x
20. Anyan ME, et al. Type IV pili interactions promote intercellular association and moderate swarming of Pseudomonas aeruginosa. *Proc Natl Acad Sci U S A*. 2014;111(50):18013–18018. https://doi.org/10.1073/pnas.1414661111
21. Radoul M, Bykov D, Rinaldo S, Cutruzzola F, Neese F, Goldfarb D, et al. Dynamic hydrogen-bonding network in the distal pocket of the nitrosyl complex of Pseudomonas aeruginosa cd1 nitrite reductase. *J Am Chem Soc*. 2011;133(9):3043–3055. https://doi.org/10.1021/ja109688w
22. Cherny KE, Sauer K. Untethering and degradation of the polysaccharide matrix are essential steps in the dispersion response of pseudomonas aeruginosa biofilms. *J Bacteriol*. 2020;202(3). https://doi.org/10.1128/JB.00575-19
23. Fleming D, Niese B, Redman W, Vanderpool E, Gordon V, Rumbaugh KP, et al. Contribution of Pseudomonas aeruginosa Exopolysaccharides Pel and Psl to wound infections. *Front Cell Infect Microbiol*. 2022;12:835754. https://doi.org/10.3389/fcimb.2022.835754
24. Ghanbari A, et al. Inoculation density and nutrient level determine the formation of mushroom-shaped structures in Pseudomonas aeruginosa biofilms. *Sci Rep*. 2016;6:32097.
25. Flemming HC, et al. Biofilms: an emergent form of bacterial life. *Nat Rev Microbiol*. 2016;14(9):563–575.
26. Lai BM, et al. Bacterial quorum sensing: cooperation and cheating. *J Appl Ecol*. 2017;28(5):1735–1742. https://doi.org/10.13287/j.1001-9332.201705.018
27. Flemming HC, et al. Microbial extracellular polymeric substances in the environment, technology and medicine. *Nat Rev Microbiol*. 2025;23:87–105. https://doi.org/10.1038/s41579-024-01098-y
28. Ban-Cucerzan A, et al. Persistent threats: a comprehensive review of biofilm formation, control, and economic implications in food processing environments. *Microorganisms*. 2025;13(8):1805. https://doi.org/10.3390/microorganisms13081805
29. Abisado RG, et al. Bacterial quorum sensing and microbial community interactions. *mBio*. 2018;9(3) https://doi.org/10.1128/mBio.02331-17 e02331-17.
30. Nadell C, et al. Spatial structure, cooperation and competition in biofilms. *Nat Rev Microbiol*. 2016;14:589–600. https://doi.org/10.1038/nrmicro.2016.84

31. Branda SS, et al. Biofilms: the matrix revisited. *Trends Microbiol.* 2005;13(1):20–26. https://doi.org/10.1016/j.tim.2004.11.006
32. Park EJ, et al. Dietary and genetic obesity promote liver inflammation and tumorigenesis by enhancing IL-6 and TNF expression. *Cell.* 2010;140(2):197–208. https://doi.org/10.1016/j.cell.2009.12.052
33. Hanahan D, Weinberg RA. Hallmarks of cancer: the next generation. *Cell.* 2011;144(5):646–674. https://doi.org/10.1016/j.cell.2011.02.013
34. Worringer KA, et al. The let-7/LIN-41 pathway regulates reprogramming to human induced pluripotent stem cells by controlling expression of prodifferentiation genes. *Cell Stem Cell.* 2014;14(1):40–52. https://doi.org/10.1016/j.stem.2013.11.001
35. Baxter KJ, et al. Biofilms: from the cradle of life to life support. *NBJ Biofilms Microbiomes.* 2026;12:11. https://doi.org/10.1038/s41522-025-00875-8
36. Montanari E, et al. Biofilm formation by the host microbiota: a protective shield against immunity and its implication in cancer. *Mol Cancer.* 2025;24:148. https://doi.org/10.1186/s12943-025-02348-0
37. Domingue JC, et al. Host responses to mucosal biofilms in the lung and gut. *Mucosal Immunol.* 2020;13:413–422. https://doi.org/10.1038/s41385-020-0270-1
38. Jensen P, et al. The immune system vs. Pseudomonas aeruginosa biofilms. *FEMS Immunol. & Med Microbiol.* 2010;59(3):292–305. https://doi.org/10.1111/j.1574-695X.2010.00706.x
39. Choi E, et al. Biofilm and cancer: interactions and future directions for cancer therapy. *Int J Mol Sci.* 2023;24(16):12836. https://doi.org/10.3390/ijms241612836
40. Queen J, et al. *Fusobacterium nucleatum* is enriched in invasive biofilms in colorectal cancer. *NPJ Biofilms Microbiomes.* 2025;11:81. https://doi.org/10.1038/s41522-025-00717-7
41. McIlvanna E, et al. *Fusobacterium nucleatum* and oral cancer: a critical review. *BMC Cancer.* 2021;21(1):1212. https://doi.org/10.1186/s12885-021-08903-4
42. Pignatelli P, et al. The role of *Fusobacterium nucleatum* in oral and colorectal carcinogenesis. *Microorganisms.* 2023;11:2358. https://doi.org/10.3390/microorganisms11092358
43. Hathroubi S, et al. Helicobacter pylori biofilm formation and its potential role in pathogenesis. *Microbiol Mol Biol Reviews: MMBR.* 2018;82(2):e00001-18. https://doi.org/10.1128/MMBR.00001-18.
44. Vieira P, et al. Helicobacter pylori and the microbiome in gastric cancer development and treatment: a year in review. *Microb Health Dis.* 2024;6:e1031. https://doi.org/10.26355/mhd_20249_1031
45. Kato M, et al. A somatic role for the histone methyltransferase Setdb1 in endogenous retrovirus silencing. *Nat Commun.* 2018;9(1):1683. https://doi.org/10.1038/s41467-018-04132-9
46. Poore GD, et al. Microbiome analyses of blood and tissues suggest cancer diagnostic approach. *Nature.* 2020;579(7800):567–574. https://doi.org/10.1038/s41586-020-2095-1
47. Keen EC, et al. Draft genome sequence of the naturally competent *Bacillus simplex* strain WY10. *Genome Announc.* 2017;5(46):e01295-17. https://doi.org/10.1128/genomeA.01295-17.
48. Riquelme E, et al. Tumor microbiome diversity and composition influence pancreatic cancer outcomes. *Cell.* 2019;178(4):795–806.e12. https://doi.org/10.1016/j.cell.2019.07.008
49. Li Z, et al. Fusobacterium in the microbiome: from health to disease across the oral–gut axis and beyond. *NPJ Biofilms Microbiomes.* 2025;11:200. https://doi.org/10.1038/s41522-025-00838-z
50. Urbaniak C, et al. The microbiota of breast tissue and its association with breast cancer. *Appl Env Microbiol.* 2016;82(16):5039–5348. https://doi.org/10.1128/AEM.01235-16

51. Geller LT, et al. Potential role of intratumor bacteria in mediating tumor resistance to the chemotherapeutic drug gemcitabine. *Science*. 2017;357(6356):1156–1160. https://doi.org/10.1126/science.aah5043
52. Routy B, et al. Gut microbiome influences efficacy of PD-1–based immunotherapy against epithelial tumors. *Science*. 2018;359:91–97. https://doi.org/10.1126/science.aan3706
53. Schneeweiss A, et al. Phase Ib study evaluating the safety and clinical activity of lumretuzumab combined with pertuzumab and paclitaxel in HER2-low metastatic breast cancer. *Cancer Res*. 2017. P6–11–13. https://doi.org/10.1158/1538-7445.SABCS16-P6-11-13.
54. Sorokin L. The impact of the extracellular matrix on inflammation. *Nat Rev Immunol*. 2010;10:712–723. https://doi.org/10.1038/nri2852

Index

Note: Page numbers followed by "*f*" indicate figures and "*t*" indicate tables.

C

R

S

Printed and bound by CPI Group (UK) Ltd, Croydon, CR0 4YY
25/09/2026
02222752-0005